Springer Series in Statistics

Springer Series in Statistics

Andersen/Borgan/Gill/Keiding: Statistical Models Based on Counting Processes.
Andrews/Herzberg: Data: A Collection of Problems from Many Fields for the Student and Research Worker.
Anscombe: Computing in Statistical Science through APL.
Berger: Statistical Decision Theory and Bayesian Analysis, 2nd edition.
Bolfarine/Zacks: Prediction Theory for Finite Populations.
Brémaud: Point Processes and Queues: Martingale Dynamics.
Brockwell/Davis: Time Series: Theory and Methods, 2nd edition.
Daley/Vere-Jones: An Introduction to the Theory of Point Processes.
Dzhaparidze: Parameter Estimation and Hypothesis Testing in Spectral Analysis of Stationary Time Series.
Fahrmeir/Tutz: Multivariate Statistical Modelling Based on Generalized Linear Models.
Farrell: Multivariate Calculation.
Federer: Statistical Design and Analysis for Intercropping Experiments.
Fienberg/Hoaglin/Kruskal/Tanur (Eds.): A Statistical Model: Frederick Mosteller's Contributions to Statistics, Science and Public Policy.
Fisher/Sen: The Collected Works of Wassily Hoeffding.
Good: Permutation Tests: A Practical Guide to Resampling Methods for Testing Hypotheses.
Goodman/Kruskal: Measures of Association for Cross Classifications.
Grandell: Aspects of Risk Theory.
Hall: The Bootstrap and Edgeworth Expansion.
Härdle: Smoothing Techniques: With Implementation in S.
Hartigan: Bayes Theory.
Heyer: Theory of Statistical Experiments.
Jolliffe: Principal Component Analysis.
Kotz/Johnson (Eds.): Breakthroughs in Statistics Volume I.
Kotz/Johnson (Eds.): Breakthroughs in Statistics Volume II.
Kres: Statistical Tables for Multivariate Analysis.
Le Cam: Asymptotic Methods in Statistical Decision Theory.
Le Cam/Yang: Asymptotics in Statistics: Some Basic Concepts.
Longford: Models for Uncertainty in Educational Testing.
Manoukian: Modern Concepts and Theorems of Mathematical Statistics.
Miller, Jr.: Simultaneous Statistical Inference, 2nd edition.
Mosteller/Wallace: Applied Bayesian and Classical Inference: The Case of *The Federalist Papers.*
Pollard: Convergence of Stochastic Processes.
Pratt/Gibbons: Concepts of Nonparametric Theory.
Read/Cressie: Goodness-of-Fit Statistics for Discrete Multivariate Data.

(continued after index)

Paul R. Rosenbaum

Observational Studies

With 14 Illustrations

Springer-Verlag
New York Berlin Heidelberg London Paris
Tokyo Hong Kong Barcelona Budapest

Paul R. Rosenbaum
Department of Statistics
University of Pennsylvania
Philadelphia, PA 19104-6302
USA

Library of Congress Cataloging-in-Publication Data
Rosenbaum, Paul R.
Observational studies/Paul R. Rosenbaum.
p. cm.—(Springer series in statistics)
Includes bibliographical references and index.
ISBN 0-387-94482-6
1. Experimental design. 2. Analysis of variance. I. Title.
II. Series.
QA279.R67 1995
001.4′2—dc20 95-2178

Printed on acid-free paper.

Production coordinated by Brian Howe and managed by Francine McNeill; manufacturing supervised by Jeff Taub.
Typeset by Asco Trade Typesetting Ltd., Hong Kong.
Printed and bound by Maple-Vail, York, PA.
Printed in the United States of America.

9 8 7 6 5 4 3 2 1

ISBN 0-387-94482-6 Springer-Verlag New York Berlin Heidelberg

For Sarah, Hannah, and Aaron

Preface

1. What the Book Is About: An Outline

An *observational study* is an empirical investigation of treatments, policies, or exposures and the effects they cause, but it differs from an experiment in that the investigator cannot control the assignment of treatments to subjects. Observational studies are common in most fields that study the effects of treatments or policies on people. Chapter 1 defines the subject more carefully, presents several observational studies, and briefly indicates some of the issues that structure the subject.

In an observational study, the investigator lacks experimental control; therefore, it is important to begin by discussing the contribution of experimental control to inference about treatment effects. The statistical theory of randomized experiments is reviewed in Chapter 2.

Analytical adjustments are widely used in observational studies in an effort to remove overt biases, that is, differences between treated and control groups, present before treatment, that are visible in the data at hand. Chapter 3 discusses the simplest of these adjustments, which do little more than compare subjects who appear comparable. Chapter 3 then examines the circumstances under which the adjustments succeed. Alas, these circumstances are not especially plausible, for they imply that the observational study differs from an experiment only in ways that have been recorded. If treated and control groups differed before treatment in ways not recorded, there would be a hidden bias. Chapter 4 discusses sensitivity analyses that ask how the findings of a study might be altered by hidden biases of various magnitudes. It turns out that observational studies vary markedly in their sensitivity to hidden bias. The degree of sensitivity to hidden bias is one important consideration in judging whether the treatment caused its ostensible effects or alternatively whether these seeming effects could merely reflect a hidden bias.

Although a sensitivity analysis indicates the degree to which conclusions could be altered by hidden biases of various magnitudes, it does not indicate whether hidden bias is present. Chapters 5 through 7 concern attempts to detect hidden biases using devices such as multiple control groups, multiple referent groups in a case-referent study, or known effects. Chapter 8 concerns coherence in observational studies, a concept that falls somewhere between attempts to detect hidden bias and sensitivity analyses.

Chapter 9 discusses methods and algorithms for matching and stratification for observed covariates. Chapters 3 and 9 both concern the control of overt biases; however, Chapter 3 assumes that matched pairs or sets or strata are exactly homogeneous in the observed covariates. When there are many covariates each taking many values, the exact matching in Chapter 3 is not practical. In contrast, the methods and algorithms in Chapter 9 will often produce matched pairs or sets or strata that balance many covariates simultaneously. Chapter 10 discusses the relationship between the design of an observation study and its intended audience.

2. Dependence Among Chapters and Sections: Numbering of Sections

Chapter 1 is motivation and Chapter 2 is largely review. Chapter 4 depends on Chapter 3. Chapters 5, 6, 7, and 8 depend strongly on Chapter 4 but only weakly on each other. Chapter 9 may be read immediately following Chapter 3. Chapter 10 depends on all the previous chapters but may be read at any time. This is summarized in Figure 1.

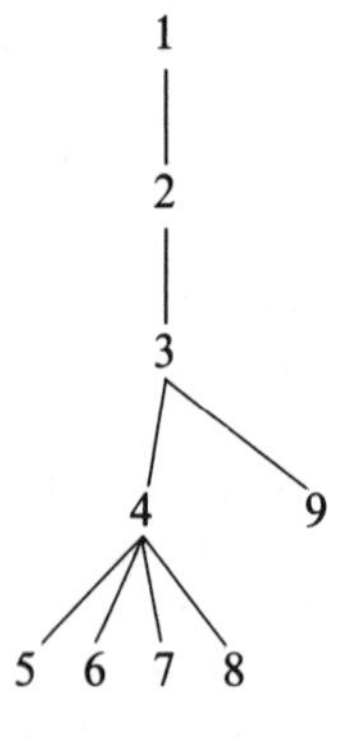

Figure 1. Dependence among chapters.

In the Contents, sections and appendices marked with an asterisk (*) may be skipped or deferred without loss of continuity. These sections discuss material that is either somewhat more technical or specialized than the remainder of the book. An asterisk warns the reader that the next section is in the logical place given its content, but it may not be the best section to read next.

Chapters, sections, and subsections are numbered as follows: Chapter 2 is as is, Section 4 in Chapter 2 is §2.4, and Subsection 3 in Section 4 of Chapter 2 is §2.4.3. Table 3.2.1 is the first table in §3.2; the table happens to be in §3.2.4, but the table number does not indicate the subsection.

There are three paths through the book:

- *The main path.* Read from the beginning to the end, skipping or deferring sections marked with an asterisk. This path is accessible with a knowledge of elementary probability and basic mathematical statistics, and it covers concepts, applications, and theory in reasonable depth.
- *A path around the technical discussion.* Many key ideas may be accessed while avoiding much of the technical discussion. To do this, follow Figure 2. This path requires some knowledge of statistical methods, but no background in mathematical statistics.
- *A quick path to what's new.* To focus on statistical methods and theory that have not previously been discussed in textbooks, glance through Chapter 2 to pick up the notation, and then follow any of the paths in Figure 3.

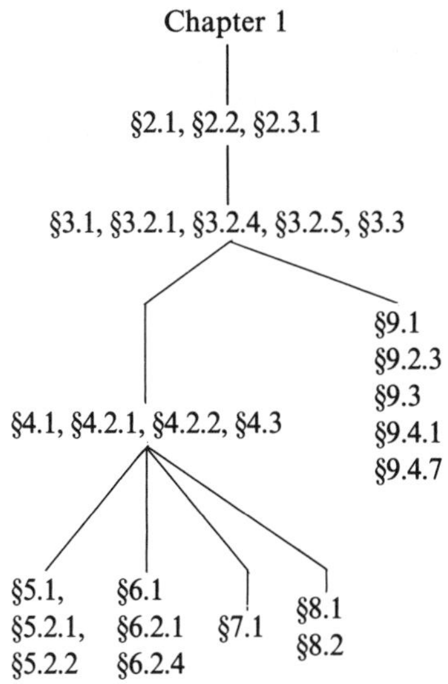

Figure 2. A path to some key ideas avoiding technical material.

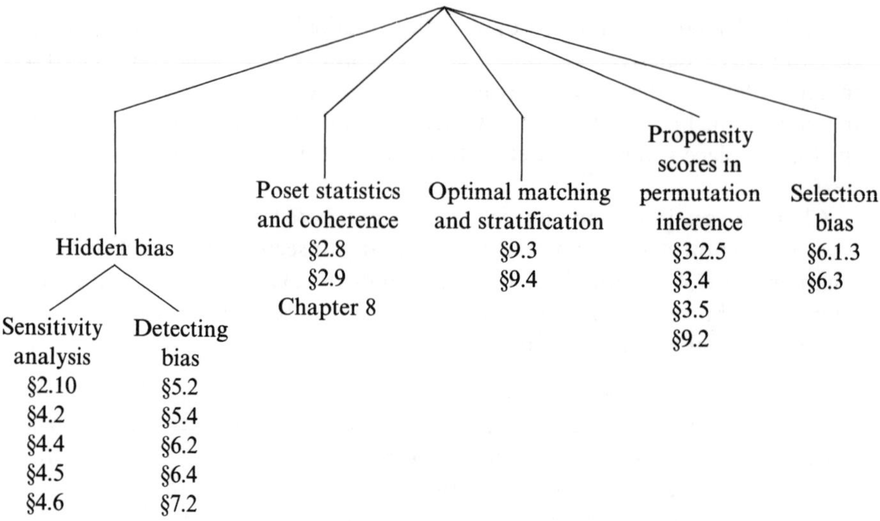

Figure 3. Quick paths to what's new.

3. Acknowledgment

My thanks go to Joe Gastwirth, Abba Krieger, and Sue Marcus for reading drafts of parts of the book. Their ideas and suggestions were a great help to me. Also, I would like to acknowledge three collaborators on journal articles that are discussed in the text, namely, Sam Gu, Abba Krieger, and Don Rubin. A paper with Sam is discussed in §9.4.7. A paper with Abba is discussed in §4.5 and §4.6.2. Two papers with Don are discussed in §9.2. Most of all, I would like to thank my wife, Karen, and my children, Sarah, Hannah, and Aaron.

Contents

* An asterisk indicates either an appendix of a section that has a footnote indicating that the section may be skipped without loss of continuity or deferred. The material in these sections is either a little more technical or a little more specialized or both. See the Preface, §2.

List of Examples

CHAPTER 1

Observational Studies

1.1. What Are Observational Studies?

William G. Cochran first presented "observational studies" as a topic defined by principles and methods of statistics. Cochran had been an author of the 1964 United States Surgeon General's Advisory Committee Report, *Smoking and Health*, which reviewed a vast literature and concluded: "Cigarette smoking is causally related to lung cancer in men; the magnitude of the effect of cigarette smoking far outweighs all other factors. The data for women, though less extensive, point in the same direction (p. 37)." Though there had been some experiments confined to laboratory animals, the direct evidence linking smoking with human health came from observational or nonexperimental studies.

In a later review, Cochran (1965) defined an observational study as an empirical investigation in which:

> ... the objective is to elucidate cause-and-effect relationships ... [in which] it is not feasible to use controlled experimentation, in the sense of being able to impose the procedures or treatments whose effects it is desired to discover, or to assign subjects at random to different procedures.

Features of this definition deserve emphasis. An observational study concerns treatments, interventions, or policies and the effects they cause, and in this respect it resembles an experiment. A study without a treatment is neither an experiment nor an observational study. Most public opinion polls, most forecasting efforts, most studies of fairness and discrimination, and many other important empirical studies are neither experiments nor observational studies.

In an experiment, the assignment of treatments to subjects is controlled by the experimenter, who ensures that subjects receiving different treatments are comparable. In an observational study, this control is absent for one of several reasons. It may be that the treatment, perhaps cigarette smoking or radon gas, is harmful and cannot be given to human subjects for experimental purposes. Or the treatment may be controlled by a political process that, perhaps quite appropriately, will not yield control merely for an experiment, as is true of much of macroeconomic and fiscal policy. Or the treatment may be beyond the legal reach of experimental manipulation even by a government, as is true of many management decisions in a private economy. Or experimental subjects may have such strong attachments to particular treatments that they refuse to cede control to an experimenter, as is sometimes true in areas ranging from diet and exercise to bilingual education. In each case, the investigator does not control the assignment of treatments and cannot ensure that similar subjects receive different treatments.

1.2. Some Observational Studies

It is encouraging to recall cases, such as *Smoking and Health*, in which observational studies established important truths, but an understanding of the key issues in observational studies begins elsewhere. Observational data have often led competent, honest scientists to false and harmful conclusions, as was the case with Vitamin C as a treatment for advanced cancer.

Vitamin C and Treatment of Advanced Cancer: An Observational Study and an Experiment Compared

In 1976, in their article in the *Proceedings of the National Academy of Sciences*, Cameron and Pauling presented observational data concerning the use of vitamin C as a treatment for advanced cancer. They gave vitamin C to 100 patients believed to be terminally ill from advanced cancer and studied subsequent survival.

For each such patient, ten historical controls were selected of the same age and gender, the same site of primary cancer, and the same histological tumor type. This method of selecting controls is called *matched sampling*—it consists of choosing controls one at a time to be similar to individual treated subjects in terms of characteristics measured prior to treatment. Used effectively, matched sampling often creates treated and control groups that are comparable in terms of the variables used in matching, though the groups may still differ in other ways, including ways that were not measured. Cameron and Pauling (1976, p. 3685) write: "Even though no formal process of randomization was carried out in the selection of our two groups, we believe that they come close to representing random subpopulations of the population of terminal cancer patients in the Vale of Leven Hospital." In a moment, we shall see whether this is so.

Patients receiving vitamin C were compared to controls in terms of time from "untreatability by standard therapies" to death. Cameron and Pauling found that, as a group, patients receiving vitamin C survived about four times longer than the controls. The difference was highly significant in a conventional statistical test, p-value < 0.0001, and so could not be attributed to "chance." Cameron and Pauling "conclude that there is strong evidence that treatment ... [with vitamin C] ... increases the survival time."

This study created interest in vitamin C as a treatment. In response, the Mayo Clinic (Moertel et al., 1985) conducted a careful randomized controlled experiment comparing vitamin C to placebo for patients with advanced cancer of the colon and rectum. In a *randomized experiment*, subjects are assigned to treatment or control on the basis of a chance mechanism, typically a random number generator, so it is only luck that determines who receives the treatment. They found no indication that vitamin C prolonged survival, with the placebo group surviving slightly but not significantly longer. Today, few scientists claim that vitamin C holds promise as a treatment for cancer.

What went wrong in Cameron and Pauling's observational study? Why were their findings so different from those of the randomized experiment? Could their mistake have been avoided in any way other than by conducting a true experiment?

Definite answers are not known, and in all likelihood will never be known. Evidently, the controls used in their observational study, though matched on several important variables, nonetheless differed from treated patients in some way that was important to survival.

The obvious difference between the experiment and the observational study was the random assignment of treatments. In the experiment, a single group of patients was divided into a treated and a control group using a random device. Bad luck could, in principle, make the treated and control groups differ in important ways, but it is not difficult to quantify the potential impact of bad luck and to distinguish it from an effect of the treatment. Common statistical tests and confidence intervals do precisely this. In fact, this is what it means to say that the difference could not reasonably be due to "chance." Chapter 2 will discuss the link between statistical inference and random assignment of treatments.

In the observational study, subjects were not assigned to treatment or control by a random device created by an experimenter. The matched sampling ensured that the two groups were comparable in a few important ways, but beyond this, there was little to ensure comparability. If the groups were not comparable before treatment, if they differed in important ways, then the difference in survival might be no more than a reflection of these initial differences.

It is worse than this. In the observational study, the control group was formed from records of patients already dead, while the treated patients were alive at the start of the study. The argument was that the treated patients were terminally in, that they would all be dead shortly, so the recent records of apparently similar patients, now dead, could reasonably be used to indicate the duration of survival absent treatment with vitamin C. Nonetheless, when the results were analyzed, some patients given vitamin C were still alive, that is, their survival times were censored. This might reflect dramatic effects of vitamin C, but it might instead reflect some imprecision in judgments about who is terminally ill and how long a patient is likely to survive, that is, imprecision about the initial prognosis of patients in the treated group. In contrast, in the control group, one can say with total confidence, without reservation or caveat, that the prognosis of a patient already dead is not good. In the experiment, all patients in both treated and control groups were initially alive.

It is worse still. While death is a relatively unambiguous event, the time from "untreatability by standard therapies" to death depends also on the time of "untreatability." In the observational study, treated patients were judged, at the start of treatment with vitamin C, to be untreatable by other therapies. For controls, a date of untreatability was determined from records. It is possible that these two different processes would produce the same number, but it is by no means certain. In contrast, in the experiment, the starting date in treated and control groups was defined in the same way for both groups, simply because the starting date was determined before a subject was assigned to treatment or control.

What do we conclude from the studies of vitamin C? First, observational studies and experiments can yield very different conclusions. When this happens, the experiments tend to be believed. Chapter 2 will develop some of the reasons why this tendency is reasonable. Second, matching and similar adjustments in observational studies, though often useful, do not ensure that treated and control groups are comparable in all relevant ways. More than this, the groups may not be comparable and yet the data we have may fail to reveal this. Methods for addressing this problem are the topic of Chapters 4 through 8. Third, while a controlled experiment uses randomization and an observational study does not, experimental control also helps in other ways. Even if we cannot randomize, we wish to exert as much experimental control as is possible, for instance, using the same eligibility criteria for treated and control groups, and the same methods for determining measurements.

Observational studies are typically conducted when experimentation is not possible. Direct comparisons of experiments and observational studies are less common, vitamin C for cancer being an exception. Another direct comparison occurred in the Salk vaccine for polio, a story that is well told by Meier (1972). Others are discussed by Chalmers, Block, and Lee (1970), LaLonde (1986), Fraker and Maynard (1987), and Zwick (1991).

Smoking and Heart Disease: An Elaborate Theory

Doll and Hill (1966) studied the mortality from heart disease of British doctors with various smoking behaviors. While dramatic associations are typically found between smoking and lung cancer, much weaker associations are found with heart disease. Still, since heart disease is a far more common cause of death, even modest increases in risk involve large numbers of deaths.

The first thing Doll and Hill did was to "adjust for age." The old are at greater risk of heart disease than the young. As a group, smokers tend to be somewhat older than nonsmokers, though of course there are many young smokers and many old nonsmokers. Compare smokers and nonsmokers directly, ignoring age, and you compare a somewhat older group to a somewhat younger group, so you expect a difference in coronary mortality even if smoking has no effect. In its essence, to "adjust for age" is to compare smokers and nonsmokers of the same age. Often results at different ages are combined into a single number called an age-adjusted mortality rate. Methods of adjustment and their properties will be discussed in Chapters 3 and 9. For now, it suffices to say that differences in Doll and Hill's age-adjusted mortality rates cannot be attributed to differences in age, for they were formed by comparing smokers and nonsmokers of the same age. Adjustments of this sort, for age or other variables, are central to the analysis of observational data.

The second thing Doll and Hill did was to consider in detail what should be seen if, in fact, smoking causes coronary disease. Certainly, increased deaths among smokers are expected, but it is possible to be more specific. Light smokers should have mortality somewhere between that of nonsmokers and heavy smokers. People who quit smoking should also have risks between those of nonsmokers and heavy smokers, though it is not clear what to expect when comparing continuing light smokers to people who quit heavy smoking.

Why be specific? Why spell out in advance what a treatment effect should look like? The importance of highly specific theories has a long history, having been advocated in general by Sir Karl Popper (1959) and in observational studies by Sir Ronald Fisher, the inventor of randomized experiments, as quoted by Cochran (1965, §5):

> About 20 years ago, when asked in a meeting what can be done in observational studies to clarify the step from association to causation, Sir Ronald Fisher replied: "Make your theories elaborate." The reply puzzled me at first, since by Occam's razor, the advice usually given is to make theories as simple as is consistent with known data. What Sir Ronald meant, as subsequent discussion showed, was that when constructing a causal hypothesis one should envisage as many different consequences of its truth as possible, and plan observational studies to discover whether each of these consequences is found to hold.

> ... this multi-phasic attack is one of the most potent weapons in observational studies.

Chapters 5 through 8 consider this advice formally and in detail.

Figure 1.2.1 gives Doll and Hill's six age-adjusted mortality rates for death from coronary disease not associated with any other specific disease. The rates are deaths per 1000 per year, so the value 3.79 means about four deaths in each 1000 doctors each year. The six groups are nonsmokers, exsmokers, and current smokers of 1–14 cigarettes, 15–24 cigarettes, and ≥ 25 cigarettes per day. Doll and Hill did not separate exsmokers by the amount they had previously smoked, though this would have been interesting and would have permitted more detailed predictions. Again, differences in age do not affect these mortality rates.

Figure 1.2.1 confirms each expectation. Mortality increases with the quantity smoked. Quitters have lower mortality than heavy smokers but higher mortality than nonsmokers. Any alternative explanation, any claim that smoking is not a cause of coronary mortality, would need to explain the entire pattern in Figure 1.2.1. Alternative explanations are not difficult to imagine, but the pattern in Figure 1.2.1 restricts their number.

DES and Vaginal Cancer: Sensitivity to Bias

Cancer of the vagina is a rare condition, particularly in young women. In 1971, Herbst, Ulfelder, and Poskanzer published a report describing eight cases of vaginal cancer in women aged 15 to 22. They were particularly interested in the possibility that a drug, diethylstilbestrol or DES, given to pregnant women, might be a cause of vaginal cancer in their daughters. Each of the eight cases was matched to four *referents*, that is, to four women who did not develop vaginal cancer. These four referents were born within five days of the birth of the case at the same hospital, and on the same

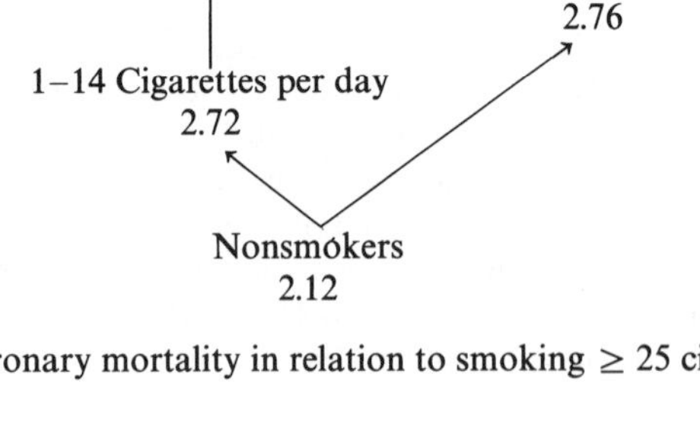

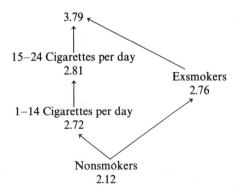

Figure 1.2.1. Coronary mortality in relation to smoking ≥ 25 cigarettes per day.

type of service, ward or private. There were then eight cases of vaginal cancer and 32 referents, and the study compared the use of DES by their mothers.

This sort of study is called a *case-referent study* or a *case-control study* or a *retrospective study*, no one terminology being universally accepted. In an experiment and in many observational studies, treated and control groups are followed forward in time to see how outcomes develop. In the current context, this would mean comparing two groups of women, a treated group whose mothers had received DES and a control group whose mothers had not. That sort of study is not practical because the outcome, vaginal cancer, is so rare—the treated and control groups would have to be enormous and continue for many years to yield eight cases of vaginal cancer. In a case-referent study, the groups compared are not defined by whether or not they received the treatment, but rather by whether or not they exhibit the outcome. The cases are compared to the referents to see if exposure to the treatment is more common among cases. Issues involved in case-referent studies will be examined in §3.3, §4.4.4–§4.4.5, and §6.

In general, the name "case-control" study is not ideal because the word "control" does not have its usual meaning of a person who did not receive the treatment. In fact, in most case-referent studies, many referents did receive the treatment. The name "retrospective" study is not ideal because there are observational studies in which data on entire treated and control groups are collected after treatments have been given and outcomes have appeared, that is collected retrospectively, and yet the groups being compared are still treated and untreated groups. See MacMahon and Pugh (1970, pp. 41–46) for some detailed discussion of this terminology.

So the study compared eight cases of vaginal cancer to 32 matched referents to see if treatment with estrogen diethylstilbestrol was more common among mothers of the cases, and indeed it was. Among the mothers of the eight cases, seven had received DES during pregnancy. Among mothers of the 32 referents, none had received DES. The association between vaginal cancer and DES appears to be almost as strong as a relationship can be, though of course only eight cases have been observed. If a conventional test designed for use in a randomized experiment is used to compare cases and referents in terms of the frequency of exposure to DES, the difference is highly significant. However, experience with the first example, vitamin C and cancer, suggests caution here.

What should be concluded from the strong association observed between DES and vaginal cancer in eight cases and 32 matched referents? Unlike the case of vitamin C and cancer, it would be neither practical nor ethical to follow up with a randomized experiment. Could such a hypothetical experiment produce very different findings? That possibility can never be entirely ruled out. Still, it is possible to ask: How severe would the unseen problems in this study have to be to produce such a strong relationship if DES did not cause vaginal cancer? How far would the observational study have to depart from an experiment to produce such a relationship if DES was harmless? How does the small size of the case group, eight cases, affect these questions? Chapter 4 will provide answers. As it turns out, only severe unseen problems and hidden biases, only dramatic departures from an experiment, could produce such a strong association in the absence of an effect of DES, the small sample size notwithstanding. In other words, this study is highly insensitive to hidden bias; its conclusions could be altered by dramatic biases, but not by small ones. This is by no means true of all observational studies. Chapter 4 concerns general methods for quantifying the sensitivity of findings to hidden biases, and it discusses the uses and limitations of sensitivity analyses.

Academic Achievement in Public and Catholic High Schools: Specific Responses to Specific Criticisms

A current controversy in the United States concerns the effectiveness of public or state-run schools, particularly as compared to existing privately operated schools. The 1985 paper by Hoffer, Greely, and Coleman is one of a series of observational studies of this question. They used data from the High School and Beyond Study (HSB), which includes a survey of US high-school students as sophomores with followup in their senior year. The HSB study provided standardized achievement test scores in several areas in sophomore and senior years, and included followup of students who dropped out of school, so as these things go, it is a rather complete and attractive source of data. Hoffer, Greely, and Coleman (1985) begin with a list of six objections made to their earlier studies, which had compared achievement test scores in public and Catholic schools, concluding that "... Catholic high schools are more effective than public high schools." As an illustration, objection #3 states: "Catholic schools seem to have an effect because they eliminate their disciplinary problems by expelling them from the school." The idea here is that Catholic schools eliminate difficult students while the public schools do not, so the students who remain in Catholic schools would be more likely to perform well even if there was no difference in the effectiveness of the two types of schools.

Criticism is enormously important to observational studies. The quality of the criticism offered in a particular field is intimately connected with the quality of the studies conducted in that field. Quality is not quantity, nor is harshness quality. Cochran (1965, §5) argues that the first critic of an observational study should be its author:

> When summarizing the results of a study that shows an association consistent with the causal hypothesis, the investigator should always list and discuss all alternative explanations of his results (including different hypotheses and biases in the results) that occur to him. This advice may sound trite, but in practice is often neglected.

Criticisms of observational studies are of two kinds, the tangible and the dismissive, objection #3 being of the tangible kind. A tangible criticism is a specific and plausible alternative interpretation of the available data; indeed, a tangible criticism is itself a scientific theory, itself capable of empirical investigation. Bross (1960) writes:

> ... a critic who objects to a bias in the design or a failure to control some established factor is, in fact, raising a counterhypothesis ... [and] has the responsibility for showing [it] is tenable. In doing so, he operates under the same ground rules as the proponent...: When a critic has shown that his counterhypothesis is tenable, his job is done, while at this point the proponent's job is just beginning. A proponent's job is not finished as long as there is a tenable hypothesis that rivals the one he asserts.

On the second page of his *The Design of Experiments*, Fisher described dismissive criticism as he argued that a theory of experimental design is needed:

> This type of criticism is usually made by what I might call a heavyweight *authority*. Prolonged experience, or at least the long possession of a scientific reputation, is almost a pre-requisite for developing successfully this line of attack. Technical details are seldom in evidence. The authoritative assertion: "His *controls* are *totally* inadequate" must have temporarily discredited many a promising line of work; and such an authoritarian method of judgement must surely continue, human nature being what it is, so long as theoretical notions of the principles of experimental design are lacking....

Dismissive criticism rests on the authority of the critic and is so broad and vague that its claims cannot be studied empirically. Judging the weight of evidence is inseparable from judging the criticisms that have been or can be raised.

Concerning objection #3, Hoffer, Greely, and Coleman respond: "... the evidence from the HSB data, although indirect, does not support this objection. Among students who reported that they had been suspended during their sophomore year, those in the Catholic sector were more likely to be in the same school as seniors than those in the public sector (63 percent to 56 percent)." In other words, difficult students, or at any rate students who were suspended, remained in Catholic school more often, not less often, than in public schools. This response to objection #3, though not decisive, does gives one pause.

1.3. Purpose of This Book

Scientific evidence is commonly and properly greeted with objections, skepticism, and doubt. Some objections come from those who simply do not like the conclusions, but setting aside such unscientific reactions, responsible scientists are responsibly skeptical. We look for failures of observation, gaps in reasoning, and alternative interpretations. We compare new evidence with past evidence. This skepticism is itself scrutinized. Skepticism must be justified, defended. One needs "grounds for doubt," in Wittgenstein's (1969, §122) phrase. The grounds for doubt are themselves challenged. Objections bring forth counterobjections and more evidence. As time passes, arguments on one side or the other become strained, fewer scientists are willing to offer them, and the arguments on that side come increasingly from individuals who seem to have some stake in the outcome. In this way, questions are settled.

Scientific questions are not settled on a particular date by a single event, nor are they settled irrevocably. We speak of the weight of evidence. Eventually, the weight is such that critics can no longer lift it, or are too weary to try. Overwhelming evidence is evidence that overwhelms responsible critics.

Experiments are better than observational studies because there are fewer grounds for doubt. The ideal experiment would leave few grounds for doubt, and at times this ideal is nearly achieved, particularly in the laboratory. Experiments often settle questions faster.

Despite this, experiments are not feasible in some settings. At times, observational studies have produced overwhelming evidence, as compelling as any in science, but at other times, observational data have misled investigators to advocate harmful policies or ineffective treatments.

A statistical theory of observational studies is a framework and a set of tools that provide measures of the weight of evidence. The purpose of this book is to give an account of statistical principles and methods for the design and analysis of observational studies. An adequate account must relate observational studies to controlled experiments, showing how uncertainty about treatment effects is greater in the absence of randomization. Analytical adjustments are common in observational studies, and the account should indicate what adjustments can and cannot do. A large literature offers many devices to detect hidden biases in observational studies, for instance, the use of several control groups, and the account must show how such devices work and when they may be expected to succeed or fail. Even when it is not possible to reduce or dispel uncertainty, it is possible to be careful in discussing its magnitude. That is, even when it is not possible to remove bias through adjustment or to detect bias through careful design, it is nonetheless possible to give quantitative expression to the magnitude of uncertainties about bias, a technique called *sensitivity analysis*. The account must indicate what can and cannot be done with a sensitivity analysis.

1.4. Bibliographic Notes

Most scientific fields that study human populations conduct observational studies. Many fields have developed a literature on the design, conduct, and interpretation of observational studies, often with little reference to related work in other fields. It is not possible to do justice to these several literatures in a short bibliographic note. There follows a short and incomplete list of fine books that contain substantial, general discussions of the methodology used for observational studies in epidemiology, public program evaluation, or the social sciences. A shared goal in these diverse works is evaluation of treatments, exposures, programs, or policies from nonexperimental data. The list is followed by references cited in Chapter 1.

Some Books and a Few Papers

Blaug, M. (1980). *The Methodology of Economics.* New York: Cambridge University Press.

Breslow, N. and Day, N. (1980, 1987). *Statistical Methods in Cancer Research*, Volumes 1 and 2. Lyon, France: International Agency for Research on Cancer.

Campbell, D. and Stanley, J. (1963). *Experimental and Quasi-Experimental Design for Research.* Chicago: Rand McNally.

Cochran, W. (1983). *Planning and Analysis of Observational Studies.* New York: Wiley.

Cook, T. and Campbell, D. (1979). *Quasi-Experimentation.* Chicago: Rand McNally.

Cox, D. R. (1992). Causality: some statistical aspects. *Journal of the Royal Statistical Society*, Series A, **155**, 291–301.

Friedman, M. (1953). *Essays in Positive Economics.* Chicago: University of Chicago Press.

Gastwirth, J. (1988). *Statistical Reasoning in Law and Public Policy.* New York: Academic Press.

Greenhouse, S. (1982). Jerome Cornfield's contributions to epidemiology. *Biometrics*, **28**, Supplement, 33–46.

Holland, P. (1986). Statistics and causal inference (with discussion). *Journal of the American Statistical Association*, **81**, 945–970.

Kelsey, J., Thompson. W., and Evans, A. (1986). *Methods in Observational Epidemiology.* New York: Oxford University Press.

Kish, L. (1987). *Statistical Design for Research.* New York: Wiley.

Lilienfeld, A. and Lilienfeld, D. (1980). *Foundations of Epidemiology.* New York: Oxford University Press.

MacMahon, B. and Pugh, T. (1970). *Epidemiology.* Boston: Little, Brown.

Mantel, N. and Haenszel, W. (1959). Statistical aspects of retrospective studies of disease. *Journal of the National Cancer Institute*, **22**, 719–748.

Meyer, M. and Fienberg, S., eds. (1992). *Assessing Evaluation Studies: The Case of Bilingual Education Strategies.* Washington, DC: National Academy Press.

Miettinen, O. (1985) *Theoretical Epidemiology.* New York: Wiley.

Rosenthal, R. and Rosnow, R., eds. (1969). *Artifact in Behavioral Research.* New York: Academic Press.

Rossi, P. and Freeman, H. (1985). *Evaluation.* Beverly Hills, CA: Russell Sage.

Rothman, K. (1986). *Modern Epidemiology.* Boston: Little, Brown.

Rubin, D. (1974). Estimating causal effects of treatments in randomized and nonrandomized studies. *Journal of Educational Psychology*, **66**, 688–701.

Schlesselman, J. (1982). *Case-Control Studies.* New York: Oxford University Press.

Suchman, E. (1967). *Evaluation Research.* New York: Russell Sage.

Susser, M. (1973). *Causal Thinking in the Health Sciences: Concepts and Strategies in Epidemiology.* New York: Oxford University Press.

Tufte, E., ed. (1970). *The Quantitative Analysis of Social Problems.* Reading, MA: Addison-Wesley.

Zellner, A. (1968). *Readings in Economic Statistics and Econometrics.* Boston: Little, Brown.

References

Bross, I.D.J. (1960). Statistical criticism. *Cancer*, **13**, 394–400. Reprinted in: *The Quantitative Analysis of Social Problems* (ed., E. Tufte). Reading, MA: Addison-Wesley, pp. 97–108.

Cameron, E. and Pauling, L. (1976). Supplemental ascorbate in the supportive treatment of cancer: Prolongation of survival times in terminal human cancer. *Proceedings of the National Academy of Sciences (USA)*, **73**, 3685–3689.

Chalmers, T., Block, J., and Lee, S. (1970). Controlled studies in clinical cancer research. *New England Journal of Medicine*, **287**, 75–78.

Cochran, W.G. (1965). The planning of observational studies of human populations (with Discussion). *Journal of the Royal Statistical Society*, Series A, **128**, 134–155. Reprinted in *Readings in Economic Statistics and Econometrics* (ed., A. Zellner, 1968). Boston: Little Brown, pp. 11–36.

Doll, R and Hill, A. (1966). Mortality of British doctors in relation to smoking: Observations on coronary thrombosis. In: *Epidemiological Approaches to the Study of Cancer and Other Chronic Diseases*, (ed., W. Haenszel). U.S. National Cancer Institute Monograph 19. Washington, DC: US Department of Health, Education, and Welfare, pp. 205–268.

Fraker, T. and Maynard, R. (1987). The adequacy of comparison group designs for evaluations of employment-related programs. *Journal of Human Resources*, **22**, 194–227.

Herbst, A., Ulfelder, H., and Poskanzer, D. (1971). Adenocarcinoma of the vagina: Association of maternal stilbestrol therapy with tumor appearance in young women. *New England Journal of Medicine*, **284**, 878–881.

Hoffer, T., Greeley, A., and Coleman, J. (1985). Achievement growth in public and Catholic schools. *Sociology of Education*, **58**, 74–97.

LaLonde, R. (1986). Evaluating the econometric evaluations of training programs with experimental data. *American Economic Review*, **76**, 604–620.

Meier, P. (1972). The biggest public health experiment ever: The 1954 field trial of the Salk poliomyelitis vaccine. In: *Statistics: A Guide to the Unknown* (ed., J. Tanur). San Francisco: Hoden Day, pp. 2–13.

Moertel, C., Fleming, T., Creagan, E., Rubin, J., O'Connell, M., and Ames, M. (1985). High-dose vitamin C versus placebo in the treatment of patients with advanced cancer who have had no prior chemotherapy: A randomized double-blind comparison. *New England Journal of Medicine*, **312**, 137–141.

Popper, K. (1959). *The Logic of Scientific Discovery*. New York: Harper & Row.

United States Surgeon General's Advisory Committee Report, *Smoking and Health*.

Wittgenstein, L. (1969). *On Certainty*. New York: Harper & Row.

Zwick, R. (1991). Effects of item order and context on estimation of NAEP reading proficiency. *Educational Measurement: Issues and Practice*, 10–16.

CHAPTER 2

Randomized Experiments

2.1. Introduction and Example: A Randomized Clinical Trial

Observational studies and controlled experiments have the same goal, inference about treatment effects, but random assignment of treatments is present only in experiments. This chapter reviews the role of randomization in experiments, and so prepares for discussion of observational studies in later chapters. A theory of observational studies must have a clear view of the role of randomization, so it can have an equally clear view of the consequences of its absence. Sections 2.1 and 2.2 give two examples, first a large controlled clinical trial, and then a small but famous example due to Sir Ronald Fisher, who is usually credited with the invention of randomization, which he called the "reasoned basis for inference" in experiments. Later sections discuss the meaning of this phrase, that is, the link between randomization and statistical methods. Most of the material in this chapter is quite old.

Randomized Trial of Coronary Surgery

The US Veterans Administration (Murphy et al., 1977) conducted a randomized controlled experiment comparing coronary artery bypass surgery with medical therapy as treatments for coronary artery disease. Bypass surgery is an attempt to repair the arteries that supply blood to the heart, arteries that have been narrowed by fatty deposits. In bypass surgery, a bypass or bridge is formed around a blockage in a coronary artery. In contrast, medical therapy uses drugs to enhance the flow of blood through nar-

Table 2.1.1. Base-Line Comparison of Coronary Patients in the Veterans Administration Randomized Trial.

Covariate	Medical %	Surgical %
New York Heart Association Class II & III	94.2	95.4
History of myocardial infarction (MI)	59.3	64.0
Definite or possible MI based on electrocardiogram	36.1	40.5
Duration of chest pain > 25 months	50.0	51.8
History of hypertension	30.0	27.6
History of congestive heart failure	8.4	5.2
History of cerebral vascular episode	3.2	2.1
History of diabetes	12.9	12.2
Cardiothoracic ratio > 0.49	10.4	12.2
Serum cholesterol > 249 mg/100 ml	31.6	20.6

rowed arteries. The study involved randomly assigning 596 patients at 13 Veterans Administration hospitals, of whom 286 received surgery and 310 received drug treatments. The random assignment of a treatment for each patient was determined by a central office after the patient had been admitted into the trial.

Table 2.1.1 is taken from their study. It compares the medical and surgical treatment groups in terms of ten important characteristics of patients measured at "base-line," that is, prior to the start of treatment. A variable measured prior to the start of treatment is called a *covariate*. Similar tables appear in reports of most clinical trials.

Table 2.1.1 shows the two groups of patients were similar in many important ways prior to the start of treatment, so that comparable groups were being compared. When the percentages for medical and surgical are compared, the difference is not significant at the 0.05 level for nine of the variables in Table 2.1.1, but is significant for serum cholesterol. This is in line with what one would expect from ten significance tests if the only differences were due to chance, that is, due to the choice of random numbers used in assigning treatments.

For us, Table 2.1.1 is important for two reasons. First, it is an example showing that randomization tends to produce relatively comparable or balanced treatment groups in large experiments. The second point is separate and more important. The ten covariates in Table 2.1.1 were not used in assigning treatments. There was no deliberate balancing of these variables. Rather the balance we see was produced by the random assignment, which made no use of the variables themselves. This gives us some reason to hope and expect that other variables, not measured, are similarly balanced. Indeed, as will be seen shortly, statistical theory supports this expectation. Had the trial not used random assignment, had it instead assigned patients one at a time to balance these ten covariates, then the balance might well have been better than in Table 2.1.1, but there would be no basis for expecting other unmeasured variables to be similarly balanced.

The VA study compared survival in the two groups 3 years after treatment. Survival in the medical group was 87% and in the surgical group 88%, both with a standard error of 2%. The 1% difference in mortality was not significant. Evidently, when comparable groups of patients received medical and surgical treatment at the VA hospitals, the outcomes were quite similar.

The statement that randomization tends to balance covariates is at best imprecise—taken too literally, it is misleading. For instance, in Table 2.1.1, the groups do differ slightly in terms of serum cholesterol. Presumably there are other variables, not measured, exhibiting imbalances similar if not greater than that for serum cholesterol. What is precisely true is that random assignment of treatments can produce some imbalances by chance, but common statistical methods, properly used, suffice to address the uncertainty introduced by these chance imbalances. To this subject, we now turn.

2.2. The Lady Tasting Tea

"A lady declares that by tasting a cup of tea made with milk she can discriminate whether the milk or the tea infusion was first added to the cup," or so begins the second chapter of Sir Ronald Fisher's book *The Design of Experiments*, which introduced the formal properties of randomization. This example is part of the tradition of statistics, and in addition it was well selected by Fisher to illustrate key points. He continues:

> Our experiment consists in mixing eight cups of tea, four in one way and four in the other, and presenting them to the subject for judgement in a random order. The subject has been told in advance of what the test will consist, namely that she will be asked to taste eight cups, that these shall be four of each kind, and that they shall be presented to her in a random order, that is in an order not determined arbitrarily by human choice, but by the actual manipulation of the physical apparatus used in games of chance, cards, dice, roulettes, etc., or more expeditiously, from a published collection of random sampling numbers purporting to give the actual results of such a manipulation. Her task is to divide the 8 cups into two sets of 4, agreeing, if possible, with the treatments received.

Fisher then asks what would be expected if the Lady was "without any faculty of discrimination," that is, if she makes no changes at all in her judgments in response to changes in the order in which tea and milk are added to the cups. To change her judgments is to have some faculty of discrimination, however slight. So suppose for the moment that she cannot discriminate at all, that she gives the same judgments no matter which four cups receive milk first. Then it is only by accident or chance that she correctly identifies the four cups in which milk was added first. Since there are $\binom{8}{4} = 70$ possible divisions of the eight cups into two groups of four, and randomization has ensured that these are equally probable, the chance of this accident is 1/70. In other words, the probability that the random ordering of the cups will yield perfect agreement with the Lady's fixed judgments is 1/70. If the Lady correctly classified the cups, this probability, $0.014 = 1/70$, is the significance level for testing the null hypothesis that she is without the ability to discriminate.

Fisher goes on to describe randomization as the "reasoned basis" for inference and "the physical basis of the validity of the test"; indeed, these phrases appear in section headings and are clearly important to Fisher. He explains:

> We have now to examine the physical conditions of the experimental technique needed to justify the assumption that, if discrimination of the kind under test is absent, the result of the experiment will be wholly governed by the laws of chance....
>
> It is [not sufficient] to insist that "all the cups must be exactly alike" in every respect except that to be tested. For this is a totally impossible requirement in our example, and equally in all other forms of experimentation....
>
> The element in the experimental procedure which contains the essential safeguard is that the two modifications of the test beverage are to be prepared "in random order." This, in fact, is the only point in the experimental procedure in which the laws of chance, which are to be in exclusive control of our frequency distribution, have been explicitly introduced.

Fisher discusses this example for 15 pages, though its formal aspects are elementary and occupy only a part of a paragraph. He is determined to establish that randomization has justified or grounded a particular inference, formed its "reasoned basis," a basis that would be lacking had the same pattern of responses, the same data, been observed in the absence of randomization.

The example serves Fisher's purpose well. The Lady is not a sample from a population of Ladies, and even if one could imagine that she was, there is but one Lady in the experiment and the hypothesis concerns her alone. Her eight judgments are not independent observations, not least because the rules require a split into four and four. Later cups differ from earlier ones, for by cup number five, the Lady has surely tasted one with milk first and one with tea first. There is no way to construe, or perhaps misconstrue, the data from this experiment as a sample from a population, or as a series of independent and identical replicates. And yet, Fisher's inference is justified, because the only probability distribution used in the inference is the one created by the experimenter.

What are the key elements in Fisher's argument? First, experiments do not require, indeed cannot reasonably require, that experimental units be homogeneous, without variability in their responses. Homogenous experimental units are not a realistic description of factory operations, hospital patients, agricultural fields. Second, experiments do not require, indeed, cannot reasonably require, that experimental units be a random sample from a population of units. Random samples of experimental units are not the reality of the industrial laboratory, the clinical trial, or the agricultural experiment. Third, for valid inferences about the effects of a treatment on the units included in an experiment, it is sufficient to require that treatments be allocated at random to experimental units—these units may be both heterogeneous in their responses and not a sample from a population. Fourth, probability enters the experiment only through the random assignment of treatments, a process controlled by the experimenter. A quantity that is not impacted by the random assignment of treatments is a fixed quantity describing the units in the experiment.

The next section repeats Fisher's argument in more general terms.

2.3. Randomized Experiments

2.3.1. Units and Treatment Assignments

There are N units available for experimentation. A unit is an opportunity to apply or withhold the treatment. Often, a unit is a person who will receive either the treatment or the control as determined by the experimenter. However, it may happen that it is not possible to assign a treatment to a single person, so a group of people form a single unit, perhaps all children in a particular classroom or school. On the other hand, a single person may present several opportunities to apply different treatments, in which case each opportunity is a unit. For instance, in §2.2, the one Lady yielded eight units.

The N units are divided into S strata or subclasses on the basis of covariates, that is, on the basis of characteristics measured prior to the assignment of treatments. The stratum to which a unit belongs is not affected by the treatment, since the strata are formed prior to treatment. There are n_s units in stratum s for $s = 1, \ldots, S$, so $N = \sum n_s$.

Write $Z_{si} = 1$ if the ith unit in stratum s receives the treatment and write $Z_{si} = 0$ if this unit receives the control. Write m_s for the number of treated units in stratum s, so $m_s = \sum_{i=1}^{n_s} Z_{si}$, and $0 \le m_s \le n_s$. Finally, write $\mathbf{Z}$ for the N-dimensional column vector containing the Z_{si} for all units in the lexical order, that is,

$$\mathbf{Z} = \begin{bmatrix} Z_{11} \\ Z_{12} \\ \vdots \\ Z_{1,n_1} \\ Z_{21} \\ \vdots \\ Z_{S,n_S} \end{bmatrix} = \begin{bmatrix} \mathbf{Z}_1 \\ \vdots \\ \mathbf{Z}_S \end{bmatrix} \quad \text{where} \quad \mathbf{Z}_s = \begin{bmatrix} Z_{s1} \\ \vdots \\ Z_{s,n_s} \end{bmatrix}. \tag{2.3.1}$$

This notation covers several common situations. If no covariates are used to divide the units, then there is a single stratum containing all units, so $S = 1$. If $n_s = 2$ and $m_s = 1$ for $s = 1, \ldots, S$, then there are S pairs of units matched on the basis of covariates, each pair containing one treated unit and one control. The situation in which $n_s \ge 2$ and $m_s = 1$ for $s = 1, \ldots, S$ is called matching with multiple controls. In this case there are S matched sets, each containing one treated unit and one or more controls.

The case of a single stratum, that is $S = 1$, is sufficiently common and important to justify slight modifications in notation. When there is only a single stratum the subscript s will be dropped, so Z_i will be written in place of Z_{1i}. The same convention will apply to other quantities which have subscripts s and i.

2.3.2. Several Methods of Assigning Treatments at Random

In a *randomized experiment*, the experimenter determines the assignment of treatments to units, that is the value of $\mathbf{Z}$, using a known random mechanism such as a table of random numbers. To say that the mechanism is known is to say that the distribution of the random variable $\mathbf{Z}$ is known because it was created by the experimenter. One requirement is placed on this random mechanism, namely, that, before treatments are assigned, every unit has a nonzero chance of receiving both the treatment and the control, or formally that $0 < \text{prob}(Z_{si} = 1) < 1$ for $s = 1, \ldots, S$ and $i = 1, \ldots, n_s$. Write Ω_0 for the set containing all possible values of $\mathbf{Z}$, that is, all values of $\mathbf{Z}$ which are given nonzero probability by the mechanism.

In practice, many different random mechanisms have been used to determine $\mathbf{Z}$. The simplest assigns treatments independently to different units, taking $\text{prob}(Z_{si} = 1) = \frac{1}{2}$ for all s, i. This method was used in the Veterans Administration experiment on coronary artery surgery in §2.1. In this case, Ω_0 is the set containing 2^N possible values of $\mathbf{Z}$, namely all N-tuples of zeros and ones, and every assignment in Ω_0 has the same probability, that is, $\text{prob}(\mathbf{Z} = \mathbf{z}) = 1/2^N$ for all $\mathbf{z} \in \Omega_0$. The number of elements in a set A will be written $|A|$, so in this case $|\Omega_0| = 2^N$. This mechanism has the peculiar property that there is a nonzero probability that all units will be assigned to the same treatment, though this probability is extremely small when N is moderately large. From a practical point of view, a more important problem with this mechanism arises when S is fairly large compared to N. In this case, the mechanism may give a high probability to the set of treatment assignments in which all units in some stratum receive the same treatment. If the strata were types of patients in a clinical trial, this would mean that all patients of some type received the same treatment. If the strata were schools in an educational experiment, it would mean that all children in some school received the same treatment. Other assignment mechanisms avoid this possibility.

The most commonly used assignment mechanism fixes the number, m_s, of treated subjects in stratum s. In other words, the only assignments $\mathbf{Z}$ with nonzero probability are those with m_s treated subjects in stratum s for $s = 1, \dots, S$. If m_s is chosen sensibly, this avoids the problem mentioned in the previous paragraph. For instance, if n_s is required to be even and m_s is required to equal $n_s/2$ for each s, then half the units in each stratum receive the treatment and half receive the control, so the final treated and control groups are exactly balanced in the sense that they contain the same number of units from each stratum.

When m_s is fixed in this way, let Ω be the set containing the $K = \prod_{s=1}^{S} \binom{n_s}{m_s}$ possible treatment assignments $\mathbf{Z} = \begin{bmatrix} \mathbf{Z}_1 \\ \vdots \\ \mathbf{Z}_S \end{bmatrix}$ in which $\mathbf{Z}_s$ is an n_s-tuple with m_s ones and $n_s - m_s$ zeros for $s = 1, \dots, S$. In the most common assignment mechanism, each of these K possible assignments is given the same probability, $\text{prob}(\mathbf{Z} = \mathbf{z}) = 1/K$ all $\mathbf{z} \in \Omega$. This type of randomized experiment, with equal probabilities and fixed m_s, will be called a *uniform randomized experiment*. When there is but a single stratum, $S = 1$, it has traditionally been called a *completely randomized experiment*, but when there are two or more strata, $S \geq 2$, it has been called a *randomized block experiment.* If the strata each contain two units, $n_s = 2$, and one receives the treatment, $m_s = 1$, then it has been called a *paired randomized experiment.*

Figure 2.3.1 shows a small illustration, a uniform randomized experiment with two strata, four units in the first stratum and two in the second. Half of the units in each stratum receive the treatment. There are $K = 12$ possible treatment assignments contained in the set Ω, and each has probability 1/12.

The following proposition is often useful. It says that in a uniform randomized experiment, the assignments in different strata are independent of each other. For the elementary proof, see the Problems in §2.12.1.

Proposition 2.1. *In a uniform randomized experiment, the* $\mathbf{Z}_1, \dots, \mathbf{Z}_S$ *are mutually independent, and* $\text{prob}(\mathbf{Z}_s = \mathbf{z}_s) = 1\Big/\binom{n_s}{m_s}$ *for each* n_s*-tuple* $\mathbf{z}_s$ *containing* m_s *ones and* $n_s - m_s$ *zeros.*

$$\begin{bmatrix} Z_{11} \\ Z_{12} \\ Z_{13} \\ Z_{14} \\ Z_{21} \\ Z_{22} \end{bmatrix}$$

$$S = 2, \quad n_1 = 4, \quad m_1 = 2, \quad n_2 = 2, \quad m_2 = 1.$$

$$K = \prod \binom{n_s}{m_s} = \binom{4}{2}\binom{2}{1} = 12.$$

$$\Omega = \left\{ \begin{matrix} \begin{bmatrix} 1 \\ 1 \\ 0 \\ 0 \\ 1 \\ 0 \end{bmatrix}, \begin{bmatrix} 1 \\ 0 \\ 1 \\ 0 \\ 1 \\ 0 \end{bmatrix}, \begin{bmatrix} 1 \\ 0 \\ 0 \\ 1 \\ 1 \\ 0 \end{bmatrix}, \begin{bmatrix} 0 \\ 1 \\ 1 \\ 0 \\ 1 \\ 0 \end{bmatrix}, \begin{bmatrix} 0 \\ 1 \\ 0 \\ 1 \\ 1 \\ 0 \end{bmatrix}, \begin{bmatrix} 0 \\ 0 \\ 1 \\ 1 \\ 1 \\ 0 \end{bmatrix}, \\ \begin{bmatrix} 1 \\ 1 \\ 0 \\ 0 \\ 0 \\ 1 \end{bmatrix}, \begin{bmatrix} 1 \\ 0 \\ 1 \\ 0 \\ 0 \\ 1 \end{bmatrix}, \begin{bmatrix} 1 \\ 0 \\ 0 \\ 1 \\ 0 \\ 1 \end{bmatrix}, \begin{bmatrix} 0 \\ 1 \\ 1 \\ 0 \\ 0 \\ 1 \end{bmatrix}, \begin{bmatrix} 0 \\ 1 \\ 0 \\ 1 \\ 0 \\ 1 \end{bmatrix}, \begin{bmatrix} 0 \\ 0 \\ 1 \\ 1 \\ 0 \\ 1 \end{bmatrix} \end{matrix} \right\}$$

Figure 2.3.1. Treatment assignments for a small hypothetical experiment.

The uniform randomized designs are by far the most common randomized experiments involving two treatments, but others are also used, particularly in clinical trials. It is useful to mention one of these methods of randomization to underscore the point that randomized experiments need not give every treatment assignment $\mathbf{z} \in \Omega_0$ the same probability. A distinguishing feature of many clinical trials is that the units are patients who arrive for treatment over a period of months or years. As a result, the number n_s of people who will fall in stratum s will not be known at the start of the experiment, so a randomized block experiment is not possible. Efron (1971) proposed the following method. Fix a probability p with $\frac{1}{2} < p < 1$. When the ith patient belonging to stratum s first arrives, calculate a current measure of imbalance in stratum s, IMBAL_{si}, defined to be the number of patients so far assigned to treatment in this stratum minus the number so far assigned to control. It is easy to check that $\text{IMBAL}_{si} = \sum_{j=1}^{i-1} (2Z_{sj} - 1)$. If $\text{IMBAL}_{si} = 0$, assign the new patient to treatment or control each with probability $\frac{1}{2}$. If $\text{IMBAL}_{si} < 0$, so there are too few treated patients in this stratum, then assign the new patient to treatment with probability p and to control with probability $1 - p$. If $\text{IMBAL}_{si} > 0$, so there are too many treated patients, then assign the new patient to treatment with probability $1 - p$ and to control with probability p. Efron examines various aspects of this method. In particular, he shows that it much better than independent assignment in producing balanced treated and control groups, that is, treated and control groups with similar numbers of patients from each stratum. He also examines potential biases due to the experimenter's knowledge of IMBAL_{si}. Zelen (1974) surveys a number of related methods with similar objectives.

2.4. Testing the Hypothesis of No Treatment Effect

2.4.1. The Distribution of a Test Statistic When the Treatment Is Without Effect

In the theory of experimental design, a special place is given to the test of the hypothesis that the treatment is entirely without effect. The reason is that, in a randomized experiment, this test may be performed virtually without assumptions of any kind, that is, relying just on the random assignment of treatments. Fisher discussed the Lady and her tea with such care to demonstrate this. Other activities, such as estimating treatment effects or building confidence intervals, do require some assumptions, often innocuous assumptions, but assumptions nonetheless. The contribution of randomization to formal inference is clearest when expressed in terms of the test of no effect. Does this mean that such tests are of greater practical importance than point or interval estimates? Certainly not. It is simply that the theory of such tests is less cluttered, and so it sets randomized and nonrandomized studies in sharper contrast. The important point is that, in the absence of difficulties such as noncompliance or loss to follow-up, assumptions play a minor role in randomized experiments, and no role at all in randomization tests of the hypothesis of no effect. In contrast, inference in a nonrandomized experiment requires assumptions that are not at all innocuous. So let us follow Fisher and develop this point with care.

Each unit exhibits a response which is observed some time after treatment. To say that the treatment has no effect on this response is to say that each unit would exhibit the same value of the response whether assigned to treatment or control. This is the definition of "no effect." If changing the treatment assigned to a unit changed that unit's response, then certainly the treatment has at least some effect.

In the traditional development of randomization inference, chance and probability enter only through the random assignment of treatments, that is, through the known mechanism that selects the treatment assignment $\mathbf{Z}$ from Ω. The only random quantities are $\mathbf{Z}$ and quantities that depend on $\mathbf{Z}$. When the treatment is without effect, the response of a unit is fixed, in the sense that this response would not change if a different treatment assignment $\mathbf{Z}$ were selected from Ω. Again, this is simply what it means for a treatment to be without effect. When testing the null hypothesis of no effect, the response of the ith unit in stratum s is written r_{si} and the N-tuple of responses for all N units is written $\mathbf{r}$. The lowercase notation for r_{si} emphasizes that, under the null hypothesis, r_{si} is a fixed quantity and not a random variable. Later on, when discussing treatments with effects, a different notation is needed.

A test statistic $t(\mathbf{Z}, \mathbf{r})$ is a quantity computed from the treatment assignment $\mathbf{Z}$ and the response $\mathbf{r}$. For instance, the treated-minus-control difference in sample means is the test statistic

$$t(\mathbf{Z}, \mathbf{r}) = \frac{\mathbf{Z}^T\mathbf{r}}{\mathbf{Z}^T\mathbf{1}} - \frac{(\mathbf{1} - \mathbf{Z})^T\mathbf{r}}{(\mathbf{1} - \mathbf{Z})^T\mathbf{1}}, \tag{2.4.1}$$

where $\mathbf{1}$ is an N-tuple of 1's. Other statistics will be discussed shortly.

Given any test statistic $t(\mathbf{Z}, \mathbf{r})$, the task is to compute a significance level for a test which rejects the null hypothesis of no treatment effect when $t(\mathbf{Z}, \mathbf{r})$ is large. More precisely:

(i) The null hypotheses of no effect is tentatively assumed to hold, so $\mathbf{r}$ is fixed.
(ii) A treatment assignment $\mathbf{Z}$ has been selected from Ω using a known random mechanism.
(iii) The observed value, say T, of the test statistic $t(\mathbf{Z}, \mathbf{r})$ has been calculated.
(iv) We seek the probability of a value of the test statistic as large or larger than that observed if the null hypothesis were true.

The significance level is simply the sum of the randomization probabilities of assignments $\mathbf{z} \in \Omega$ that lead to values of $t(\mathbf{z}, \mathbf{r})$ greater than or equal to the observed value T, namely,

$$\text{prob}\{t(\mathbf{Z}, \mathbf{r}) \geq T\} = \sum_{\mathbf{z} \in \Omega} [t(\mathbf{z}, \mathbf{r}) \geq T] \cdot \text{prob}(\mathbf{Z} = \mathbf{z}), \tag{2.4.2}$$

$$\text{where} \quad [\text{event}] = \begin{cases} 1 & \text{if event occurs,} \\ 0 & \text{otherwise,} \end{cases} \tag{2.4.3}$$

and $\text{prob}(\mathbf{Z} = \mathbf{z})$ is determined by the known random mechanism that assigned treatments. This is a direct calculation, though not always a straightforward one when Ω is extremely large.

In the case of a uniform randomized experiment, there is a simpler expression for the significance level (2.4.2) since $\text{prob}(\mathbf{Z} = \mathbf{z}) = 1/K = 1/|\Omega|$. It is the proportion of treatment assignments $\mathbf{z} \in \Omega$ giving values of the test statistic $t(\mathbf{z}, \mathbf{r})$ greater than or equal to T, namely,

$$\text{prob}\{t(\mathbf{Z}, \mathbf{r}) \geq T\} = \frac{|\{\mathbf{z} \in \Omega: t(\mathbf{z}, \mathbf{r}) \geq T\}|}{|\Omega|} = \frac{|\{\mathbf{z} \in \Omega: t(\mathbf{z}, \mathbf{r}) \geq T\}|}{K}. \tag{2.4.4}$$

To illustrate, consider again Fisher's example of the Lady who tastes $N = 8$ cups of tea, all in a single stratum, so $S = 1$. A treatment assignment is an 8-tuple containing four 1's and four 0's. For instance, the assignment

$\mathbf{Z} = (1, 0, 0, 1, 1, 0, 0, 1)^T$ would signify that cups 1, 4, 5, and 8 had milk added first and the other cups had tea added first. The set of treatment assignments Ω contains all possible 8-tuples containing four 1's and four 0's, so Ω contains $|\Omega| = K = \binom{8}{4} = 70$ such 8-tuples. The actual assignment was selected at random in the sense that $\text{prob}(\mathbf{Z} = \mathbf{z}) = 1/K = 1/70$ for all $\mathbf{z} \in \Omega$. Notice that $\mathbf{z}^T\mathbf{1} = 4$ for all $\mathbf{z} \in \Omega$.

The Lady's response for cup i is either $r_i = 1$ signifying that she classifies this cup as milk first or $r_i = 0$ signifying that she classifies it as tea first. Then $\mathbf{r} = (r_1, \ldots, r_8)^T$. Recall that she must classify exactly four cups as milk first, so $\mathbf{1}^T\mathbf{r} = 4$. The test statistic is the number of cups correctly identified, and this is written formally as $t(\mathbf{Z}, \mathbf{r}) = \mathbf{Z}^T\mathbf{r} + (\mathbf{1} - \mathbf{Z})^T(\mathbf{1} - \mathbf{r}) = 2\mathbf{Z}^T\mathbf{r}$, where the second equality follows from $\mathbf{1}^T\mathbf{1} = 8$, $\mathbf{Z}^T\mathbf{1} = 4$, and $\mathbf{1}^T\mathbf{r} = 4$. To make this illustration concrete, suppose that $\mathbf{r} = (1, 1, 0, 0, 0, 1, 1, 0)$, so the Lady classifies the first, second, sixth, and seventh cups as milk first. To say that the treatment has no effect is to say that she would give this classification no matter how milk was added to the cups, that is no matter how treatments were assigned to cups. If changing the cups to which milk is added first changes her responses, then she is discerning something, and the treatment has some effect, however slight or erratic.

There is only one treatment assignment $\mathbf{z} \in \Omega$ leading to perfect agreement with the Lady's responses, namely, $\mathbf{z} = (1, 1, 0, 0, 0, 1, 1, 0)$, so if $t(\mathbf{Z}, \mathbf{r}) = 8$ the significance level (2.4.4) is $\text{prob}\{t(\mathbf{Z}, \mathbf{r}) \geq 8\} = 1/70$. This says that the chance of perfect agreement by accident is $1/70 = 0.014$, a small chance. In other words, if the treatment is without effect, the chance that a random assignment of treatments will just happen to produce perfect agreement is 1/70.

It is not possible to have seven agreements since to err once is to err twice. How many assignments $\mathbf{z} \in \Omega$ lead to exactly $t(\mathbf{Z}, \mathbf{r})$ = six agreements? One such assignment with six agreements is $\mathbf{z} = (1, 0, 1, 0, 0, 1, 1, 0)$. Starting with perfect agreement, $\mathbf{z} = (1, 1, 0, 0, 0, 1, 1, 0)$, any one of the four 1's may be made a 0 and any of the four 0's may be made a 1, so there are $16 = 4 \times 4$ assignments with exactly $t(\mathbf{Z}, \mathbf{r}) = 6$ agreements. Hence, there are 17 assignments leading to six or more agreements, so six agreements yields the significance level (2.4.4) equal to $\text{prob}\{t(\mathbf{Z}, \mathbf{r}) \geq 6\} = 17/70 = 0.24$, no longer a small probability. It would not be surprising to see six or more agreements if the treatment were without effect—it happens by chance as frequently as seeing two heads when flipping two coins.

The key point deserves repeating. Probability enters the calculation only through the random assignment of treatments. The needed probability distribution is known, not assumed. The resulting significance level does not depend upon assumptions of any kind. If the same calculation were performed in a nonrandomized study, it would require an assumption that the distribution of treatment assignments, $\text{prob}(\mathbf{Z} = \mathbf{z})$, is some particular distribution, perhaps the assumption that all assignments are equally probable,

$\text{prob}(\mathbf{Z} = \mathbf{z}) = 1/K$. In a nonrandomized study, there may be little basis on which to ground or defend this assumption, it may be wrong, and it will certainly be open to responsible challenge and debate. In other words, the importance of the argument just considered is that it is one way of formally expressing the claim that randomized experiments are not open to certain challenges that can legitimately be made to nonrandomized studies.

2.4.2. Some Common Randomization Tests

Many commonly used tests are randomization tests in that their significance levels can be calculated using (2.4.4), though the tests are sometimes derived in other ways as well. This section briefly recalls and reviews some of these tests. The purpose of the section is to provide a reference for these methods in a common terminology so they may be discussed and used at later stages. Though invented at different times, it is natural to see the methods as members of a few classes whose properties are similar, and this is done beginning in §2.4.3. In most cases, the methods described have various optimality properties which are not discussed here; see Cox (1970) for the optimality properties of the procedures for binary outcomes and Lehmann (1975) for optimality properties of the nonparametric procedures. In all cases, the experiment is the uniform randomized experiment in §2.3.2 with $\text{prob}(\mathbf{Z} = \mathbf{z}) = 1/K$ for all $\mathbf{z} \in \Omega$.

Fisher's exact test for a 2×2 contingency table is, in fact, the test just used for the example of the Lady and her tea. Here, there is one stratum, $S = 1$, the outcome r_i is *binary*, that is, $r_i = 1$ or $r_i = 0$, and the test statistic is the number of responses equal to 1 in the treated group, that is, $t(\mathbf{Z}, \mathbf{r}) = \mathbf{Z}^T\mathbf{r}$. The 2×2 contingency table records the values of Z_i and r_i, as shown in Table 2.4.1 for Fisher's example. Notice that the marginal totals in this table are fixed since $N = 8$ cups, $\mathbf{1}^T\mathbf{r} = 4$ and $\mathbf{1}^T\mathbf{Z} = 4$ are fixed in this experiment. Under the hypothesis of no effect, the randomization distribution of the test statistic $\mathbf{Z}^T\mathbf{r}$ is the hypergeometric distribution. The usual *chi-square test* for a 2×2 table is an approximation to the randomization significance level when N is large.

Table 2.4.1. The 2×2 Table for Fisher's Exact Test for the Lady Tasting Tea.

		Response, r_i		
		1	0	Total
Treatment or control, Z_i	1	$\mathbf{Z}^T\mathbf{r}$	$4 - \mathbf{Z}^T\mathbf{r}$	4
	0	$4 - \mathbf{Z}^T\mathbf{r}$	$\mathbf{Z}^T\mathbf{r}$	4
	Total	4	4	8

The *Mantel–Haenszel* (1959) *statistic* is the analog of Fisher's exact test when there are two or more strata, $S \geq 2$, and the outcome r_{si} is binary. It is extensively used in epidemiology and certain other fields. The data may be recorded in a $2 \times 2 \times S$ contingency table giving treatment Z by outcome r by stratum s. The test statistic is again the number of 1 responses among treated units, $t(\mathbf{Z}, \mathbf{r}) = \mathbf{Z}^T\mathbf{r} = \sum_{s=1}^{S} \sum_{i=1}^{n_s} Z_{si} r_{si}$. Under the null hypothesis, the contribution from stratum s, that is, $\sum_{i=1}^{n_s} Z_{si} r_{si}$, again has a hypergeometric distribution, and (2.4.4) is the distribution of the sum of S independent hypergeometric variables. The Mantel–Haenszel statistic yields an approximation to the distribution of $\mathbf{Z}^T\mathbf{r}$ based on its expectation and variance, as described in more general terms in the next section. One technical attraction of this statistic is that the large sample approximation tends to work well for a $2 \times 2 \times S$ table with large N even if S is also large, so there may be few subjects in each of the S tables. In particular, the statistic is widely used in matching with multiple controls, in which case $m_s = 1$ for each s.

McNemar's (1947) *test* is for paired binary data, that is, for S pairs with $n_s = 2$, $m_s = 1$, and $r_{si} = 1$ or $r_{si} = 0$. The statistic is, yet again, the number of 1 responses among treated units, that is, $t(\mathbf{Z}, \mathbf{r}) = \mathbf{Z}^T\mathbf{r}$. McNemar's statistic is, in fact, a special case of the Mantel–Haenszel statistic, though the $2 \times 2 \times S$ table now describes S pairs and certain simplifications are possible. In particular, the distribution of $\mathbf{Z}^T\mathbf{r}$ in (2.4.4) is that of a constant plus a certain binomial random variable.

Developing these methods for $2 \times 2 \times S$ tables in a different way, Birch (1964) and Cox (1966, 1970) show that these three tests with binary responses possess an optimality property, so there is a sense in which Fisher's exact test, the Mantel–Haenszel test, and McNemar's test are the best tests for the problems they address. Specifically, they show that the test statistic $t(\mathbf{Z}, \mathbf{r}) = \mathbf{Z}^T\mathbf{r}$ together with the significance level (2.4.4) is a uniformly most powerful unbiased test against alternatives defined in terms of constant odds ratios.

Mantel's (1963) *extension* of the Mantel–Haenszel test is for responses r_{si} that are confined to a small number of values representing a numerical scoring of several ordered categories. As an example of such an outcome, the New York Heart Association classifies coronary patients into one of four categories based on the degree to which the patient is limited in physical activity by coronary symptoms such as chest pain. The categories are:

(1) no limitation of physical activity;
(2) slight limitation, comfortable at rest, but ordinary physical activity results in pain or other symptoms;
(3) marked limitation, minor activities results in coronary symptoms; and
(4) unable to carry on any physical activity without discomfort, which may be present even at rest.

The outcome r_{si} for a patient is then one of the integers 1, 2, 3, or 4. In this case the data might be recorded as a $2 \times 4 \times S$ contingency table for $Z \times r \times s$. Mantel's test statistic is the sum of the response scores for treated units, that is, $t(\mathbf{Z}, \mathbf{r}) = \mathbf{Z}^T\mathbf{r}$. Birch (1965) shows that the test is optimal in a certain sense.

In the case of a single stratum, $S = 1$, *Wilcoxon's* (1945) *rank sum test* is commonly used to compare outcomes taking many numerical values. In this test, the responses are ranked from smallest to largest. If all N responses were different numbers, the ranks would be the numbers $1, 2, \ldots, N$. If some of the responses were equal, then the average of their ranks would be used. Write q_i for the rank of r_i, and write $\mathbf{q} = (q_1, \ldots, q_N)^T$. For instance, if $N = 4$, and $r_1 = 2.3$, $r_2 = 1.1$, $r_3 = 2.3$, and $r_4 = 7.9$, then $q_1 = 2.5$, $q_2 = 1$, $q_3 = 2.5$, and $q_4 = 4$, since r_2 is smallest, r_4 is largest, and r_1 and r_3 share the ranks 2 and 3 whose average rank is $2.5 = (2 + 3)/2$. Note that the ranks $\mathbf{q}$ are a function of the responses $\mathbf{r}$ which are fixed if the treatment has no effect, so $\mathbf{q}$ is also fixed. The rank sum statistic is the sum of the ranks of the treated observations, $t(\mathbf{Z}, \mathbf{r}) = \mathbf{Z}^T\mathbf{q}$, and its significance level is determined from (2.4.4). The properties of the rank sum test have been extensively studied; for instance, see Lehmann (1975, §1) or Hettmansperger (1984, §3). Wilcoxon's rank sum test is equivalent to the *Mann and Whitney* (1947) *test*.

In the case of S matched pairs with $n_s = 2$ and $m_s = 1$ for $s = 1, \ldots, S$, Wilcoxon's (1945) signed rank test is commonly used for responses taking many values. Here, $(Z_{s1}, Z_{s2}) = (1, 0)$ if the first unit in pair s received the treatment or $(Z_{s1}, Z_{s2}) = (0, 1)$ if the second unit received the treatment. In this test, the absolute differences in responses within pairs $|r_{s1} - r_{s2}|$ are ranked from 1 to S, with average ranks used for ties. Let d_s be the rank of $|r_{s1} - r_{s2}|$ thus obtained. The signed rank statistic is the sum of the ranks for pairs in which the treated unit had a higher response than the control unit. To write this formally, let $c_{s1} = 1$ if $r_{s1} > r_{s2}$ and $c_{s1} = 0$ otherwise, and similarly, let $c_{s2} = 1$ if $r_{s2} > r_{s1}$ and $c_{s2} = 0$ otherwise, so $c_{s1} = c_{s2} = 0$ if $r_{r1} = r_{s2}$. Then $Z_{s1}c_{s1} + Z_{s2}c_{s2}$ equals 1 if the treated unit in pair s had a higher response than the control unit, and equals zero otherwise. It follows that the signed rank statistic is $\sum_{s=1}^{S} d_s \sum_{i=1}^{2} c_{si} Z_{si}$. Note that d_s and c_{si} are functions of $\mathbf{r}$ and so are fixed under the null hypothesis of no treatment effect. Also, if $r_{s1} = r_{s2}$, then pair s contributes zero to the value of the statistic no matter how treatments are assigned. As with the rank sum test, the signed rank test is widely used and has been extensively studied; for instance, see Lehmann (1975, §3) or Hettmansperger (1984, §2). Section 3.2.4 below contains a numerical example using the sign-rank statistic in an observational study.

For stratified responses, a method that is sometimes used involves calculating the rank sum statistic separately in each of the S strata, and taking the sum of these S rank sums as the test statistic. This is the *stratified rank sum statistic*. It is easily checked that this statistic has the form $t(\mathbf{Z}, \mathbf{r}) = \mathbf{Z}^T\mathbf{q}$ resembling the rank sum statistic; however, the ranks in $\mathbf{q}$ are no longer a permutation of the numbers $1, 2, \ldots, N$, but rather of the numbers $1, \ldots,$

$n_1, 1, \ldots, n_2, \ldots, 1, \ldots, n_S$, with adjustments for ties if needed. Also Ω has changed.

Hodges and Lehmann (1962) find the stratified rank sum statistic to be inefficient when S is large compared to N. In particular, for paired data with $S = N/2$, the stratified rank test is equivalent to the sign test, which in turn is substantially less efficient than the signed rank test for data from short tailed distributions such as the Normal. They suggest as an alternative the method of aligned ranks: the mean in each stratum is subtracted from the responses in that stratum creating aligned responses that are ranked from 1 to N, momentarily ignoring the strata. Writing $\mathbf{q}$ for these aligned ranks, the *aligned rank statistic* is the sum of the aligned ranks in the treated group, $t(\mathbf{Z}, \mathbf{r}) = \mathbf{Z}^{\mathrm{T}}\mathbf{q}$. See also Lehmann (1975, §3.3).

Another statistic is the median test. Let $c_{si} = 1$ if r_{si} is greater than the median of the responses in stratum s and let $c_{si} = 0$ otherwise, and let $\mathbf{c}$ be the N-tuple containing the c_{si}. Then $t(\mathbf{Z}, \mathbf{r}) = \mathbf{Z}^{\mathrm{T}}\mathbf{c}$ is the number of treated responses that exceed their stratum medians. With a single stratum, $S = 1$, the median test is quite good in large samples if the responses have a double exponential distribution, a distribution with a thicker tail than the normal; see, for instance, Hettmansperger (1984, §3.4, p. 146). In this test, the median is sometimes replaced by other quantiles or other measures of location.

Start with any statistic $t(\mathbf{Z}, \mathbf{r})$ and the randomization distribution of $t(\mathbf{Z}, \mathbf{r})$ may be determined from (2.4.4). This is true even of statistics that are commonly referred to a theoretical distribution instead, for instance, the *two-sample or paired t-tests*, among others. Welch (1937) and Wilk (1955) studied the relationship between the randomization distribution and the theoretical distribution of statistics that were initially derived from assumptions of Normally and independently distributed responses. They suggest that the theoretical distribution may be viewed as a computationally convenient approximation to the desired but computationally difficult randomization distribution. That is, they suggest that t-tests, like rank tests or Mantel–Haenszel tests, may be justified solely on the basis of the use of randomization in the design of an experiment, without reference to Normal, independent errors. These findings depend on the good behavior of moments of sums of squares of responses over the randomization distribution; therefore, they depend on the absence of extreme responses. Still, the results are important as a conceptual link between Normal theory and randomization inference.

2.4.3. Classes of Test Statistics

The similarity among the commonly used test statistics in §2.4.2 is striking but not accidental. In this book, these statistics will not be discussed individually, except when used in examples. The important properties of the methods are shared by large classes of statistics, so it is both simpler and less repetitive to discuss the classes.

Though invented by different people at different times for different purposes, the commonly used statistics in §2.4.2 are similar for good reason. As the sample size N increases, the number K of treatment assignments in Ω grows very rapidly, and the direct calculation in (2.4.4) becomes very difficult to perform, even with the fastest computers. To see why this is true, take the simplest case consisting of one stratum, $S = 1$, and an equal division of the n subjects into $m = n/2$ treated subjects and $m = n/2$ controls. Then there are $K = \binom{n}{n/2}$ treatment assignments in Ω. If one more unit is added to the experiment, increasing the sample size to $n + 1$, then K is increased by a factor of $\frac{n+1}{(n/2)+1}$, that is, K nearly doubles. Roughly speaking, if the fastest computer can calculate (2.4.4) directly for at most a sample of size n, and if computing power doubles every year for ten years, then ten years hence computing power will be $2^{10} = 1024$ times greater than today and it will be possible to handle a sample of size $n + 10$. Direct calculation of (2.4.4) is not practical for large n.

The usual solution to this problem is to approximate (2.4.4) using a large sample or asymptotic approximation. The most common approximations use the moments of the test statistic, its expectation and variance, and sometimes higher moments. The needed moments are easily derived for certain classes of statistics, including all those in §2.4.2.

As an alternative to asymptotic approximation, there are several proposals for computing (2.4.4) exactly, but they are not, as yet, commonly used. One is to compute (2.4.4) exactly but indirectly using clever computations that avoid working with the set Ω. For some statistics this can be done by calculating the characteristic function of the test statistic and inverting it using the fast Fourier transform; see Pagano and Tritchler (1983). A second approach is to design experiments differently so that Ω is a much smaller set, perhaps containing 10,000 or 100,000 treatment assignments. In this case, direct calculation is possible and any test statistic may be used; see Tukey (1985) for discussion.

The first class of statistics will be called *sum statistics* and they are of the form $t(\mathbf{Z}, \mathbf{r}) = \mathbf{Z}^{\mathrm{T}}\mathbf{q}$ where $\mathbf{q}$ is some function of $\mathbf{r}$. A sum statistic sums the scores q_{si} for treated units. All of the statistics in §2.4.3 are sum statistics for suitable choices of $\mathbf{q}$. In Fisher's exact test, the Mantel–Haenszel test and McNemar's test, $\mathbf{q}$ is simply equal to $\mathbf{r}$. In the rank sum test, $\mathbf{q}$ contains the ranks of $\mathbf{r}$. In the median test, $\mathbf{q}$ is the vector of ones and zeros identifying responses r_{si} that exceed stratum medians. In the signed rank statistic, $q_{si} = d_s c_{si}$.

Simple formulas exist for the moments of sum statistics under the null hypothesis that the treatment is without effect. In this case, $\mathbf{r}$ is fixed, so $\mathbf{q}$ is also fixed. The moment formulas use the properties of simple random sampling without replacement. Recall that a simple random sample without replacement of size m from a population of size n is a random subset of m

elements from a set with n elements where each of the $\binom{n}{m}$ subsets of size m has the same probability $1\Big/\binom{n}{m}$. Cochran (1963) discusses simple random sampling. In a uniform randomized experiment, the m_s treated units in stratum s are a simple random sample without replacement from the n_s units in stratum s. The following proposition is proved in the problems in §2.12.2.

Proposition 2.2. *In a uniform randomized experiment, if the treatment has no effect, the expectation and variance of a sum statistic* $\mathbf{Z}^{\mathrm{T}}\mathbf{q}$ *are*

$$E(\mathbf{Z}^{\mathrm{T}}\mathbf{q}) = \sum_{s=1}^{S} m_s \bar{q}_s,$$

and

$$\operatorname{var}(\mathbf{Z}^{\mathrm{T}}\mathbf{q}) = \sum_{s=1}^{S} \frac{m_s(n_s - m_s)}{n_s(n_s - 1)} \sum_{i=1}^{n_s} (q_{si} - \bar{q}_s)^2,$$

where

$$\bar{q}_s = \frac{1}{n_s} \sum_{i=1}^{n_s} q_{si}.$$

Moments are easily determined for sum statistics, but other classes of statistics have other useful properties. The first such class, the *sign-score statistics*, is a subset of the sum statistics. A statistic is a sign-score statistic if it is of the form $t(\mathbf{Z}, \mathbf{r}) = \sum_{s=1}^{S} d_s \sum_{i=1}^{n_s} c_{si} Z_{si}$ where c_{si} is binary, $c_{si} = 1$ or $c_{si} = 0$, and both d_s and c_{si} are functions of $\mathbf{r}$. Fisher's exact test, the Mantel–Haenszel test and McNemar's test are sign-score statistics with $d_s = 1$ and $c_{si} = r_{si}$. The signed rank and median test statistics are also sign-score statistics, but with c_{si} and d_s defined differently. A sign-score statistic is a sum statistic with $q_{si} = d_s c_{si}$, but many sum statistics, including the rank sum statistic, are not sign-score statistics. In Chapter 4, certain calculations are simpler for sign-score statistics than for certain other sum statistics, and this motivates the distinction.

Another important class of statistics are *arrangement-increasing functions* of $\mathbf{Z}$ and $\mathbf{r}$, which will be defined in a moment. Informally, a statistic $t(\mathbf{Z}, \mathbf{r})$ is arrangement-increasing if it increases in value as the coordinates of $\mathbf{Z}$ and $\mathbf{r}$ are rearranged into an increasingly similar order within each stratum. In fact, all of the statistics in §2.4.2 are arrangement-increasing, so anything that is true of arrangement-increasing statistics is true of all the commonly used statistics in §2.3.2. Hollander, Proschan, and Sethuraman (1977) discuss many properties of arrangement increasing functions.

A few preliminary terms are useful. The numbers S and n_s, $s = 1, \ldots, S$ with $N = \sum n_s$, are taken as given. A *stratified N-tuple* $\mathbf{a}$ is an N-tuple in which the N coordinates are divided into S strata with n_s coordinates in stratum s, where a_{si} is the ith of the n_s coordinates in stratum s. For instance, $\mathbf{Z}$ and $\mathbf{r}$ are each stratified N-tuples. If $\mathbf{a}$ is a stratified N-tuple, and if i and j are different positive integers less than or equal to n_s, then let $\mathbf{a}_{sij}$ be the stratified N-tuple formed from $\mathbf{a}$ by interchanging a_{si} and a_{sj}, that is, by placing the value a_{sj} in the ith position in stratum s and placing the value a_{si} in the jth position in stratum s. To avoid repetition, whenever the symbol $\mathbf{a}_{sij}$ appears, it is assumed without explicit mention that the subscripts are appropriate, so s is a positive integer between 1 and S and i and j are different positive integers less than or equal to n_s. A function $f(\mathbf{a}, \mathbf{b})$ of two stratified N-tuples is *invariant* if $f(\mathbf{a}, \mathbf{b}) = f(\mathbf{a}_{sij}, \mathbf{b}_{sij})$ for all s, i, j, so renumbering units in the same stratum does not change the value of $f(\mathbf{a}, \mathbf{b})$. For instance, the function $\mathbf{z}^T\mathbf{q}$ is an invariant function of $\mathbf{z}$ and $\mathbf{q}$.

Definition. An invariant function $f(\mathbf{a}, \mathbf{b})$ of two stratified N-tuples is *arrangement-increasing* (or AI) if $f(\mathbf{a}, \mathbf{b}_{sij}) \geq f(\mathbf{a}, \mathbf{b})$ whenever $(a_{si} - a_{sj}) \cdot (b_{si} - b_{sj}) \leq 0$.

Notice what this definitions says. Consider the ith and jth unit in stratum s. If $(a_{si} - a_{sj})(b_{si} - b_{sj}) < 0$, then of these two units, the one with the higher value of a has the lower value of b, so these two coordinates are out of order. However, in $\mathbf{a}$ and $\mathbf{b}_{sij}$, these two coordinates are in the same order, for b_{si} and b_{sj} have been interchanged. The definition says that an arrangement increasing function will be larger, or at least no smaller, when these two coordinates are switched into the same order.

Notice also what the definition says when $(a_{si} - a_{sj})(b_{si} - b_{sj}) = 0$. In this case, either $a_{si} = a_{sj}$ or $b_{si} = b_{sj}$ or both. In this case, $f(\mathbf{a}, \mathbf{b}_{sij}) = f(\mathbf{a}, \mathbf{b})$.

Consider some examples. The function $\mathbf{z}^T\mathbf{q}$ is arrangement-increasing as a function of $\mathbf{z}$ and $\mathbf{q}$. To see this, note that $\mathbf{z}^T\mathbf{q}_{sij} - \mathbf{z}^T\mathbf{q} = (z_{si}q_{sj} + z_{sj}q_{si}) - (z_{si}q_{si} + z_{sj}q_{sj}) = -(z_{si} - z_{sj})(q_{si} - q_{sj})$, so if $(z_{si} - z_{sj})(q_{si} - q_{sj}) \leq 0$ then $\mathbf{z}^T\mathbf{q}_{sij} - \mathbf{z}^T\mathbf{q} \geq 0$. This shows $\mathbf{z}^T\mathbf{q}$ is arrangement-increasing.

Table 2.4.2 is a small illustration for the rank sum statistic with a single stratum, $S = 1$, $n = 4$ units, of whom $m = 2$ received the treatment. Here,

Table 2.4.2. A Hypothetical Example Showing an Arrangement-Increasing Statistic.

i		$\mathbf{z}$	$\mathbf{q}$	$\mathbf{q}_{23}$
1	Treated	1	4	4
2	Treated	1	2	3
3	Control	0	3	2
4	Control	0	1	1
Rank sum			6	7

$(z_2 - z_3)(q_2 - q_3) = (1 - 0)(2 - 3) = -1 \leq 0$, and the rank sum $\mathbf{z}^T\mathbf{q} = 6$ is increased to $\mathbf{z}^T\mathbf{q}_{23} = 7$ by interchanging q_2 and q_3.

As a second example, consider the function $t(\mathbf{z}, \mathbf{r}) = \mathbf{z}^T\mathbf{q}$ where $\mathbf{q}$ is a function of $\mathbf{r}$, which may be written explicitly as $\mathbf{q}(\mathbf{r})$. Then $t(\mathbf{z}, \mathbf{r})$ may or may not be arrangement-increasing in $\mathbf{z}$ and $\mathbf{r}$ depending upon how $\mathbf{q}(\mathbf{r})$ varies with $\mathbf{r}$. The common statistics in §2.4.2 all have the following two properties:

(i) permute $\mathbf{r}$ within strata and $\mathbf{q}$ is permuted in the same way; and
(ii) within each stratum, larger r_{si} receive larger q_{si}.

One readily checks that $t(\mathbf{z}, \mathbf{r}) = \mathbf{z}^T\mathbf{q}$ is arrangement increasing if $\mathbf{q}(\mathbf{r})$ has these two properties, because the first property insures that $t(\mathbf{z}, \mathbf{r})$ is invariant, and the second insures that $r_{si} - r_{sj} \geq 0$ implies $q_{si} - q_{sj} \geq 0$, so $(z_{si} - z_{sj})(r_{si} - r_{sj}) \leq 0$ implies $(z_{si} - z_{sj})(q_{si} - q_{sj}) \leq 0$, and the argument of the previous paragraph applies. The important conclusion is that all of the statistics in §2.4.2 are arrangement-increasing.

In describing the behavior of a statistic when the null hypothesis does not hold and instead the treatment has an effect, a final class of statistics is useful. Many statistics that measure the size of the difference between treated and control groups would tend to increase in value if responses in the treated group were increased and those in the control group were decreased. Let us express this formally. A treated unit has $2Z_{si} - 1 = 1$, since $Z_{si} = 1$, and a control unit has $2Z_{si} - 1 = -1$ since $Z_{si} = 0$. Let $\mathbf{z} \in \Omega$ be a possible treatment assignment and let $\mathbf{r}$ and $\mathbf{r}^*$ be two possible values of the N-tuple of responses such that $(r_{si}^* - r_{si})(2z_{si} - 1) \geq 0$ for all s, i. With treatments given by $\mathbf{z}$, this says that $r_{si}^* \geq r_{si}$ for every treated unit and $r_{si}^* \leq r_{si}$ for every control unit. In words, if higher responses indicated favorable outcomes, then every treated unit does better with $\mathbf{r}^*$ than with $\mathbf{r}$, and every control does worse with $\mathbf{r}^*$ than with $\mathbf{r}$. That is, the difference between treated and control groups looks larger with $\mathbf{r}^*$ than with $\mathbf{r}$. The test statistic is *order-preserving* if $t(\mathbf{z}, \mathbf{r}) \leq t(\mathbf{z}, \mathbf{r}^*)$ whenever $\mathbf{r}$ and $\mathbf{r}^*$ are two possible values of the response such that $(r_{si}^* - r_{si})(2z_{si} - 1) \geq 0$ for all s, i. All of the commonly used statistics in §2.4.2 are order-preserving.

Table 2.4.3 contains a small hypothetical example to illustrate the idea of an order-preserving statistic. Here there is a single stratum, $S = 1$, four subjects, $n = 4$, of whom $m = 2$ received the treatment. Notice that when r_i and r_i^* are compared, treated subjects have $r_i^* \geq r_i$ while controls have $r_i^* \leq r_i$. If the responses are ranked 1, 2, 3, 4, and the ranks in the treated group are summed to give Wilcoxon's rank sum statistic, then the rank sum is larger for r_i^* than for r_i.

In summary, this section has considered four classes of statistics:

(i) the sum statistics;
(ii) the arrangement-increasing statistics;
(iii) the order-preserving statistics; and
(iv) the sign-score statistics.

Table 2.4.3. Hypothetical Example of an Order-Preserving Statistic.

i		z_i	$2z_i - 1$	r_i	r_i^*
1	Treated	1	1	5	6
2	Treated	1	1	2	4
3	Control	0	−1	3	2
4	Control	0	−1	1	1
Rank sum				6	7

All of the commonly used statistics in §2.4.2 are members of the first three classes, and most are sign-score statistics; however, the rank sum statistic, the stratified rank sum statistic, and Mantel's extension are not sign-score statistics.

2.5. Models for Treatment Effects

2.5.1. Responses When the Treatment Has an Effect

If the treatment has an effect, then the observed N-tuple of responses for the N units will be different for different treatment assignments $\mathbf{z} \in \Omega$—this is what it means to say the treatment has an effect. In earlier sections, the null hypothesis of no treatment effect was assumed to hold, so the observed responses were fixed, not varying with $\mathbf{z}$, and the response was written $\mathbf{r}$. When the null hypothesis of no effect is not assumed to hold, the response changes with $\mathbf{z}$, and the response observed when the treatment assignment is $\mathbf{z} \in \Omega$ will be written $\mathbf{r}_\mathbf{z}$. The null hypothesis of no treatment effect says that $\mathbf{r}_\mathbf{z}$ does not vary with $\mathbf{z}$, and instead $\mathbf{r}_\mathbf{z}$ is a constant the same for all $\mathbf{z}$; in this case, $\mathbf{r}$ was written for this constant. Notice that, for each $\mathbf{z} \in \Omega$, the response $\mathbf{r}_\mathbf{z}$ is some nonrandom N-tuple—probability has not yet entered the discussion. Write $r_{si\mathbf{z}}$ for the (s, i) coordinate of $\mathbf{r}_\mathbf{z}$, that is, for the response of the ith unit in stratum s when the N units receive the treatment assignment $\mathbf{z}$.

To make this definite, return for a moment to Fisher's Lady tasting tea. If the Lady could not discriminate at all, then no matter how milk is added to the cup—that is, no matter what $\mathbf{z}$ is—she will classify the cups in the same way—that is, she will give the same binary 8-tuple of responses $\mathbf{r}$. On the other hand, if she discriminates perfectly, always classifying cups correctly, then her 8-tuple of responses will vary with $\mathbf{z}$; indeed, the responses will match the treatment assignments so that $\mathbf{r}_\mathbf{z} = \mathbf{z}$.

If treatments are randomly assigned, then the treatment assignment $\mathbf{Z}$ is a random variable, so the observed responses are also random variables as they depend on $\mathbf{Z}$. Specifically, the observed response is the random variable $\mathbf{r}_\mathbf{Z}$, that is, one of the many possible $\mathbf{r}_\mathbf{z}$, $\mathbf{z} \in \Omega$, selected by picking a treatment assignment $\mathbf{Z}$ by the random mechanism that governs the experiment. Write $\mathbf{R} = \mathbf{r}_\mathbf{Z}$ for the observed response, where $\mathbf{R}$ like $\mathbf{Z}$ is a random variable.

In principle, each possible treatment assignment $\mathbf{z} \in \Omega$ might yield a pattern of responses $\mathbf{r}_\mathbf{z}$ that is unrelated to the pattern observed with another $\mathbf{z}$. For instance, in a completely randomized experiment with 50 subjects divided into two groups of 25, there might be $|\Omega| = \binom{50}{25} \doteq 1.3 \times 10^{14}$ different and unrelated 50-tuples $\mathbf{r}_\mathbf{z}$. Since it is difficult to comprehend a treatment effect in such terms, we look for regularities, patterns or models of the behavior of $\mathbf{r}_\mathbf{z}$ as $\mathbf{z}$ varies over Ω. The remainder of §2.5 discusses several models for $\mathbf{r}_\mathbf{z}$ as $\mathbf{z}$ varies over Ω.

2.5.2. No Interference Between Units

A first model is that of "no interference between units" which means that "the observation on one unit should be unaffected by the particular assignment of treatments to the other units (Cox, 1958a, §2.4)." Rubin (1986) calls this SUTVA for the "stable unit treatment value assumption." Formally, no interference means that $r_{si\mathbf{z}}$ varies with z_{si} but not with the other coordinates of $\mathbf{z}$. In other words, the response of the ith unit in stratum s depends on the treatment assigned to this unit, but not on the treatments assigned to other units, so this unit has only two possible values of the response rather than $|\Omega|$ possible values. When this model is assumed, write r_{Tsi} and r_{Csi} for the responses of the ith unit in stratum s when assigned, respectively, to treatment or control; that is, r_{Tsi} is the common value of $r_{si\mathbf{z}}$ for all $\mathbf{z} \in \Omega$ with $z_{si} = 1$, and r_{Csi} is the common value of $r_{si\mathbf{z}}$ for all $\mathbf{z} \in \Omega$ with $z_{si} = 0$. Then the observed response from the ith unit in stratum s is $R_{si} = r_{Tsi}$ if $Z_{si} = 1$ or $R_{si} = r_{Csi}$ if $Z_{si} = 0$, which may also be written $R_{si} = Z_{si} r_{Tsi} + (1 - Z_{si}) r_{Csi}$. This model, with one potential response for each unit under each treatment, has been important both to experimental design—see Neyman (1923), Welch (1937), Wilk (1955), Cox (1958b, §5), and Robinson (1973)—and to causal inference more generally—see Rubin (1974, 1977) and Holland (1986). When there is no interference between units, write $\mathbf{r}_T$ for $(r_{T11}, \ldots, r_{TSn_S})^\mathrm{T}$ and $\mathbf{r}_C$ for $(r_{C11}, \ldots, r_{CSn_S})^\mathrm{T}$.

"No interference between units" is a model and it can be false. The model is not appropriate for the eight units or cups in the case of the Lady tasting tea. The Lady must classify four cups each way, so changing the classification of one cup necessitates a change in the classification of another. If the treatment has an effect, there will be interference between the eight cups; an exception is the effect $\mathbf{r_z} = \mathbf{z}$. However, when the units are different people and the treatment is a medical intervention, the model is often reasonable.

The problems in §2.12.3 consider models with interference between units. Interference arises naturally when the units involve multiple measurements on the same person at different times under different treatments.

2.5.3. The Model of an Additive Effect, and Related Models

The model of an additive treatment effect assumes units do not interfere with each other, and the administration of the treatment raises the response of a unit by a constant amount τ, so that $r_{Tsi} = r_{Csi} + \tau$ for each s, i. The principal attraction of the model is that there is a definite parameter to estimate, namely, the additive treatment effect τ. As seen in §2.7, in a uniform randomized experiment, many estimators do indeed estimate τ when this model holds.

Under the additive model, the observed response from the ith unit in stratum s is $R_{si} = r_{Csi} + \tau Z_{si}$, or $\mathbf{R} = \mathbf{r}_C + \tau \mathbf{Z}$. It follows that the *adjusted responses*, $\mathbf{R} - \tau\mathbf{Z} = \mathbf{r}_C$ are fixed, not varying with the treatment assignment $\mathbf{Z}$, so the adjusted responses satisfy the null hypothesis of no effect. This fact will be useful in drawing inferences about τ.

There are many similar models, including the model of a multiplicative effect, $r_{Tsi} = \sigma r_{Csi}$, the model of a linear effect, $r_{Tsi} = \sigma r_{Csi} + \tau$ with a two-dimensional parameter (τ, σ), and models with parameters linking the magnitude of the effect $r_{Tsi} - r_{Csi}$ to the values of covariates. See also Problem 6 for discussion of a model of an additive first difference effect.

2.5.4. Positive Effects and Larger Effects

The model of an additive effect assumes a great deal about the relationship between r_{TSi} and r_{Csi}. At times, it is desirable to describe the behavior of statistical procedures while assuming much less. When there is no interference between units, an effect is a pair $(\mathbf{r}_T, \mathbf{r}_C)$ giving the responses of each unit under each treatment. Two useful concepts are positive effects and larger effects. Unlike the model of an additive treatment effect, positive effects and larger effects are meaningful not just for continuous responses, but also for binary responses, for ordinal responses, and as seen later in §2.8, for censored responses and multivariate responses.

A treatment has a *positive effect* if $r_{Tsi} \geq r_{Csi}$ for all units (s, i) with strict inequality for at least one unit. A more compact way of writing this is that $(\mathbf{r}_T, \mathbf{r}_C)$ is a positive effect if $\mathbf{r}_T \geq \mathbf{r}_C$ with $\mathbf{r}_T \neq \mathbf{r}_C$. This says that application of the treatment never decreases a unit's response and sometimes increases it. For instance, there is a positive effect if the effect is additive and $\tau > 0$. Hamilton (1979) discusses this model in detail when the outcome is binary.

Consider two possible effects, say $(\mathbf{r}_T, \mathbf{r}_C)$ and $(\mathbf{r}_T^*, \mathbf{r}_C^*)$. Then $(\mathbf{r}_T^*, \mathbf{r}_C^*)$ is a *larger effect* than $(\mathbf{r}_T, \mathbf{r}_C)$ if $r_{Tsi}^* \geq r_{Tsi}$ and $r_{Csi}^* \leq r_{Csi}$ for all s, i. For instance, the simplest example occurs when the treatment effect is additive, $\mathbf{r}_C^* = \mathbf{r}_C$, $\mathbf{r}_T = \mathbf{r}_C + \tau\mathbf{1}$, and $\mathbf{r}_T^* = \mathbf{r}_C^* + \tau^*\mathbf{1}$, for in this case $(\mathbf{r}_T^*, \mathbf{r}_C^*)$ exhibits a *larger effect* than $(\mathbf{r}_T, \mathbf{r}_C)$ if $\tau^* \geq \tau$. Write $\mathbf{R}$ and $\mathbf{R}^*$ for the observed responses from, respectively, the effects $(\mathbf{r}_T, \mathbf{r}_C)$ and $(\mathbf{r}_T^*, \mathbf{r}_C^*)$, so $R_{si}^* = r_{Tsi}^*$ if $Z_{si} = 1$ and $R_{si}^* = r_{Csi}^*$ if $Z_{si} = 0$.

If a statistical test rejects the null hypothesis 5% of the time when it is true, one would hope that it would reject at least 5% of the time when it is false in the anticipated direction. Recall that a statistical test is *unbiased* against a collection of alternative hypotheses if the test is at least as likely to reject the null hypothesis when one of the alternatives is true as when the null hypothesis is true. The next proposition says that all of the common tests in §2.4.2 are unbiased tests against positive treatment effects, and the test statistic is larger when the effect is larger. The proposition is proved in somewhat more general terms in the Appendix, §2.9.

Proposition 2.3. *In a randomized experiment, a test statistic that is order-preserving yields an unbiased test of no effect against the alternative of a positive effect, and if* $(\mathbf{r}_T^*, \mathbf{r}_C^*)$ *is a larger effect than* $(\mathbf{r}_T, \mathbf{r}_C)$, *then* $t(\mathbf{Z}, \mathbf{R}^*) \geq t(\mathbf{Z}, \mathbf{R})$.

2.6. Confidence Intervals

2.6.1. Testing General Hypotheses

So far, the test statistic $t(\mathbf{Z}, \mathbf{R})$ has been used to test the null hypothesis of no treatment effect. There is an extension to test hypotheses that specify a particular treatment effect. In §2.6.2, this extension is used to construct confidence intervals. As always, the confidence interval is the set of hypotheses not rejected by a test.

Consider testing the hypothesis $H_0\colon \tau = \tau_0$ in the model of an additive effect, $\mathbf{R} = \mathbf{r}_C + \tau\mathbf{Z}$. The idea is as follows. If the null hypothesis $H_0\colon \tau = \tau_0$ were true, then $\mathbf{r}_C = \mathbf{R} - \tau_0\mathbf{Z}$, so testing $H_0\colon \tau = \tau_0$ is the same as testing that $\mathbf{R} - \tau_0\mathbf{Z}$ satisfies the null hypothesis of no treatment effect.

More precisely, if $\mathbf{r}_C$ were known, the probability, say α, that $t(\mathbf{Z}, \mathbf{r}_C)$ is greater than or equal to some fixed number T could be determined from (2.4.2). If the null hypothesis were true, then $\mathbf{r}_C$ would equal the *adjusted responses*, $\mathbf{R} - \tau_0\mathbf{Z}$, so under the null hypothesis, $\mathbf{r}_C$ can be calculated from τ_0 and the observed data. If the hypothesis $H_0\colon \tau = \tau_0$ is true, then the chance that $t(\mathbf{Z}, \mathbf{R} - \tau_0\mathbf{Z}) \geq T$ is α, where α is calculated as described above with $\mathbf{r}_C = \mathbf{R} - \tau_0\mathbf{Z}$.

Table 2.6.1. Example of Confidence Interval Computations.

Unit	Response under control	Treatment	Observed response	Adjusted response	Ranks of adjusted responses
i	r_{Ci}	Z_i	$R_i = r_{Ci} + \tau Z_i$	$R_i - \tau_0 Z_i$	q_i
1	2	1	9	8	7
2	1	0	1	1	1
3	3	0	3	3	2
4	4	0	4	4	3
5	0	1	7	6	5
6	4	1	11	10	8
7	1	1	8	7	6
8	5	0	5	5	4
			$\tau = 7, \tau_0 = 1$		

Now, suppose the null hypothesis is not true, say instead $\tau > \tau_0$, and consider the behavior of the above test. In this case, $\mathbf{R} = \mathbf{r}_C + \tau\mathbf{Z}$ and the adjusted responses $\mathbf{R} - \tau_0\mathbf{Z}$ equal $\mathbf{r}_C + (\tau - \tau_0)\mathbf{Z}$, so the adjusted responses will vary with the assigned treatment $\mathbf{Z}$. If a unit receives the treatment, it will have an adjusted response that is $\tau - \tau_0$ higher than if this unit receives the control. If the test statistic is order-preserving, as is true of all the statistics in §2.4.2, then $t(\mathbf{Z}, \mathbf{R} - \tau_0\mathbf{Z}) = t\{\mathbf{Z}, \mathbf{r}_C + (\tau - \tau_0)\mathbf{Z}\} \geq t(\mathbf{Z}, \mathbf{r}_C) = t(\mathbf{Z}, \mathbf{R} - \tau\mathbf{Z})$, where the inequality follows from the definition of an order-preserving statistic. In words, if the null hypothesis is false and instead $\tau > \tau_0$, then an order-preserving test statistic $t(\mathbf{Z}, \mathbf{R} - \tau_0\mathbf{Z})$ will be larger with τ_0 than it would have been had we tested the correct value τ.

Table 2.6.1 illustrates these computations with a rank sum test. It is a hypothetical uniform randomized experiment with $N = 8$ units, all in a single stratum $S = 1$, with $m = 4$ units assigned to treatment, and an additive treatment effect $\tau = 7$, though the null hypothesis incorrectly says H_0: $\tau = \tau_0 = 1$. The rank sum computed from the adjusted responses $\mathbf{R} - 1\mathbf{Z}$ is $7 + 5 + 8 + 6 = 26$, which is the largest possible rank sum for $N = 8$, $m = 4$, and the one-sided significance level is $\binom{8}{4}^{-1} = 1/70 = 0.014$. The two-sided significance level is $2 \times 0.014 = 0.028$. After removing the hypothesized $\tau_0 = 1$ from treated units, the treated units continue to have higher responses than the controls. The same approach is used to test hypotheses about other models in §2.5.3

2.6.2. Confidence Intervals

Under the model of an additive treatment effect, $\mathbf{R} = \mathbf{r}_C + \tau\mathbf{Z}$, a $1 - \alpha$ confidence set for τ is obtaining by testing each value of τ as in §2.6.1 and collecting all values not rejected at level α into a set A. More precisely, A is the set of values of τ which, when tested, yield significance levels or P-values greater than or equal to α. For instance, in the example in Table 2.6.1, the value $\tau = 1$ would not be contained in a 95% confidence set. When the true value τ is tested, it is rejected with probability no greater than α, so the random set A contains the true τ with probability at least $1 - \alpha$. This is called "inverting" a test, and it is the standard way of obtaining a confidence set from a test; see, for instance, Cox and Hinkley (1974, §7.2). For many test statistics, a two-sided test yields a confidence set that is an interval, whose endpoints may be determined by a line search, as illustrated in §4.3.6. Section 3.2.4 uses this confidence interval in an observational study of lead in the blood of children.

2.7. Point Estimates

2.7.1. Unbiased Estimates of the Average Effect

The most quoted fact about randomized experiments is that they lead to unbiased estimates of the average treatment effect. Take the simplest case, a uniform randomized experiment with a single stratum, with no interference between units. In this case, there are m treated units, $N - m$ control units, $E(Z_i) = m/N$, $R_i = r_{Ti}$ if $Z_i = 1$, and $R_i = r_{Ci}$ if $Z_i = 0$. The difference between the mean response in the treated group, namely $1/m \sum Z_i R_i$, and the mean response in the control group, namely $1/(N - m) \sum (1 - Z_i) R_i$, has expectation

$$E\left\{\sum \frac{Z_i R_i}{m} - \frac{(1 - Z_i) R_i}{N - m}\right\} = E\left\{\sum \frac{Z_i r_{Ti}}{m} - \frac{(1 - Z_i) r_{Ci}}{N - m}\right\}$$

$$= \sum \frac{(m/N) r_{Ti}}{m} - \frac{(1 - m/N) r_{Ci}}{N - m} = \frac{1}{N} \sum r_{Ti} - r_{Ci},$$

and the last term is the average of the N treatment effects $r_{Ti} - r_{Ci}$ for the N experimental units. In words, the difference in sample means is unbiased for the average effect of the treatment. Notice carefully that this is true assuming only that there is no interference between units. There is no assumption that the treatment effect $r_{Ti} - r_{Ci}$ is constant from unit to unit, no assumptions about interactions.

The estimate is unbiased for the average effect on the N units in this study, namely, $1/N \sum r_{Ti} - r_{Ci}$, but this says nothing about the effect on other units not in the study. Campbell and Stanley (1963) say that a randomized experiment has *internal validity* in permitting inferences about effects for the N units in the study, but it need not have *external validity* in that there is no guarantee that the treatment will be equally effective for other units outside the study.

The difference in sample means may be biased when there are two or more strata and the experimenter assigns disproportionately more subjects to the treatment in some strata than in others. However, there is an unbiased estimate that corrects the imbalance. It consists of calculating, within stratum s, the difference between the average response in treated group, namely $1/m_s \sum_i Z_{si} R_{si}$, and the average response in the control group, namely $1/(n_s - m_s) \sum_i (1 - Z_{si}) R_{si}$, and weighting this difference by the proportion of units in stratum s, namely, n_s/N. The estimate, called *direct adjustment*, is then:

$$\sum_{s=1}^{S} \frac{n_s}{N} \sum_{i=1}^{n_s} \left\{ \frac{Z_{si} R_{si}}{m_s} - \frac{(1 - Z_{si}) R_{si}}{n_s - m_s} \right\}. \tag{2.7.1}$$

To check that (2.7.1) is unbiased, recall that, in a uniform randomized experiment, Z_{si} has expectation m_s/n_s. It follows that (2.7.1) has expectation

$$\begin{aligned} E&\left[\sum_{s=1}^{S} \frac{n_s}{N} \sum_{i=1}^{n_s} \left\{ \frac{Z_{si} R_{si}}{m_s} - \frac{(1 - Z_{si}) R_{si}}{n_s - m_s} \right\} \right] \\ &= E\left[\sum_{s=1}^{S} \frac{n_s}{N} \sum_{i=1}^{n_s} \left\{ \frac{Z_{si} r_{Tsi}}{m_s} - \frac{(1 - Z_{si}) r_{Csi}}{n_s - m_s} \right\} \right] \\ &= \sum_{s=1}^{S} \frac{n_s}{N} \sum_{i=1}^{n_s} \left\{ \frac{(m_s/n_s) r_{Tsi}}{m_s} - \frac{(1 - m_s/n_s) r_{Csi}}{n_s - m_s} \right\} \\ &= \frac{1}{N} \sum r_{Ti} - r_{Ci}, \end{aligned}$$

so direct adjustment does indeed give an unbiased estimate of the average effect. Rubin (1977) does calculations of this kind.

In effect, direct adjustment views the treated units and the control units as two stratified random samples from the N units in the experiment. Then (2.7.1) is the usual stratified estimate of mean response to treatment in the population of N units minus the usual estimate of the mean response to control in the population of N units. Notice again that direct adjustment is unbiased for the average treatment effect even if that effect varies from unit to unit or from stratum to stratum. On the other hand, the average effect is but a summary of the effects, and not a complete description, when the effect varies from one stratum to another.

2.7.2. Hodges–Lehmann Estimates of an Additive Effect

Under the model of an additive effect, $\mathbf{R} = \mathbf{r}_C + \tau \mathbf{Z}$, there are many estimates of τ. One due to Hodges and Lehmann (1963) is closely tied to the test in §2.4 and the confidence interval in §2.6. Recall that $H_0: \tau = \tau_0$ is tested using $t(\mathbf{Z}, \mathbf{R} - \tau_0 \mathbf{Z})$, that is, by subtracting the hypothesized treatment effect $\tau_0 \mathbf{Z}$ from the observed responses $\mathbf{R}$, and asking whether the adjusted responses $\mathbf{R} - \tau_0 \mathbf{Z}$ appear to be free of a treatment effect. The Hodges–Lehmann estimate of τ is that value $\hat{\tau}$ such that the adjusted responses $\mathbf{R} - \hat{\tau} \mathbf{Z}$ appear to be exactly free of a treatment effect. Consider this in detail. Throughout this section, the experiment is a uniform randomized experiment.

Suppose that we can determine the expectation, say $\bar{\bar{t}}$, of the statistic $t(\mathbf{Z}, \mathbf{R} - \tau\mathbf{Z})$ when calculated using the correct τ, that is, when calculated from responses $\mathbf{R} - \tau\mathbf{Z}$ that have been adjusted so they are free of a treatment effect. For instance, in an experiment with a single stratum, the rank sum statistic has expectation $\bar{\bar{t}} = m(N + 1)/2$ if the treatment has no effect. This is true because, in the absence of a treatment effect, the rank sum statistic is the sum of m scores randomly selected from N scores whose mean is $(N + 1)/2$. In the same way, in a stratified experiment, the stratified rank sum statistic has expectation $\bar{\bar{t}} = \frac{1}{2}\sum m_s(n_s + 1)$ in the absence of a treatment effect. In a paired experiment, in the absence of a treatment effect, the expectation of the signed rank statistic is $\bar{\bar{t}} = (N + 1)/4$, since we expect to sum $\frac{1}{2}$ of N scores which average $(N + 1)/2$. In the absence of an effect, in an experiment with a single stratum, the difference in sample means (2.4.1) has expectation $\bar{\bar{t}} = 0$. In each of these cases, $\bar{\bar{t}}$ may be determined without knowing τ, so there is a Hodges–Lehmann estimate.

Roughly speaking, the Hodges–Lehmann estimate is the solution $\hat{\tau}$ of the equation $\bar{\bar{t}} = t(\mathbf{Z}, \mathbf{R} - \hat{\tau}\mathbf{Z})$. In other words, $\hat{\tau}$ is the value such that the adjusted responses $\mathbf{R} - \hat{\tau}\mathbf{Z}$ appear to be entirely free of a treatment effect, in the sense that the test statistic $t(\mathbf{Z}, \mathbf{R} - \hat{\tau}\mathbf{Z})$ exactly equals its expectation in the absence of an effect.

If $t(\cdot, \cdot)$ is an order-preserving statistic, as is true of all of the statistics in §2.3, then $t(\mathbf{Z}, \mathbf{R} - \hat{\tau}\mathbf{Z})$ is monotone decreasing as a function of $\hat{\tau}$ with $\mathbf{Z}$ and $\mathbf{R}$ fixed. This says: The larger the treatment effect $\hat{\tau}\mathbf{Z}$ removed from the observed responses $\mathbf{R}$, the smaller the statistic becomes. This is useful in solving $\bar{\bar{t}} = t(\mathbf{Z}, \mathbf{R} - \hat{\tau}\mathbf{Z})$. If a $\hat{\tau}$ has been tried such that $\bar{\bar{t}} < t(\mathbf{Z}, \mathbf{R} - \hat{\tau}\mathbf{Z})$, then a larger $\hat{\tau}$ will tend to make $t(\mathbf{Z}, \mathbf{R} - \hat{\tau}\mathbf{Z})$ smaller, moving it toward $\bar{\bar{t}}$. Similarly, if $\bar{\bar{t}} > t(\mathbf{Z}, \mathbf{R} - \hat{\tau}\mathbf{Z})$, then a smaller $\hat{\tau}$ is needed.

Problems arise immediately. For rank statistics, such as the rank sum and the signed rank, $t(\mathbf{Z}, \mathbf{R} - \hat{\tau}\mathbf{Z})$ varies in discrete jumps as $\hat{\tau}$ is varied, so there may be no value $\hat{\tau}$ such that $\bar{\bar{t}} = t(\mathbf{Z}, \mathbf{R} - \hat{\tau}\mathbf{Z})$. To see this, take a trivial case, a uniform experiment in one stratum, sample size $N = 2$, one treated unit $m = 1$. Then the rank sum statistic is either 1 or 2 depending upon which of the two units receive the treatment, but $\bar{\bar{t}} = 1\frac{1}{2}$, so it is not possible to find a $\hat{\tau}$ such that $\bar{\bar{t}} = t(\mathbf{Z}, \mathbf{R} - \hat{\tau}\mathbf{Z})$.

Not only may $\bar{\bar{t}} = t(\mathbf{Z}, \mathbf{R} - \hat{\tau}\mathbf{Z})$ have no solution $\hat{\tau}$, but it may have infinitely many solutions. If $t(\mathbf{Z}, \mathbf{R} - \hat{\tau}\mathbf{Z})$ varies in discrete jumps, it will be constant for intervals of values of $\hat{\tau}$.

Hodges and Lehmann resolve these problems in the following way. They define the solution of an equation $\bar{\bar{t}} = t(\mathbf{Z}, \mathbf{R} - \hat{\tau}\mathbf{Z})$ as $\text{SOLVE}\{\bar{\bar{t}} = t(\mathbf{Z}, \mathbf{R} - \hat{\tau}\mathbf{Z})\}$ defined by

$$\hat{\tau} = \text{SOLVE}\{\bar{\bar{t}} = t(\mathbf{Z}, \mathbf{R} - \hat{\tau}\mathbf{Z})\}$$
$$= \frac{\inf\{\tau\colon \bar{\bar{t}} > t(\mathbf{Z}, \mathbf{R} - \tau\mathbf{Z})\} + \sup\{\tau\colon \bar{\bar{t}} < t(\mathbf{Z}, \mathbf{R} - \tau\mathbf{Z})\}}{2}.$$

Table 2.7.1. Computing a Hodges–Lehmann Estimate.

τ	4.9999	5	5.0001	5.9999	6	6.0001
$t(\mathbf{Z}, \mathbf{R} - \tau\mathbf{Z})$	20	19	18	18	17	15

This defines the Hodges–Lehmann estimate. Roughly speaking, if there is no exact solution, then average the smallest τ that is too large and the largest τ that is too small.

Consider the small example in Table 2.6.1. Under the null hypothesis of no effect, the rank sum statistic has expectation $\bar{\bar{t}} = m(N + 1)/2 = 4(8 + 1)/2 = 18$, that is, half of the sum of all eight ranks, $36 = 1 + 2 + \cdots + 8$. Table 2.7.1 gives values of $t(\mathbf{Z}, \mathbf{R} - \tau\mathbf{Z})$ for several values of τ. As noted, since $t(\cdot, \cdot)$ is order-preserving, in Table 2.7.1, $t(\mathbf{Z}, \mathbf{R} - \tau\mathbf{Z})$ decreases in τ. We want as our estimate a value $\hat{\tau}$ such that $18 = t(\mathbf{Z}, \mathbf{R} - \hat{\tau}\mathbf{Z})$, but the table indicates that any value between 5 and 6 will do. As the table suggests, $\inf\{\tau: \bar{\bar{t}} > t(\mathbf{Z}, \mathbf{R} - \tau\mathbf{Z})\} = 6$ and $\sup\{\tau: \bar{\bar{t}} < t(\mathbf{Z}, \mathbf{R} - \tau\mathbf{Z})\} = 5$, so the Hodges–Lehmann estimate is $\hat{\tau} = (6 + 5)/2 = 5.5$.

For particular test statistics, there are other ways of computing $\hat{\tau}$. This is true, for instance, for a single stratum using the rank sum test. In this case, it may be shown $\hat{\tau}$ is the median of the $m(N - m)$ pairwise differences formed by taking each of the m treated responses and subtracting each of the $N - m$ control responses.

The Hodges–Lehmann estimate $\hat{\tau}$ inherits properties from the test statistic $t(\cdot, \cdot)$. Consistency is one such property. Recall that a test is *consistent* if the probability of rejecting each false hypothesis tends to one as the sample size increases. Recall that an estimate is consistent if the probability that it is close to the true value tends to one as the sample size increases. As one would expect, these ideas are interconnected. A test that rejects incorrect values of τ leads to an estimate that moves away from these incorrect values. In other words, under mild conditions, consistent tests lead to consistent Hodges–Lehmann estimates; see Maritz (1981, §1.4) for some details.

*2.8. More Complex Outcomes

2.8.1. Partially Ordered Outcomes

So far, the outcome R_{si} has been a number, possibly a continuous measurement, possibly a binary event, possibly a discrete score, but always a single number. However, for more complex responses, much of the earlier discus-

* The material in this section is used only in Chapter 8.

sion continues to apply with little or no change. The purpose of §2.8 is to discuss issues that arise with certain complex outcomes, including multivariate responses and censored observations.

When the outcome R_{si} is a single number, it is clear what it means to speak of a high or low response, and it is clear what it means to ask whether responses are typically higher among treated units than among controls. For more complex responses, it may happen that some responses are higher than some others; and yet not every pair of possible responses can be ordered. For example, unit 1 may have a more favorable outcome than units 2 and 3, but units 2 and 3 may have different outcomes neither of which can be described as entirely more favorable than the other. For instance, patient 1 may live longer and have a better quality of life than patients 2 and 3, but patient 2 may outlive patient 3 though patient 3 had a better quality of life than patient 2. In this case, outcomes may be partially ordered rather than totally ordered, an idea that will be formalized in a moment. Common examples are given in §2.8.2 and §2.8.3.

A *partially ordered set or poset* is a set A together with a relation $\lesssim$ on A such that three conditions hold:

(i) $a \lesssim a$ for all $a \in A$;
(ii) $a \lesssim b$ and $b \lesssim a$ implies $a = b$ for all $a, b \in A$; and
(iii) if $a \lesssim b$ and $b \lesssim c$ then $a \lesssim c$ for all $a, b, c \in A$.

There is *strict inequality* between a and b if $a \lesssim b$ and $a \neq b$. A poset A is *totally ordered* if $a \lesssim b$ or $b \lesssim a$ for every $a, b \in A$. The real numbers with conventional inequality $\leq$ are totally ordered. If A is partially ordered but not totally ordered, then for some $a, b \in A$, $a \neq b$, neither a nor b is higher than the other, that is, neither $a \lesssim b$ nor $b \lesssim a$. Sections 2.8.2 and 2.8.3 discuss two common examples of partially ordered outcomes, namely censored and multivariate outcomes. Following this, in §2.8.4, general methods for partially ordered outcomes are discussed.

2.8.2. Censored Outcomes

In some experiments, an outcome records the time to some event. In a clinical trial, the outcome may be the time between a patient's entry into the trial and the patient's death. In a psychological experiment, the outcome may be the time lapse between administration of a stimulus by the experimenter and the production of a response by an experimental subject. In a study of remedial education, the outcome may be the time until a certain level of proficiency in reading is reached.

Times may be censored in the sense that, when data analysis begins, the event may not yet have occurred. The patient may be alive at the close of the study. The stimulus may never elicit a response. The student may not develop proficiency in reading during the period under study.

If the event occurs for a unit after, say, 3 months, the unit's response is written 3. If the unit entered the study 3 months ago, the event has not yet occurred, and the analysis is done today, the unit's response is written 3+ signifying that the event has not occurred in the initial 3 months.

Censored times are partially ordered. To see this, consider a simple illustration. In a clinical trial, patient 1 died at 3 months, patient 2 died at 12 months, and patient 3 entered the study 6 months ago and is alive today yielding a survival of 6+ months. Then patient 1 had a shorter survival than patients 2 and 3, but it is not possible to say whether patient 2 survived longer than patient 3 because we do not know whether patient 3 will survive for a full year.

The set A of censored survival times contains the nonnegative real numbers together with the nonnegative real numbers with a plus appended. Define the partial order $\lesssim$ on A as follows: if a and b are nonnegative real numbers, then:

(i) $a \lesssim b$ if and only if $a \le b$;
(ii) $a \lesssim b+$ if and only if $a \le b$; and
(iii) $a \lesssim a$ and $a+ \lesssim a+$.

Here, (i) indicates that "a" and "b" are both deaths and "a" died first. In (ii), "a" died before "b" was censored, so "b" certainly outlived "a." Of course, (iii) is just the case of equality—every censored time is equal to itself, and so is less than or equal to itself. It is easy to check that this is indeed a partial order, and that strict inequality indicates certainty about who died first.

2.8.3. Multivariate Outcomes and Other Partially Ordered Outcomes

Quite often, a single number is not enough to describe the outcome for a unit. In an educational intervention, there may be test scores in several areas, such as reading and mathematics. In a clinical trial, the outcome may involve both survival and quality of life. A multivariate response is a p-tuple of outcomes describing an individual. If the p components are numbers, then the multivariate response inherits a partial order as follows: $(a_1, \ldots, a_p) \lesssim (b_1, \ldots, b_p)$ if and only if $a_1 \le b_1$, $a_2 \le b_2, \ldots,$ and $a_p \le b_p$. It is easy to check that this defines a partial order. As an example, if the outcome is the 2-tuple consisting of a reading score and a mathematics score, then one student has a higher multivariate response than another only if the first student did at least as well as the second student on both tests.

In fact, the components of the p-tuple need not be numbers—rather they may be any partially ordered outcomes. In the same way, the p-tuple inherits a partial order from the partial orders of individual outcomes. For instance, the outcome might be a 2-tuple consisting of a censored survival time and a number measuring quality of life. The censored survival times are partially but not totally ordered. In this case, a patient who died early with a poor quality of life would have a lower outcome than a patient who was censored late with a good quality of life.

Multivariate responses may be given other partial orders appropriate to particular contexts. Here is one which gives greatest emphasis to the first coordinate and about equal emphasis to the other two: $(a_1, a_2, a_3) \precsim (b_1, b_2, b_3)$ if $a_1 \leq b_1$ or if $\{a_1 = b_1$ and $a_2 \leq b_2$ and $a_3 \leq b_3\}$. In an educational setting, this might say that a student who graduates had a better outcome than one who did not regardless of test scores, but among those who graduate, one student is better than another only if both reading and math scores are as good or better.

2.8.4. A Test Statistic for Partially Ordered Outcomes

The task is to test the null hypothesis of no treatment effect against the alternative that treated units tend to have higher responses than controls in the sense of a partial order $\precsim$ on the outcomes. For this purpose, define indicators L_{sij} for $s = 1, \ldots, S$, $i = 1, \ldots, n_s$, $j = 1, \ldots, n_s$, as follows:

$$L_{sij} = \begin{cases} 1 & \text{if } R_{sj} \precsim R_{si} \text{ with } R_{si} \neq R_{sj}, \\ -1 & \text{if } R_{si} \precsim R_{sj} \text{ with } R_{si} \neq R_{sj}, \\ 0 & \text{otherwise.} \end{cases} \tag{2.8.1}$$

In words, L_{sij} compares the ith and jth units in stratum s, and L_{sij} is 1 if the ith is strictly greater than the jth, is -1 if the ith is strictly smaller than the jth, and is zero in all other cases. The statistic is

$$t(\mathbf{Z}, \mathbf{R}) = \sum_{s=1}^{S} \sum_{i=1}^{n_s} \sum_{j=1}^{n_s} Z_{si}(1 - Z_{sj}) L_{sij}. \tag{2.8.2}$$

Consider the statistic in detail. The term $Z_{si}(1 - Z_{sj}) L_{sij}$ equals 1 if, in stratum s, the ith unit received the treatment, the jth unit received the control, and these two units had unequal responses with the treated unit having a higher response, $R_{sj} \precsim R_{si}$. Similarly, $Z_{si}(1 - Z_{sj}) L_{sij}$ equals -1 if, in stratum s, the ith unit is treated, the jth is a control, and the control had the higher response, $R_{si} \precsim R_{sj}$. In all other cases, $Z_{si}(1 - Z_{sj}) L_{sij}$ equals zero. So the test statistic is the number of comparisons of treated and control units in the same stratum in which the treated unit had the higher response minus the number in which the control unit had the higher response.

This statistic generalizes several familiar statistics. If the outcome is a single number and the partial order $\precsim$ is ordinary inequality $\leq$, then (2.8.2) is equivalent to the Mann-Whitney (1947) statistic and the Wilcoxon (1945) rank sum statistic. If the outcome is censored and $\precsim$ is the partial order in §2.8.2, then the statistic is Gehan's (1965) statistic.

A device due to Mantel (1967) shows that (2.8.2) is a sum statistic. The steps are as follows. First note that, for any subset A of $\{1, 2, \ldots, n_s\}$,

$$\sum_{i \in A} \sum_{j \in A} L_{sij} = 0 \tag{2.8.3}$$

since L_{sij} and L_{sji} both appear in the sum, with $L_{sij} = -L_{sji}$ and they cancel. Using this fact with $A = \{i: 1 \le i \le n_s \text{ with } Z_{si} = 1\}$ yields

$$0 = \sum_{i \in A} \sum_{j \in A} L_{sij} = \sum_{i=1}^{n_s} \sum_{j=1}^{n_s} Z_{si} Z_{sj} L_{sij},$$

which permits the test statistic to be rewritten as the sum statistic

$$t(\mathbf{Z}, \mathbf{R}) = \sum_{s=1}^{S} \sum_{i=1}^{n_s} Z_{si} \sum_{j=1}^{n_s} L_{sij} = \sum_{s=1}^{S} \sum_{i=1}^{n_s} Z_{si} q_{si} \quad \text{with} \quad q_{si} = \sum_{j=1}^{n_s} L_{sij}.$$

As a result, the expectation and variance of the test statistic under the null hypothesis are given by Proposition 2.2. In fact, in that Proposition, $\bar{q}_s = 0$ for each s using (2.8.3).

The score q_{si} has an interpretation. It is the number of units in stratum s with outcomes less that unit i minus the number with outcomes greater than i. The score q_{si} is large if unit i has a response larger than that of most units in stratum s. For instance, in Gehan's statistic for censored outcomes, the score q_{si} is the number of patients in stratum s who definitely died before patient i minus the number who definitely died after patient i. A numerical example using the scores q_{si} is given in §8.1.3.

2.8.5. Order-Preserving Statistics, Positive Effects, Larger Effects

In §2.4 and §2.5, three terms were discussed, namely, order-preserving statistics, positive effects, and larger effects. These terms apply to partially ordered outcomes with virtually no change, as will be seen in a moment. In each case, the definitions in §2.4 and §2.5 are the special case of the definitions in this section with the partial order $\lesssim$ given by ordinary inequality $\le$ of real numbers.

Let $\mathbf{r}$ and $\mathbf{r}^*$ be two possible values of the N-tuple of partially ordered outcomes. If $r_{si} \lesssim r_{si}^*$ for every treated unit and $r_{si}^* \lesssim r_{si}$ for every control unit, then the treated and control groups appear further apart for outcome $\mathbf{r}^*$ than for outcome $\mathbf{r}$. A test statistic $t(\cdot, \cdot)$ is *order-preserving* if $t(\mathbf{z}, \mathbf{r}) \le t(\mathbf{z}, \mathbf{r}^*)$ whenever $\mathbf{r}$ and $\mathbf{r}^*$ are two possible values of the response such $r_{si} \lesssim r_{si}^*$ if $z_{si} = 1$ and $r_{si}^* \lesssim r_{si}$ if $z_{si} = 0$ for all s, i. In words, the statistic is larger when the outcomes in treated and control groups are further apart. The statistic in §2.8.4 is order-preserving; see the problems in §8.5.2 for proof.

If there is no interference between units, then $(\mathbf{r}_T, \mathbf{r}_C)$ is a *positive effect* if $\mathbf{r}_T \ne \mathbf{r}_C$ and $r_{Csi} \lesssim r_{Tsi}$ for every s, i. In the case of censored survival times, this would mean that each patient would definitely survive at least as long under the treatment as under the control, or else would continue to be censored at the same time due to the end of the study. An effect $(\mathbf{r}_T^*, \mathbf{r}_C^*)$ is a *larger effect* than $(\mathbf{r}_T, \mathbf{r}_C)$ if $r_{Tsi} \lesssim r_{Tsi}^*$ and $r_{Csi}^* \lesssim r_{Csi}$ for all s, i, that is, if the treated responses are higher and the control responses are lower.

The following proposition is the extension of Proposition 2.3 to partially ordered responses. Again, the proof is given in the Appendix, §2.9.

Proposition 2.4. *In a randomized experiment, a test statistic that is order-preserving yields an unbiased test of no effect against the alternative of a positive effect, and if* $(\mathbf{r}_T^*, \mathbf{r}_C^*)$ *is a larger effect than* $(\mathbf{r}_T, \mathbf{r}_C)$, *then* $t(\mathbf{Z}, \mathbf{R}^*) \geq t(\mathbf{Z}, \mathbf{R})$.

*2.9. Appendix: Order-Preserving Tests Under Alternatives

This appendix proves Propositions 2.3 and 2.4 which describe the behavior of order-preserving test statistics under the alternative hypotheses of positive effects or larger effects. It may be of interest to contrast these propositions with a result in Lehmann (1959, §5.8, Lemma 2) which is similar in spirit though quite different in detail. It suffices to prove Proposition 2.4 since Proposition 2.3 is the special case of the former in which the partial order is ordinary inequality. The proof depends of the following lemma:

Lemma 2.5. *Let* $t(\cdot, \cdot)$ *be order-preserving. If* $(\mathbf{r}_T, \mathbf{r}_C)$ *is a positive effect, then* $t(\mathbf{z}, \mathbf{r}_\mathbf{z}) \geq t(\mathbf{z}, \mathbf{r}_\mathbf{a})$ *for all* $\mathbf{z}, \mathbf{a} \in \Omega$. *If* $(\mathbf{r}_T^*, \mathbf{r}_C^*)$ *is a larger effect than* $(\mathbf{r}_T, \mathbf{r}_C)$, *then* $t(\mathbf{z}, \mathbf{r}_\mathbf{z}^*) \geq t(\mathbf{z}, \mathbf{r}_\mathbf{z})$ *for all* $\mathbf{z} \in \Omega$.

Proof of Lemma. Let $(\mathbf{r}_T, \mathbf{r}_C)$ be a positive effect, let $\mathbf{z}, \mathbf{a} \in \Omega$, and consider $\mathbf{r}_\mathbf{z}$ and $\mathbf{r}_\mathbf{a}$. If $z_{si} = 1$, then $r_{si\mathbf{z}} = r_{Tsi}$ while $r_{si\mathbf{a}}$ may equal either r_{Tsi} or r_{Csi} depending on a_{si}, but in either case, $r_{si\mathbf{a}} \lesssim r_{si\mathbf{z}}$ since $(\mathbf{r}_T, \mathbf{r}_C)$ is a positive effect. Similarly, if $z_{si} = 0$, then $r_{si\mathbf{z}} = r_{Csi} \lesssim r_{si\mathbf{a}}$. Since $t(\cdot, \cdot)$ is order-preserving, this implies $t(\mathbf{z}, \mathbf{r}_\mathbf{z}) \geq t(\mathbf{z}, \mathbf{r}_\mathbf{a})$, proving the first part of the lemma.

Now let $\mathbf{z} \in \Omega$, let $(\mathbf{r}_T^*, \mathbf{r}_C^*)$ be a larger effect than $(\mathbf{r}_T, \mathbf{r}_C)$, and consider $\mathbf{r}_\mathbf{z}^*$ and $\mathbf{r}_\mathbf{z}$. If $z_{si} = 1$, then $r_{si\mathbf{z}} = r_{Tsi} \lesssim r_{Tsi}^* = r_{si\mathbf{z}}^*$. If $z_{si} = 0$, then $r_{si\mathbf{z}}^* = r_{Csi}^* \lesssim r_{Csi} = r_{si\mathbf{z}}$. Hence $t(\mathbf{z}, \mathbf{r}_\mathbf{z}^*) \geq t(\mathbf{z}, \mathbf{r}_\mathbf{z})$ since $t(\cdot, \cdot)$ is order-preserving, completing the proof. □

Proof of Proposition 2.4. The lemma directly shows that if $(\mathbf{r}_T^*, \mathbf{r}_C^*)$ is a larger effect than $(\mathbf{r}_T, \mathbf{r}_C)$, then $t(\mathbf{Z}, \mathbf{R}^*) \geq t(\mathbf{Z}, \mathbf{R})$. To prove unbiasedness, let $\mathbf{Z}$ be randomly selected from Ω where $\text{prob}(\mathbf{Z} = \mathbf{z})$ is known but need not be uniform. If the random treatment assignment turns out to be $\mathbf{Z} = \mathbf{a}$, then the observed outcome is $\mathbf{R} = \mathbf{r}_\mathbf{a}$. If the null hypothesis were true, if the treatment had no effect, the observed response would be the same $\mathbf{r_a}$ no matter how treatments were assigned, that is, the observed response would be $\mathbf{R} = \mathbf{r_a}$ no matter what value $\mathbf{Z}$ assumed. If the null hypothesis were false and the treatment had a positive effect, the observed response would vary depending upon the treatment assignment, $\mathbf{R} = \mathbf{r_z}$ if $\mathbf{Z} = \mathbf{z}$. For any fixed number T

$$\begin{aligned}\text{prob}\{t(\mathbf{Z}, \mathbf{R}) \geq T\} \\ &= \sum_{\mathbf{z} \in \Omega} [t(\mathbf{z}, \mathbf{r_z}) \geq T] \, \text{prob}(\mathbf{Z} = \mathbf{z}) \\ &\geq \sum_{\mathbf{z} \in \Omega} [t(\mathbf{z}, \mathbf{r_a}) \geq T] \, \text{prob}(\mathbf{Z} = \mathbf{z}) \qquad \text{for every } \mathbf{a} \in \Omega \text{ by the lemma.}\end{aligned}$$

In other words, the chance that the test statistic $t(\mathbf{Z}, \mathbf{R})$ exceeds any number T is at least as great under the alternative hypothesis of a positive effect as under the null hypothesis of no effect, proving unbiasedness. □

*2.10. Appendix: The Set of Treatment Assignments

2.10.1. Outline and Motivation: The Special Structure of Ω

The set Ω of treatment assignments plays an important role both in randomized experiments and in the discussion of observational studies in later chapters. This set Ω possess a special structure, first noted by Savage (1964). Using this structure, a single theorem may refer to a large classes of test statistics and to all of the simple designs, including matched pairs, matching with multiple controls, two-group comparisons, and stratified comparisons. The purpose of this section is to describe the special structure of Ω. Appendices in later chapters refer back to this Appendix.

Savage (1964) observed that the set Ω is a finite distributive lattice. This is useful because there are tidy theorems about probability distributions on a finite distributive lattice, including the FKG inequality and Holley's inequality. This section will:

(i) offer a little motivation;
(ii) review the definition of a distributive lattice;
(iii) show that Ω is indeed such a lattice; and
(iv) discuss the relevant probability inequalities.

The material in this Appendix may be read without previous experience with lattices.

For motivation, consider a simple case. There is a single stratum, $S = 1$, so the s subscript is dropped in this example, and there are $n = 4$ units of which $m = 2$ receive the treatment. Then Ω contains $\binom{4}{2} = 6$ possible treat-

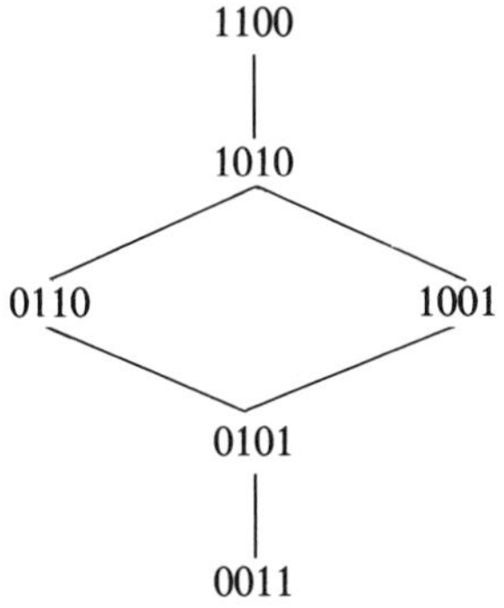

Figure 2.10.1. The lattice Ω of treatment assignments, $S = 1, n = 4, m = 2$.

ment assignments. Assume for this motivating example that the null hypothesis of no treatment effect holds, and renumber the four subjects so their observed responses are in decreasing order, $r_1 \geq r_2 \geq r_3 \geq r_4$. Since no quantity we calculate ever depends on the numbering of subjects, this renumbering changes nothing, but it is notationally convenient. The six possible treatment assignments appear in Figure 2.10.1.

The treatment assignment $\mathbf{z} = (1, 1, 0, 0)$ at the top in Figure 2.10.1 is the one that would suggest the largest positive treatment effect, since this assignment places the two largest responses, r_1 and r_2, in the treated group. The assignment below this, namely, $\mathbf{z} = (1, 0, 1, 0)$ would suggest a smaller treatment effect than (1, 1, 0, 0), since r_3 has replaced r_2, but it would suggest a larger treatment effect than any other assignment. The assignments (0, 1, 1, 0) and (1, 0, 0, 1) are not directly comparable to each other, since the latter places the largest and smallest responses in the treated group while the former places the two middle responses in the treated group; however, both are lower than (1, 0, 1, 0) and both are higher than (0, 1, 0, 1).

Consider the behavior of a test statistic $t(\mathbf{z}, \mathbf{r})$ as we move through Figure 2.10.1. Suppose, for instance, there are no ties among the responses, $r_1 > r_2 > r_3 > r_4$, and $t(\mathbf{z}, \mathbf{r})$ is the rank sum statistic. Then $t(\mathbf{z}, \mathbf{r}) = 7$ for $\mathbf{z} =$ 1100, $t(\mathbf{z}, \mathbf{r}) = 6$ for 1010, $t(\mathbf{z}, \mathbf{r}) = 5$ for both 1001 and 0110, $t(\mathbf{z}, \mathbf{r}) = 4$ for 0101, and $t(\mathbf{z}, \mathbf{r}) = 3$ for 0011, so $t(\mathbf{z}, \mathbf{r})$ increases steadily along upward paths in Figure 2.10.1. If, instead, $t(\mathbf{z}, \mathbf{r})$ were the difference between the mean response in treated and control groups, it would again be increasing along upward paths.

Suppose, instead, that r_2 and r_3 were tied, so $r_1 > r_2 = r_3 > r_4$. In this case, the rank sum statistic would give average rank 2.5 to both r_2 and r_3, so moving from 1100 to 1010 would not change $t(\mathbf{z}, \mathbf{r})$. Notice, however, that even with ties, $t(\mathbf{z}, \mathbf{r})$ is monotone increasing (i.e., nondecreasing) along upward paths.

Actually, the order in Figure 2.10.1 applies to many statistics whether ties are present or not. If $t(\mathbf{z}, \mathbf{r})$ is any arrangement-increasing statistic, then

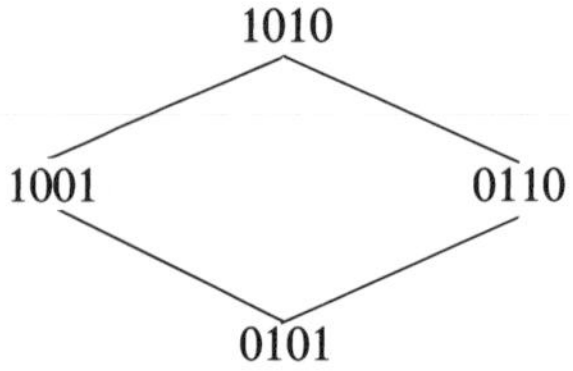

Figure 2.10.2. The lattice Ω of treatment assignments, $S = 2$, $n_1 = n_2 = 2$, $m_1 = m_2 = 1$.

$t(\mathbf{z}, \mathbf{r})$ is monotone-increasing on upward paths in Figure 2.10.1. One might say that all reasonable statistics will be larger for certain $\mathbf{z}$'s than for certain others, that reasonable statistics can differ only in how they order assignments that are not comparable like 1001 and 0110. This is not quite true, however; see Problem 8.

Take a look at a second example, the case of $S = 2$ matched pairs, so $n_s = 2$ and $m_s = 1$ for $s = 1, 2$. Then Ω contains $2^2 = 4$ treatment assignments $\mathbf{z} = (z_{11}, z_{12}, z_{21}, z_{22})$. Again, assume the null hypothesis of no treatment effect and renumber the units in each pair so that in the first pair $r_{11} \geq r_{12}$, and in the second pair $r_{21} \geq r_{22}$. The set Ω appears in Figure 2.10.2.

The assignment $\mathbf{z}$ in Figure 2.10.2 suggesting the largest positive treatment effect is $\mathbf{z} = (1, 0, 1, 0)$ since in both pairs the treated unit had a higher response than the control. For $\mathbf{z} = 1001$ and $\mathbf{z} = 0110$, the treated unit had the higher response in one pair and the lower response in the other. In the assignment $\mathbf{z} = 0101$ the treated unit had a lower response than the control in both pairs.

Once again, common statistics are monotone-increasing along upward paths in Figure 2.10.2. For instance, this is true of the signed rank statistic, which equals zero at the bottom of Figure 2.10.2, equals one or two in the middle, and equals three at the top. Indeed, all arrangement-increasing functions are monotone-increasing along upward paths in Figure 2.10.2.

What does all this suggest? There are certain treatment assignments $\mathbf{z} \in \Omega$ that are higher than others, and this is true without reference to the nature of the response $\mathbf{r}$ or the specific test statistic $t(\mathbf{z}, \mathbf{r})$. The responses might be continuous or they might be discrete scores or they might be binary. The test statistic might be the signed rank statistic or the McNemar statistic. In all these cases, $\mathbf{z} = 1010$ is higher than $\mathbf{z} = 1001$ in Figure 2.10.2. Certain statements about treatment assignments $\mathbf{z} \in \Omega$ should be true generally, without reference to the specific nature of the outcome or the test statistic.

2.10.2. A Brief Review of Lattices

Briefly, a lattice is a partially ordered set in which each pair of elements has a greatest lower bound and a least upper bound. This terminology will be discussed formally in a moment, but first consider what this means in Figure 2.10.1. A point $\mathbf{z}$ in Figure 2.10.1 is below another $\mathbf{z}^*$ if there is a path up from $\mathbf{z}$ to $\mathbf{z}^*$, for instance, 0110 is below 1100. The points 1001 and 0110 are not comparable—there is not a path up from one to the other—so Ω is partially but not totally ordered. The least upper bound of 0110 and 1001 is 1010, for it is the smallest element above both of them. The least upper bound of 1010 and 1100 is 1100. A nice introduction to lattices is given by MacLane and Birkoff (1988).

A set Ω is *partially ordered* by a relation $\lesssim$ if for all $\mathbf{z}, \mathbf{z}^*, \mathbf{z}^{**} \in \Omega$:

(i) $\mathbf{z} \lesssim \mathbf{z}$;
(ii) $\mathbf{z} \lesssim \mathbf{z}^*$ and $\mathbf{z}^* \lesssim \mathbf{z}$ implies $\mathbf{z} = \mathbf{z}^*$; and
(iii) $\mathbf{z} \lesssim \mathbf{z}^*$ and $\mathbf{z}^* \lesssim \mathbf{z}^{**}$ implies $\mathbf{z} \lesssim \mathbf{z}^{**}$.

An *upper bound* for $\mathbf{z}, \mathbf{z}^* \in \Omega$ is an element $\mathbf{z}^{**}$ such that $\mathbf{z} \lesssim \mathbf{z}^{**}$ and $\mathbf{z}^* \lesssim \mathbf{z}^{**}$. A *least upper bound* $\mathbf{z}^{**}$ for $\mathbf{z}, \mathbf{z}^*$ is an upper bound which is below all other upper bounds for $\mathbf{z}, \mathbf{z}^*$, that is, if $\mathbf{z}^{***}$ is any upper bound for $\mathbf{z}, \mathbf{z}^*$, then $\mathbf{z}^{**} \lesssim \mathbf{z}^{***}$. If a least upper bound for $\mathbf{z}, \mathbf{z}^*$ exists, then it is unique by (ii). Lower bound and *greatest lower bound* are defined similarly. A *lattice* is a partially ordered set Ω in which every pair $\mathbf{z}, \mathbf{z}^*$ of elements has a least upper bound, written $\mathbf{z} \vee \mathbf{z}^*$, and a greatest lower bound, written $\mathbf{z} \wedge \mathbf{z}^*$. A lattice Ω is *finite* if the set Ω contain only finitely many elements. In Figure 2.10.1, both 1010 and 1100 are upper bounds for the pair 1001 and 0110, but the least upper bound is $1001 \vee 0110 = 1010$.

The partial order $\lesssim$ and the operations $\vee$ and $\wedge$ are tied together by the following relationship: $\mathbf{z} \lesssim \mathbf{z}^*$ if and only if $\mathbf{z} \vee \mathbf{z}^* = \mathbf{z}^*$ and $\mathbf{z} \wedge \mathbf{z}^* = \mathbf{z}$. In fact, using this relationship, a lattice may be defined beginning with the operations $\vee$ and $\wedge$ rather beginning with the partial order $\lesssim$, that is, defining the partial order in terms of the operations. The following theorem is well known; see MacLane and Birkoff (1988, §XIV, 2) for proof.

Theorem 2.6. *A set Ω with operations $\vee$ and $\wedge$ is a lattice if and only if for all $\mathbf{z}, \mathbf{z}^*, \mathbf{z}^{**} \in \Omega$:*

L1. $\mathbf{z} \vee \mathbf{z} = \mathbf{z}$ *and* $\mathbf{z} \wedge \mathbf{z} = \mathbf{z}$;
L2. $\mathbf{z} \vee \mathbf{z}^* = \mathbf{z}^* \vee \mathbf{z}$ *and* $\mathbf{z} \wedge \mathbf{z}^* = \mathbf{z}^* \wedge \mathbf{z}$;
L3. $\mathbf{z} \vee (\mathbf{z}^* \vee \mathbf{z}^{**}) = (\mathbf{z} \vee \mathbf{z}^*) \vee \mathbf{z}^{**}$ *and* $\mathbf{z} \wedge (\mathbf{z}^* \wedge \mathbf{z}^{**}) = (\mathbf{z} \wedge \mathbf{z}^*) \wedge \mathbf{z}^{**}$; *and*
L4. $\mathbf{z} \wedge (\mathbf{z} \vee \mathbf{z}^*) = \mathbf{z} \vee (\mathbf{z} \wedge \mathbf{z}^*) = \mathbf{z}$.

Here, L2 and L3 are the commutative and associate laws, L1 is called idempotence, and L4 is called absorption. A lattice is *distributive* if the distributive law also holds,

$$\mathbf{z} \vee (\mathbf{z}^* \wedge \mathbf{z}^{**}) = (\mathbf{z} \vee \mathbf{z}^*) \wedge (\mathbf{z} \vee \mathbf{z}^{**}) \qquad \text{for all} \quad \mathbf{z}, \mathbf{z}^*, \mathbf{z}^{**} \in \Omega.$$

2.10.3. The Set of Treatment Assignments Is a Distributive Lattice

This section gives Savage's (1964) demonstration that Ω is a distributive lattice. With each N-dimensional $\mathbf{z} \in \Omega$, associate a vector $\mathbf{c}$ of dimension $\sum m_s$, as follows. The vector $\mathbf{c}$ is made up of S pieces, where piece s has m_s coordinates. It is suggestive and almost accurate to say that $\mathbf{c}$ contains the ranks of the responses of treated units, each stratum being ranked separately, the ranks being arranged in decreasing order in each stratum. This would be exactly true if there were no ties, but it is not exactly true in the case of ties. Here is the exact definition, with or without ties. If $z_{s1} = 0$, $z_{s2} = 0, \ldots, z_{s,i-1} = 0$, $z_{si} = 1$, then $c_{s1} = n_s - i + 1$. Continuing, if $z_{s,i+1} = 0, \ldots, z_{s,j-1} = 0$, $z_{sj} = 1$, then $c_{s2} = n_s - j + 1$, and so on. In terms of the $\mathbf{c}$, Figure 2.10.1 becomes Figure 2.10.3, and Figure 2.10.2 becomes Figure 2.10.4. For instance, in Figure 2.10.1, $\mathbf{z} = 1100$ becomes $\mathbf{c} = 43$, since the first 1 in $\mathbf{z}$ appears in position $i = 1$, so $n - i + 1 = 4 - 1 + 1 = 4$ and the second 1 in $\mathbf{z}$ appears in position $j = 2$, so $n - j + 1 = 4 - 2 + 1 = 3$.

If there are ties among the responses in a stratum, then $\mathbf{c}$ is no longer a collection of ranks, because $\mathbf{c}$ distinguishes units with the same tied response. In the end, this is not a problem. The lattice order makes a few distinctions among treatment assignments that statistical procedures will ignore.

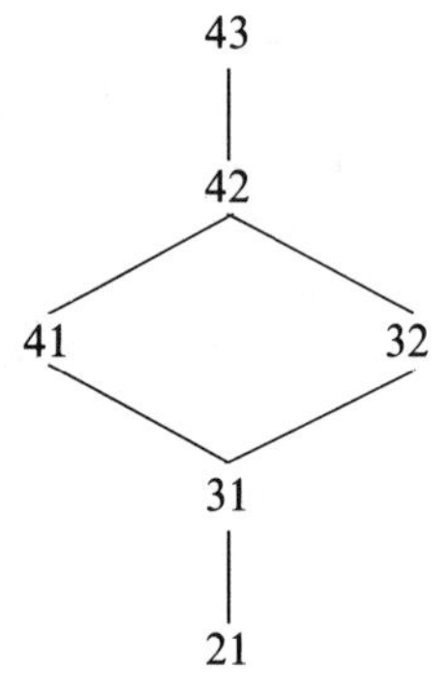

Figure 2.10.3. The lattice in terms of $\mathbf{c}$ for $S = 1$, $n = 4$, $m = 2$.

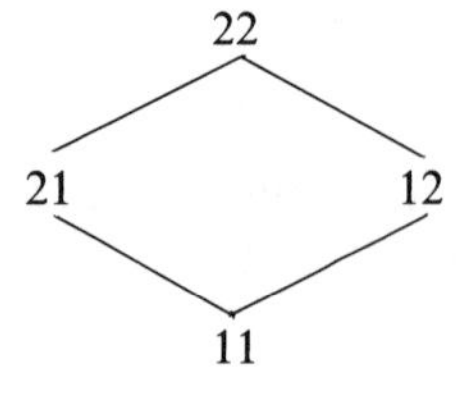

Figure 2.10.4. The lattice in terms of $\mathbf{c}$ for $S = 2$, $n_1 = n_2 = 2$, $m_1 = m_2 = 1$.

It is readily checked that each $\mathbf{z}$ has one and only one corresponding $\mathbf{c}$. Given $\mathbf{z}, \mathbf{z}^* \in \Omega$, with corresponding $\mathbf{c}$ and $\mathbf{c}^*$, the operations $\vee$ and $\wedge$ operations are defined as follows. Define $\mathbf{c} \vee \mathbf{c}^*$ and $\mathbf{c} \wedge \mathbf{c}^*$ as the vectors containing, respectively, $\max(c_{si}, c_{si}^*)$ and $\min(c_{si}, c_{si}^*)$. Define $\mathbf{z} \vee \mathbf{z}^*$ and $\mathbf{z} \wedge \mathbf{z}^*$ as the elements of Ω corresponding to $\mathbf{c} \vee \mathbf{c}^*$ and $\mathbf{c} \wedge \mathbf{c}^*$. It is readily checked that this definition makes sense, that is, that $\mathbf{c} \vee \mathbf{c}^*$ and $\mathbf{c} \wedge \mathbf{c}^*$ always correspond to elements of Ω. For instance, in Figure 2.10.1, $\mathbf{z} = 0110$ and $\mathbf{z}^* = 1001$ correspond to $\mathbf{c} = 32$ and $\mathbf{c}^* = 41$, so $\mathbf{c} \vee \mathbf{c}^* = 42$ and $\mathbf{c} \wedge \mathbf{c}^* = 31$, so $\mathbf{z} \vee \mathbf{z}^* = 1010$ and $\mathbf{z} \wedge \mathbf{z}^* = 0101$, as is consistent with Figure 2.10.1. Notice carefully that the coordinate (s, i) of $\mathbf{z} \vee \mathbf{z}^*$ is not generally equal to $\max(z_{si}, z_{si}^*)$.

To show that Ω is a lattice with these operations, one needs to check L1–L4 in Theorem 2.6, but L1–L3 hold trivially for $\max(c_{si}, c_{si}^*)$ and $\min(c_{si}, c_{si}^*)$. To show $\mathbf{z} \wedge (\mathbf{z} \vee \mathbf{z}^*) = \mathbf{z}$ in L4, it suffices to show $\mathbf{c} \wedge (\mathbf{c} \vee \mathbf{c}^*) = \mathbf{c}$. If $c_{si} \geq c_{si}^*$, then $\min\{c_{si}, \max(c_{si}, c_{si}^*)\} = \min(c_{si}, c_{si}) = c_{si}$, while if $c_{si} < c_{si}^*$, $\min\{c_{si}, \max(c_{si}, c_{si}^*)\} = \min(c_{si}, c_{si}^*) = c_{si}$, so $\mathbf{c} \wedge (\mathbf{c} \vee \mathbf{c}^*) = \mathbf{c}$ as required. The second part of L4 is proved in the same way. So Ω is a lattice.

More than this, Ω is a distributive lattice. As proof, it suffices to show $\mathbf{c} \vee (\mathbf{c}^* \wedge \mathbf{c}^{**}) = (\mathbf{c} \vee \mathbf{c}^*) \wedge (\mathbf{c} \vee \mathbf{c}^{**})$, that is, to show $\max\{c_{si}, \min(c_{si}^*, c_{si}^{**})\} = \min\{\max(c_{si}, c_{si}^*), \max(c_{si}, c_{si}^{**})\}$. There are two cases. If $c_{si} \geq \min(c_{si}^*, c_{si}^{**})$, then $\max\{c_{si}, \min(c_{si}^*, c_{si}^{**})\} = c_{si}$, but also c_{si} is less than or equal to both $\max(c_{si}, c_{si}^*)$ and $\max(c_{si}, c_{si}^{**})$ yet it equals one of them, so $\min\{\max(c_{si}, c_{si}^*), \max(c_{si}, c_{si}^{**})\} = c_{si}$. On the other hand, if $c_{si} < \min(c_{si}^*, c_{si}^{**})$, then $\max\{c_{si}, \min(c_{si}^*, c_{si}^{**})\} = \min(c_{si}^*, c_{si}^{**})$, but $\max(c_{si}, c_{si}^*) = c_{si}^*$, and $\max(c_{si}, c_{si}^{**}) = c_{si}^{**}$, so $\min\{\max(c_{si}, c_{si}^*), \max(c_{si}, c_{si}^{**})\} = \min(c_{si}^*, c_{si}^{**})$, as required to complete the proof.

2.10.4. Inequalities for Probability Distributions on a Lattice

This section discusses two inequalities for probability distributions on a finite distributive lattice, namely the FKG inequality and Holley's inequality. These inequalities are the principal tool that makes use of the lattice properties of Ω. The original proofs of these inequalities are somewhat involved, but Ahlswede and Daykin (1978) developed a simpler proof involving nothing more than elementary probability. Their proof is nicely presented in several recent texts (Anderson 1987, §6; Bollobas, 1986, §19), to which the reader may refer.

A real-valued function on Ω, $f: \Omega \to \mathbb{R}$ is isotonic if $\mathbf{z} \lesssim \mathbf{z}^*$ implies $f(\mathbf{z}) \leq f(\mathbf{z}^*)$. Throughout this appendix, $\mathbf{r}$ has been sorted into order within each stratum, $r_{si} \geq r_{s,i+1}$ for each s, i. With this order, the arrangement-increasing statistics $t(\mathbf{z}, \mathbf{r})$ are some of the isotonic functions on Ω. Actually, the arrangement-increasing statistics are the interesting isotonic functions, for they are the isotonic functions that are unchanged by interchanging tied responses in the same stratum. If there are ties, that is if $r_{si} = r_{s,i+1}$ for some s and i, then there are isotonic functions that are not arrangement-increasing, specifically functions that increase when $z_{si} = 0$, $z_{s,i+1} = 1$ is replaced by $z_{si} = 1$, $z_{s,i+1} = 0$; however, these functions are not interesting as test statistics $t(\mathbf{z}, \mathbf{r})$ because they distinguish between people who gave identical responses. From a practical point of view, the important point is that a property of all isotonic functions on Ω is automatically a property of all arrangement-increasing functions, and all of the statistics in §2.4.2 are arrangement-increasing.

The first inequality is due to Fortuin, Kasteleyn, and Ginibre (1971).

Theorem 2.7 (The FKG Inequality). *Let $f(\cdot)$ and $g(\cdot)$ be isotonic functions on a finite distributive lattice Ω. If a random element $\mathbf{Z}$ of Ω is selected by a probability distribution satisfying*

$$\text{prob}(\mathbf{Z} = \mathbf{z} \vee \mathbf{z}^*) \cdot \text{prob}(\mathbf{Z} = \mathbf{z} \wedge \mathbf{z}^*) \geq \text{prob}(\mathbf{Z} = \mathbf{z}) \cdot \text{prob}(\mathbf{Z} = \mathbf{z}^*) \quad \textit{for all} \quad \mathbf{z}, \mathbf{z}^* \in \Omega,$$

then

$$\text{cov}\{f(\mathbf{Z}), g(\mathbf{Z})\} \geq 0.$$

For example, randomization gives equal probabilities to all elements of Ω, so the randomization distribution satisfies the condition for the FKG inequality. Hence, under the null hypothesis of no effect in a randomized experiment, any two arrangement-increasing statistics have a nonnegative correlation.

The next theorem is due to Holley (1974).

Theorem 2.8 (Holley's Inequality). *Let $f(\cdot)$ be an isotonic function on a finite distributive lattice Ω. If $\mathbf{Z}$ and $\tilde{\mathbf{Z}}$ are random elements of Ω selected by two probability distributions satisfying*

$$\text{prob}(\mathbf{Z} = \mathbf{z} \vee \mathbf{z}^*) \cdot \text{prob}(\tilde{\mathbf{Z}} = \mathbf{z} \wedge \mathbf{z}^*) \geq \text{prob}(\mathbf{Z} = \mathbf{z}) \cdot \text{prob}(\tilde{\mathbf{Z}} = \mathbf{z}^*) \quad \textit{for all} \quad \mathbf{z}, \mathbf{z}^* \in \Omega,$$

then

$$E\{f(\mathbf{Z})\} \geq E\{f(\tilde{\mathbf{Z}})\}.$$

In other words, the premise of Holley's inequality is a sufficient condition for $\mathbf{Z}$ to be stochastically larger than $\tilde{\mathbf{Z}}$, in the sense that for every increasing function $f(\cdot)$, $f(\mathbf{Z})$ is expected to exceed $f(\tilde{\mathbf{Z}})$. Holley's inequality will help later in comparing a nonrandom assignment of treatments to a random assignment.

2.10.5. An Identity in Ω

There is a useful identity in the set Ω of treatment assignments. The identity links $\vee$ and $\wedge$ to the addition of vectors, and therefore it will be useful in verifying the conditions of the FKG inequality and Holley's inequality. It is true for this lattice, but not true generally for all lattices.

Lemma 2.9. *For all* $\mathbf{z}, \mathbf{z}^* \in \Omega$,

$$\mathbf{z} \vee \mathbf{z}^* + \mathbf{z} \wedge \mathbf{z}^* = \mathbf{z} + \mathbf{z}^*.$$

Proof. Fix a coordinate (s, i), so the task is to show $z_{si} + z_{si}^* = z_{\wedge si} + z_{\vee si}$ where $z_{\wedge si}$ and $z_{\vee si}$ are the (s, i) coordinates of $\mathbf{z} \wedge \mathbf{z}^*$ and $\mathbf{z} \vee \mathbf{z}^*$, respectively. Let $\mathbf{c}$ and $\mathbf{c}^*$ correspond with $\mathbf{z}$ and $\mathbf{z}^*$, respectively. There are three cases, depending upon the value of $z_{si} + z_{si}^*$.

(1) If $z_{si} + z_{si}^* = 0$, then $c_{sj} \neq n_s - i + 1$ and $c_{sj}^* \neq n_s - i + 1$ for $j = 1, \ldots, m_s$, so $\max(c_{sj}, c_{sj}^*) \neq n_s - i + 1$ and $\min(c_{sj}, c_{sj}^*) \neq n_s - i + 1$ for $j = 1, \ldots, m_s$, so $z_{\wedge si} = z_{\vee si} = 0$ as required.

(2) If $z_{si} + z_{si}^* = 2$, then there is a j and a k such that $c_{sj} = n_s - i + 1$ and $c_{sk}^* = n_s - i + 1$. If $j = k$, then $\max(c_{sj}, c_{sj}^*) = \min(c_{sj}, c_{sj}^*) = n_s - i + 1$, so $z_{\wedge si} = z_{\vee si} = 1$, as required. If $j < k$, then $n_s - i + 1 = c_{sj} > c_{sk}$ and $c_{sj}^* > c_{sk}^* = n_s - i + 1$, so $\min(c_{sj}, c_{sj}^*) = c_{sj} = n_s - i + 1$ and $\max(c_{sk}, c_{sk}^*) = c_{sk}^* = n_s - i + 1$, so $z_{\wedge si} = z_{\vee si} = 1$, as required. The case $j > k$ is similar.

(3) If $z_{si} = 1$ and $z_{si}^* = 0$, so $z_{si} + z_{si}^* = 1$, then there is a j such that $c_{sj} = n_s - i + 1$ but $c_{sk}^* \neq n_s - i + 1$ for $k = 1, \ldots, m_s$. In this case, either $n_s - i + 1 = \max(c_{sj}, c_{sj}^*)$ or $n_s - i + 1 = \min(c_{sj}, c_{sj}^*)$ but not both, and moreover $n_s - i + 1 \neq \max(c_{sk}, c_{sk}^*)$ and $n_s - i + 1 \neq \min(c_{sk}, c_{sk}^*)$ for $k \neq j$, so $z_{\wedge si} + z_{\vee si} = 1$, as required. The case $z_{si} = 0$ and $z_{si}^* = 1$ is similar. □

If there were no ties, so $\mathbf{c}$ and $\mathbf{c}^*$ are ranks, then Proposition 2.9 has the following interpretation. Within each stratum, the operations $\vee$ and $\wedge$ take the ranks in $\mathbf{c}$ and $\mathbf{c}^*$ and apportion them in forming $\mathbf{c} \vee \mathbf{c}^*$ and $\mathbf{c} \wedge \mathbf{c}^*$, but in this process they do not create or delete ranks that appear in $\mathbf{c}$ and $\mathbf{c}^*$.

2.11. Bibliographic Notes

Fisher is usually credited with the invention of randomized experiments. See, in particular, his important and influential book, *The Design of Experiments*, first published in 1935. Randomization is discussed in many articles and textbooks. In particular, see Kempthorne (1952) and Cox (1958a, §5) for discussion of randomization in experimental design, and see Lehmann (1975) and Maritz (1981) for discussion of its role in nonparametrics.

Mantel's (1963) paper was significant not just for the method he proposed, but also for its strengthening of the link between nonparametric methods and contingency table methods. The model for a treatment effect in §2.5.2 in which each unit has two potential responses, one under treatment and the other under control, has a long history. In an article first published in Polish and recently translated into English, Neyman (1923) used it to study the behavior of statistical tests under random assignment of treatments. Related work was done by Welch (1937), Wilk (1955), Cox (1958, §5), Robinson (1973), among others. Rubin (1974, 1977) first used the model in observational studies. In particular, he discussed the conditions under which matching, stratification, and covariance adjustment all estimate the same treatment effect. See also Hamilton (1979) and Holland (1986). Arrangement-increasing functions have been studied under various names by Eaton (1967), Hollander, Proshan, and Sethuraman (1977), and Marshall and Olkin (1979, §6F); see also Savage (1957). Sign-score statistics are discussed in Rosenbaum (1988) in connection with sensitivity analysis where these statistics permit certain simplifications. The discussion of complex outcomes in §2.8 draws from Mann and Whitney (1947), Gehan (1965), Mantel (1967), and Rosenbaum (1991, 1994). The material in §2.10 uses ideas from Savage (1964) and Rosenbaum (1989).

2.12. Problems

2.12.1. Proof of Proposition 2.1: Independence in Uniform Randomized Experiments

1. Let A and B be two finite, disjoint sets, $|A| = \alpha < \infty$, $|B| = \beta < \infty$, $A \cap B = \emptyset$, and let $A \times B$ be the set of all ordered pairs (a, b), $a \in A$, $b \in B$. Show that if (a, b) is selected at random from $A \times B$, each element having probability $1/(\alpha\beta)$, then a and b are independent.

2. Use Problem 1 to prove Proposition 2.1. Prove Proposition 2.1 first for $S = 2$. Then prove Proposition 2.1 for $S + 1$ assuming it is true for S.

2.12.2. Proof of Proposition 2.2: Expectation and Variance of a Sum Statistic

3. Problems 3, 4, and 5 prove Proposition 2.2, and the notation of that the proposition is used. Pick one (s, i). In a uniform randomized experiment, what is the expectation of Z_{si}? What is the expectation of $Z_{si}q_{si}$ where q_{si} is fixed? What is the expectation of $\mathbf{Z}^{\mathrm{T}}\mathbf{q}$?

4. In a uniform randomized experiment with q_{si} fixed, why is it true that:

$$\operatorname{var}\left(\sum_{s=1}^{S}\sum_{i=1}^{n_s} Z_{si}q_{si}\right) = \sum_{s=1}^{S} \operatorname{var}\left(\sum_{i=1}^{n_s} Z_{si}q_{si}\right)?$$

Why is it true that

$$\operatorname{var}\left(\sum_{i=1}^{n_s} Z_{si} q_{si}\right) = \operatorname{var}\left(\sum_{i=1}^{n_s} Z_{si}(q_{si} - \bar{q}_s)\right)?$$

5. Pick an (s, i) and an (s, j) with $i \neq j$. In a uniform randomized experiment with q_{si} fixed, what is the variance of $Z_{si}(q_{si} - \bar{q}_s)$? What is the expectation of $Z_{si}Z_{sj}$? What is the covariance of $Z_{si}(q_{si} - \bar{q}_s)$ and $Z_{sj}(q_{sj} - \bar{q}_s)$? What is the variance of $\sum_{i=1}^{n_s} Z_{si}(q_{si} - \bar{q}_s)$? Complete the proof of Proposition 2.2.

2.12.3. Interference Between Units

6. In §2.5.1, it was noted that the model of no interference between units is not reasonable for Fisher's lady tasting tea, because to change the classification of one cup requires a change in the classification of another. Consider the following model. The lady correctly classifies cups once she has tasted at least one cup with milk first and one with tea first, continuing correctly until forced to err by the constraint $4 = \mathbf{r}^T\mathbf{1}$. In what way does the model fail to determine $\mathbf{r}_\mathbf{z}$, $\mathbf{z} \in \Omega$? Is this model compatible with "no interference between units"? Why or why not?

7. Suppose that the n_s units in stratum s are consecutive weekly measurements on a person s, and that weeks are randomly selected for treatment according to the procedures of a uniform randomized experiment. Consider the following model for the responses, $r_{s,1,\mathbf{z}} = \eta_{s1} + \Delta z_{s1}$ and $r_{s,i,\mathbf{z}} - r_{s,i-1,\mathbf{z}} = \eta_{si} + \Delta z_{si}$ for $i \geq 2$, where η_{si} and Δ are unknown parameters. This model says that the increase in response from week $i - 1$ to week i is a constant depending upon the person s and the week i, plus an impact Δ if the treatment was given to person s in week i. Show that the null hypothesis of no treatment effect holds if $\Delta = 0$. Deduce that (2.4.4) may be used to test the null hypothesis that $\Delta = 0$. Show that this model is incompatible with "no interference between units" if $\Delta \neq 0$.

2.12.4. Arrangement-Increasing Functions

8. Find a plausible statistic $t(\mathbf{z}, \mathbf{r})$ for comparing two groups that is *not* arrangement increasing. (*Hint*: Consider a test statistic that is the difference of a measure of the locations in the treated and control group. Suppose this measure of location "rejects" extreme observations, either explicitly by testing for outliers or implicitly as in a redescending m-estimate. Consider the following artificial data:

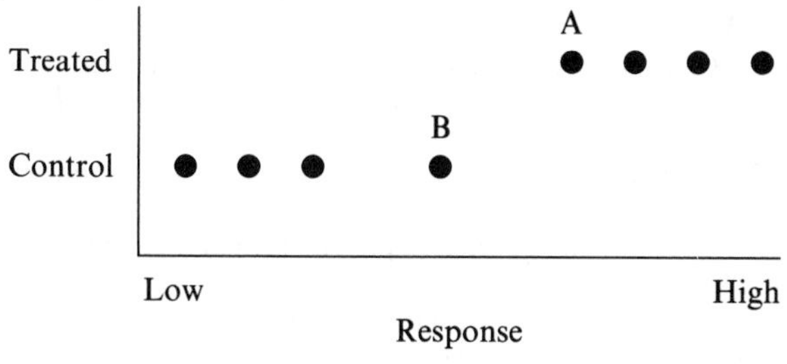

What can happen when point A is interchanged with point B?)

9. Let $\mathbf{a}$ be an N-dimensional vector such that $\mathbf{a}^T\mathbf{a} = 1$. Show that the matrix $\mathbf{I} - 2\mathbf{a}\mathbf{a}^T$ is an orthogonal matrix. Show that $\mathbf{I} - 2\mathbf{a}\mathbf{a}^T$ reflects points through the hyperplane orthogonal to $\mathbf{a}$. Find a vector $\mathbf{a}$ such that $(\mathbf{I} - 2\mathbf{a}\mathbf{a}^T)\mathbf{y}$ interchanges the ith and jth coordinates of $\mathbf{y}$. Express the definition of an arrangement-increasing function in terms of reflections. (In fact, arrangement-increasing functions are a special case of a larger class of functions defined in terms of reflections; see Eaton (1982, 1987) and the references given there.)

2.12.5. Statistics That Yield the Same Test or Estimate

10. Let $f(\cdot)$ be a strictly increasing function, that is, $x < y$ implies $f(x) < f(y)$. Consider two test statistics, $t(\mathbf{Z}, \mathbf{r})$ and $f\{t(\mathbf{Z}, \mathbf{r})\}$. Show that these two test statistics give the same numerical value of the significance level (2.4.2). Show that in a uniform randomized experiment, the statistics $\mathbf{Z}^T\mathbf{r}$ and the difference in sample means (2.4.1) give the same numerical value of the significance level (2.4.4).

11. Consider a uniform randomized experiment with a single stratum, $S = 1$, in which the test statistic $t(\mathbf{Z}, \mathbf{r})$ is the treated-minus-control difference in sample means (2.4.1). What is the Hodges–Lehmann estimate of an additive treatment effect that is obtained from this test statistic $t(\mathbf{Z}, \mathbf{r})$?

References

Ahlswede, R. and Daykin, D. (1978). An inequality for the weights of two families of sets, their unions, and intersections. *Z. Wahrsch. Verus Gebiete*, **43**, 183–185.

Anderson, I. (1987). *Combinatorics of Finite Sets*. New York: Oxford University Press.

Birch, M.W. (1964). The detection of partial association, I: The 2 × 2 case. *Journal of the Royal Statistical Society*, Series B, **26**, 313–324.

Birch, M.W. (1965). The detection of partial association, II: The general case. *Journal of the Royal Statistical Society*, Series B, **27**, 111–124.

Bollobas, B. (1986). *Combinatorics*. New York: Cambridge University Press.

Campbell, D. and Stanley, J. (1963). *Experimental and Quasi-Experimental Designs for Research*. Chicago: Rand McNally.

Cochran, W.G. (1963). *Sampling Techniques*. New York: Wiley.

Cox, D.R. (1958a). *Planning of Experiments*. New York: Wiley.

Cox, D.R. (1958b). The interpretation of the effects of non-additivity in the Latin square. *Biometrika*, **45**, 69–73.

Cox, D.R. (1966). A simple example of a comparison involving quantal data. *Biometrika*, **53**, 215–220.

Cox, D.R. (1970). *The Analysis of Binary Data*. London: Methuen.

Cox, D.R. and Hinkley, D.V. (1974). *Theoretical Statistics*. London: Chapman & Hall.

Eaton, M. (1967). Some optimum properties of ranking procedures. *Annals of Mathematical Statistics*, **38**, 124–137.

Eaton, M. (1982). A review of selected topics in probability inequalities. *Annals of Statistics*, **10**, 11–43.

Eaton, M. (1987). *Lectures on Topics in Probability Inequalities*. Amsterdam: Centrum voor Wiskunde en Informatica.

Efron, B. (1971). Forcing a sequential experiment to be balanced. *Biometrika*, **58**, 403–417.
Fisher, R.A. (1935, 1949). *The Design of Experiments*. Edinburgh: Oliver & Boyd.
Fortuin, C., Kasteleyn, P., and Ginibre, J. (1971). Correlation inequalities on some partially ordered sets. *Communications in Mathematical Physics*, **22**, 89–103.
Gehan, E. (1965). A generalized Wilcoxon test for comparing arbitrarily singly-censored samples. *Biometrika*, **52**, 203–223.
Hamilton, M. (1979). Choosing a parameter for 2×2 table or $2 \times 2 \times 2$ table analysis. *American Journal of Epidemiology*, **109**, 362–375.
Hettmansperger, T. (1984). *Statistical Inference Based on Ranks*. New York: Wiley.
Hodges, J. and Lehmann, E. (1962). Rank methods for combination of independent experiments in the analysis of variance. *Annals of Mathematical Statistics*, **33**, 482–497.
Hodges, J. and Lehmann, E. (1963). Estimates of location based on rank tests. *Annals of Mathematical Statistics*, **34**, 598–611.
Holland, P. (1986). Statistics and causal inference (with Discussion). *Journal of the American Statistical Association*, **81**, 945–970.
Hollander, M., Proschan, F., and Sethuraman, J. (1977). Functions decreasing in transposition and their applications in ranking problems. *Annals of Statistics*, **5**, 722–733.
Hollander, M. and Wolfe, D. (1973). *Nonparametric Statistical Methods*. New York: Wiley.
Holley, R. (1974). Remarks on the FKG inequalities. *Communications in Mathematical Physics*, **36**, 227–231.
Kempthorne, O. (1952). *The Design and Analysis of Experiments*. New York: Wiley.
Lehmann, E.L. (1959). *Testing Statistical Hypotheses*. New York: Wiley.
Lehmann, E.L. (1975). *Nonparametrics: Statistical Methods Based on Ranks*. San Francisco: Holden-Day.
MacLane, S. and Birkoff, G. (1988). *Algebra*. New York: Chelsea.
Mann, H. and Whitney, D. (1947). On a test of whether one of two random variables is stochastically larger than the other. *Annals of Mathematical Statistics*, **18**, 50–60.
Mantel, N. (1963). Chi-square tests with one degree of freedom: Extensions of the Mantel–Haenszel procedure. *Journal of the American Statistical Association*, **58**, 690–700.
Mantel, N. and Haenszel, W. (1959). Statistical aspects of retrospective studies of disease. *Journal of the National Cancer Institute*, **22**, 719–748.
Mantel, N. (1967). Ranking procedures for arbitrarily restricted observations. *Biometrics*, **23**, 65–78.
Maritz, J. (1981). *Distribution-Free Statistical Methods*. London: Chapman & Hall.
Marshall, A. and Olkin, I. (1979). *Inequalities: Theory of Majorization and Its Applications*. New York: Academic Press.
McNemar, Q. (1947). Note on the sampling error of the differences between correlated proportions or percentage. *Psychometrika*, **12**, 153–157.
Murphy, M., Hultgren, H., Detre, K., Thomsen, J., and Takaro, T. (1977). Treatment of chronic stable angina: A preliminary report of survival data of the randomized Veterans Administration Cooperative study. *New England Journal of Medicine*, **297**, 621–627.
Neyman, J. (1923). On the application of probability theory to agricultural experiments. Essay on principles. Section 9. (In Polish) *Roczniki Nauk Roiniczych, Tom X*, pp. 1–51. Reprinted in *Statistical Science* 1990, **5**, 463–480, with discussion by T. Speed and D. Rubin.
Pagano, M. and Tritchler, D. (1983). Obtaining permutation distributions in polynomial time. *Journal of the American Statistical Association*, **78**, 435–440.

Robinson, J. (1973). The large sample power of permutation tests for randomization models. *Annals of Statistics*, **1**, 291–296.

Rosenbaum, P.R. (1988). Sensitivity analysis for matching with multiple controls. *Biometrika*, **75**, 577–581.

Rosenbaum, P.R. (1989). On permutation tests for hidden biases in observational studies: An application of Holley's inequality to the Savage lattice. *Annals of Statistics*, **17**, 643–653.

Rosenbaum, P.R. (1991). Some poset statistics. *Annals of Statistics*, **19**, 1091–1097.

Rosenbaum, P.R. (1994). Coherence in observational studies. *Biometrics*, **50**, 368–374.

Rubin, D.B. (1974). Estimating the causal effects of treatments in randomized and nonrandomized studies. *Journal of Educational Psychology*, **66**, 688–701.

Rubin, D.B. (1977). Assignment to treatment group on the basis of a covariate. *Journal of Educational Statistics*, **2**, 1–26.

Rubin, D.B. (1986). Which ifs have causal answers? *Journal of the American Statistical Association*, **81**, 961–962.

Savage, I.R. (1957). Contributions to the theory of rank order statistics: The trend case. *Annals of Mathematical Statistics*, **28**, 968–977.

Savage, I.R. (1964). Contributions to the theory of rank order statistics: Applications of lattice theory. *Review of the International Statistical Institute*, **32**, 52–63.

Tukey, J.W. (1985). Improving crucial randomized experiments—especially in weather modification—by double randomization and rank combination. In: *Proceedings of the Berkeley Conference in Honor of Jerzy Neyman and Jack Kiefer* (eds., L. Le Cam and R. Olshen), Volume 1, Belmont, CA: Wadsworth, pp. 79–108.

Welch, B.L. (1937). On the z-test in randomized blocks and Latin squares. *Biometrika*, **29**, 21–52.

Wilcoxon, F. (1945). Individual comparisons by ranking methods. *Biometrics*, **1**, 80–83.

Wilk, M.B. (1955). The randomization analysis of a generalized randomized block design. *Biometrika*, **42**, 70–79.

Zelen, M. (1974). The randomization and stratification of patients to clinical trials. *Journal of Chronic Diseases*, **27**, 365–375.

CHAPTER 3

Overt Bias in Observational Studies

3.1. Introduction: An Example and Planning Adjustments

3.1.1. Outline: When Can Methods for Randomized Experiments Be Used?

An observational study is biased if the treated and control groups differ prior to treatment in ways that matter for the outcomes under study. An overt bias is one that can be seen in the data at hand—for instance, prior to treatment, treated subjects are observed to have lower incomes than controls. A hidden bias is similar but cannot be seen because the required information was not observed or recorded. Overt biases are controlled using adjustments, such as matching or stratification. In other words, treated and control subjects may be seen to differ in terms of certain observed covariates, but these visible differences may be removed by comparing treated and control subjects with the same values of the observed covariates, that is, subjects in the same matched set or stratum defined by the observed covariates. It is natural to ask when the standard methods for randomized experiments may be applied to matched or stratified data from an observational study. This chapter discusses a model for an observational study in which there is overt bias but no hidden bias. The model is, at best, one of many plausible models, but it does clarify when methods for randomized experiments may be used in observational studies, and so it becomes the starting point for thinking about hidden biases. Dealing with hidden bias is the focus of Chapters 4 through 8. To permit discussion of conceptual issues in this chapter, Chapter 9 will discuss the algorithmic issues that arise in constructing matched sets or strata with many covariates. The remainder of §3.1 considers an example and then discusses some of the planning steps that precede adjustments for covariates.

3.1.2. An Example with a Single Covariate

Cochran (1968) presents three stark examples of overt biases and their removal through adjustments. We will look at one of these. The data are from a study by Best and Walker of mortality in three groups of men: nonsmokers, cigarette smokers, and cigar and pipe smokers. Nonsmokers had a mortality rate of 20.2 deaths per 1000 people per year, cigarette smokers had 20.5 deaths, and cigar and pipe smokers had 35.5 deaths. The naive interpretation would be that cigarettes are harmless, but either cigars or pipes or both are dangerous. Cochran then gives the mean age in each group: 54.9 years for nonsmokers, 50.5 years for smokers, and 65.9 for cigar and pipe smokers. Clearly, the cigar and pipe smokers are older, so their higher death rate is not surprising, and may not reflect an effect of cigars or pipes. On the other hand, the cigarette smokers are the youngest group, and yet there mortality rate is slightly higher than the somewhat older nonsmokers. Perhaps cigarettes are not harmless.

Cochran then adjusts mortality for age, that is, removes an overt bias in the outcome by adjusting for an imbalance in a covariate. He uses age to divide the men into three strata or subclasses so that men in the same stratum have similar ages. Nonsmokers, cigarette smokers, and cigar and pipe smokers of roughly the same age are then compared to each other within each stratum, and the results are combined into a single rate using direct adjustment, essentially as described in §2.7.1. The adjusted mortality rate is 20.3 deaths per 1000 per year for nonsmokers, 28.3 for cigarette smokers, and 21.2 for cigar and pipe smokers. Now it is cigarettes that appear dangerous.

Which rates should be trusted, unadjusted or adjusted? Neither. The unadjusted rates are clearly wrong as a basis for estimating the effects of smoking, for they compare men who are not comparable in terms of one of the most important features of human mortality, namely, age. The adjusted rates are not clearly wrong. They might estimate the effects of smoking. However, it is possible that there is another covariate that was not recorded that has an impact similar to age; in this case, there would be a hidden bias. The current chapter discusses the conditions under which the methods in Chapter 2 for randomized experiments successfully estimate treatment effects in observational studies. These conditions become the basis in later chapters for thinking about hidden biases.

Cochran used three age strata. One might reasonably ask whether three strata are sufficient, whether such broad age groups suffice to remove the overt bias due to age, and indeed this is the main question in Cochran's paper. If instead of three-age strata, 12 strata are used, then the adjusted rates are 20.2 for nonsmokers, 29.5 for cigarette smokers, and 19.8 for cigar and pipe smokers. Three-age strata and twelve-age strata produce similar adjusted rates, both of which are very different from the rates prior to adjustment. Cochran presents a theoretical argument concluding that five strata, each containing 20% of the subjects, will remove about 90% of the bias in a single continuous covariate such as age.

3.1.3. Planning Adjustments for Overt Biases

Options narrow as an investigation proceeds. What is easy early on may become difficult or impossible later. This section discusses the earliest stages of planning and data collection, as they relate to adjustments for bias. The points raised are elementary, but at times ignored. When ignored, the problems created can be far from elementary, at times insurmountable.

The control of overt biases begins before the study is designed. A first step in planning an observational study is to determine what treatments will be studied, and in the process to distinguish outcomes from covariates. Outcomes measure quantities that may be affected by the treatment, while covariates are not affected; see §2.5. Cox (1958, §4.2) uses the term concomitant observations in place of covariate and writes:

> The essential point in our assumptions about these observations is that the value for any unit must be unaffected by the particular assignment of treatments to units actually used. In practice this means that either: (a) the concomitant observations are taken before the assignment of treatments to units is made; or (b) the concomitant observations are made after the assignment of treatments, but before the effect of treatments has had time to develop ...; or (c) we can assume from our knowledge of the nature of the concomitant observations concerned, that they are unaffected by the treatment.

As an example of type (c), Cox mentions the covariate which records the relative humidity in a textile factory, where it is known that the treatments under study could not possibly affect the relative humidity.

If adjustments are not confined to covariates, then adjustments may remove part or all of the effect of the treatment. To illustrate, consider an extreme, hypothetical example. Imagine a study comparing a placebo and a drug intended to reduce blood pressure, the outcome being the incidence of stroke. If the groups were compared after adjustment for blood pressure levels 6 months after the start of treatment, then the adjusted incidence of stroke might be similar in drug and placebo groups, not because the drug has failed to work, but rather because the drug reduces the risk of stroke by reducing blood pressure. If the effect of the drug on blood pressure is removed, the effect on stroke is removed with it.

While adjustments for an outcome can remove part of the treatment effect, adjustments of this sort are occasionally performed. It may be suspected that the treatment has only slight effects on a particular outcome, but this outcome may be strongly related to an important covariate that was not measured. An example occurred in the studies by Coleman, Hoffer, and Kilgore (1982) and Goldberger and Cain (1982) of the effects of Catholic versus public high schools. These studies compared cognitive test scores in the senior year of high school adjusting for various covariates, but the studies also adjusted for an outcome, namely, cognitive test scores in the sophomore year. The sophomore year test scores may already be affected by the difference between Catholic and public high schools, so they are, in principle, outcomes, not covariates. Still, it is natural to suspect that any effect of Catholic versus public high schools is produced gradually and cumulatively, and that only a part of the effect is present in the sophomore year. These studies used this outcome as a surrogate for an important covariate that was unavailable, namely, cognitive test scores prior to the start of high school. There are, then, two hazards. First, adjusting for sophomore test scores can remove part of the difference between Catholic and public schools. Second, failing to adjust for an early test score may yield a comparison of students who were not comparable in terms of their cognitive abilities prior to the start of high school. Notice that the second hazard is not present in a randomized experiment, so in an experiment, it is possible to give unequivocal advice that adjustments for outcomes should be avoided when estimating treatment effects. In an observational study, both hazards are present, and must be weighed; see Rosenbaum (1984a, §4) for discussion of alternative methods of analysis. The important point for the initial planning of observational studies is the distinction between outcomes and covariates, and their different status in adjustments.

The next step in planning is to list the covariates that will be measured. It is at this stage that biases become either overt or hidden. Since there is no way to completely address a hidden bias, a small change in this list may determine whether or not the study is convincing. A small oversight, easily corrected in the planning stage, may be an insurmountable problem at a later stage. In the design of randomized clinical trials, the standard practice is to begin with a written protocol that describes the data that will be collected and the main analyses that will be performed. Before the trial starts, the protocol is circulated for critical comment. Observational studies would, I believe, benefit from a written protocol and critical commentary.

Adjustments for overt biases may begin with data collection rather than with data analysis. Often treated subjects are matched to controls to form pairs or matched sets of subjects who are comparable in terms of observed covariates, and matching may take place before outcomes are measured. Chapter 9 discusses matching methods. Here, three points should be mentioned. First, unlike analytical adjustments, adjustments that are built into the study design are irrevocable. In the hypothetical example above concerning drug versus placebo to prevent stroke, it would be a mistake to adjust for blood pressure after treatment. If this mistake were made using an analytical method such as in §3.1.2, then it could be corrected by performing a different analysis, but if the mistake were made by matched sampling then it would be difficult to correct.

Second, certain covariates are more easily controlled through matching in the design than through analytical adjustments. Typically, these are covariates which classify subjects into many small categories. Matching can insure that treated and control subjects belong to the same categories, but if matching is not used in the design of the study, some categories may have treated subjects and no controls or controls and no treated subjects. For instance, consider a study (Rosenbaum, 1986) that compared cognitive skills in what would be the senior year of high school for sophomores who dropped out of school and similar sophomores who remained in school. This was done with a national sample of students, and the high school was an important covariate which takes many values. The study used matched pairs of students *from the same school* having similar test scores, academic performance, and disciplinary records in the sophomore year, before the dropout left school.

Cost is an important consideration in deciding whether or not to match. If some covariate information is readily available, but other data is difficult or expensive to obtain, then matching becomes more attractive, but if data comes with negligible costs, then matching during the design becomes somewhat less attractive. The reason is that, in many studies, some controls will be so different from treated subjects that they are of little use for comparisons. In the example above, many high-school students look very different from most dropouts in terms of test scores, academic performance, and disciplinary problems, so these students are of limited use in trying to determine how students who drop out would have performed had they remained in school. Matching may avoid collecting data on controls who will later be of little use.

A compromise between selecting matched pairs and using all potential controls is to match each treated subject to several controls. Ury (1975) studies the efficiency of studies that match several controls to each treated subject, finding that there is little to be gained from having more than four controls per treated subject. Chapter 9 discusses the construction of matched sets with equal and variable numbers of controls per treated subject.

Having collected the data on covariates, the question arises: Should adjustments be made for all observed covariates? If not, how should covariates be selected for adjustment? These questions are somewhat controversial, not so much because the issues involved are unclear, but rather because there is no fully satisfactory answer. In principle, there is little or no reason to avoid adjustment for a true covariate, a variable describing subjects before treatment. There is little harm in comparing subjects who were comparable before treatment in ways that are not relevant for the outcomes of interest. In experiments, randomization tends to make treated and control groups comparable in terms of all covariates, relevant and irrelevant. In practice, the situation is often more involved, and increasing the number of covariates used in adjustments increases costs and complexities, and may make it more difficult to adjust for the most important covariates. In part, there are issues of data quality and completeness. As more covariates are collected and analyzed, it becomes increasingly difficult to ensure that all covariates meet high standards of accuracy and completeness, and increasingly difficult to ensure that each covariate receives the needed attention when used in modeling or matching. If there are many covariates, each with some missing data, there may be few subjects with complete data on all covariates, and this may make the analysis more difficult than it would otherwise be. This considerations weigh most heavily on covariates having doubtful relevance to outcomes of interest.

Perhaps the most common method for selecting covariates is also the most widely criticized. It entails comparing treated and control groups with respect to a long list of covariates, say using a t-test, and adjusting only for those covariates for which significant differences are found. There are three problems with this. First, the process does not consider the relationship between covariate and outcome. Second, there is no reason to believe that the absence of statistical significance implies the imbalance in the covariate is small enough to be ignored. Third, the process considers covariates one at a time, while the adjustments will control the covariates simultaneously. Addressing the first two problems, Cochran (1965, §3.1) studied this technique under a simple linear regression model in which all quantities are Normally distributed and a single covariate is the only source of bias. He looked at the coverage probability of the 95% confidence interval for the effect of the treatment on an outcome when no adjustment had been made for the covariate. This coverage probability was 90% or more providing the t-statistic for the covariate was less than 1.5 in absolute value and providing the squared correlation between the covariate and the outcome was 0.5 or less. This limitation on the square correlation is often reasonable for a covariate whose relevance is in doubt. He concluded: "If a single [covariate] shows a value of t above 1.5, these results suggest that we have another look at this [covariate] when the values of the [outcome] become known." Canner (1984, 1991) discusses closely related issues.

The following approach is often reasonable and practical. Begin by selecting a tentative list of covariates for adjustments using scientific knowledge of the relevant covariates together with exploratory comparisons of covariates in the treated and control groups, perhaps including some version of the technique evaluated by Cochran (1965, §3.1). With this tentative list, determine the tentative method of adjustment, that is, select the matched pairs or sets, define the strata, and determine whatever modeling technique will be used. Apply this method of adjustment to the covariates excluded from the tentative list, identifying any covariates exhibiting a large imbalance after adjustment. Reconsider the tentative list of covariates in light of this analysis. This approach addresses, at least in part, each of the three problems in the previous paragraph. The focus is on covariates known to be relevant since they are included in the initial list. At the same time, the data are given several opportunities to call attention to imbalances that might not be anticipated. An example along these lines is discussed in Rosenbaum and Rubin (1984).

3.2. Adjustments by Exact Stratification and Matching

3.2.1. Treatment Assignment with Unknown Probabilities

When is an observational study free of hidden bias? When do adjustments such as matching and stratification remove all of the bias? This section describes a model for an observational study with overt bias but no hidden bias. In most observational studies, this model is, at best, one of many plausible models—hidden biases are possible. The model is a start, indicating the inferences that would be appropriate were hidden biases absent. Chapters 4 through 8 try to determine whether hidden biases are present and ask how inferences might change if they are.

Initially, there are M units available for study, and each has a value of an observed covariate $\mathbf{x}$, which may contain several variables. Often the covariates $\mathbf{x}$ are used to reorganize the data prior to analysis, for instance, by matching or stratifying on $\mathbf{x}$. Number the M units $j = 1, \ldots, M$, so $\mathbf{x}_{[j]}$ is the covariate for the jth unit and the treatment assignment for this unit is $Z_{[j]}$. The bracketed subscript $[j]$ signifies the numbering of units before they are reorganized. After reorganization, a unit will have a different subscript without a bracket.

As a model for an observational study, imagine that unit j is assigned to treatment with probability $\pi_{[j]} = \text{prob}(Z_{[j]} = 1)$ and to control with probability $(1 - \pi_{[j]})$, with assignments for distinct units being independent, and with $0 < \pi_{[j]} < 1$. The model says that treatments were assigned by flipping biased coins, possibly a different coin with a different bias for each unit, where the biases of the coins or the π's are unknown. The model says:

$$\text{prob}(Z_{[1]} = z_1, \ldots, Z_{[M]} = z_M) = \prod_{j=1}^{M} \pi_{[j]}^{z_j}\{1 - \pi_{[j]}\}^{1-z_j}. \tag{3.2.1}$$

In an observational study, since $\pi_{[j]}$ is unknown, the distribution of $Z_{[1]}, \ldots, Z_{[M]}$ is unknown, and it is not possible to draw inferences as in Chapter 2 where randomization created a known distribution of treatment assignments.

Consider now the model for an observational study with overt biases but no hidden biases. An observational study is *free of hidden bias* if the π's, though unknown, are known to depend only on the observed covariates $\mathbf{x}_{[j]}$, so two units with the same value of $\mathbf{x}$ have the same chance π of receiving the treatment. Formally, the study is free of hidden bias if there exists a function $\lambda(\cdot)$, whose form will typically be unknown, such that $\pi_{[j]} = \lambda(\mathbf{x}_{[j]})$ for $j = 1, \ldots, M$. If the study is free of hidden bias, then (3.2.1) becomes

$$\operatorname{prob}(Z_{[1]} = z_1, \ldots, Z_{[M]} = z_M) = \prod_{j=1}^{M} \lambda(\mathbf{x}_{[j]})^{z_j} \{1 - \lambda(\mathbf{x}_{[j]})\}^{1-z_j}. \quad (3.2.2)$$

In short, an observational study is *free of hidden bias* when (3.2.2) holds. Rubin (1977) calls (3.2.2) "randomization on the basis of a covariate."

When the study is free of hidden bias, the function $\lambda(\mathbf{x})$ is called the propensity score. In §9.2, the propensity score $\lambda(\mathbf{x})$ will be defined so as to also be meaningful when hidden biases are present; however, in that case, $\pi_{[j]} \neq \lambda(\mathbf{x}_{[j]})$, and (3.2.2) does not follow from (3.2.1). A study is free of hidden bias when the treatment assignment probabilities $\pi_{[j]}$ are given by the propensity score $\lambda(\mathbf{x}_{[j]})$ which is always a function of the observed covariates $\mathbf{x}_{[j]}$. Chapter 4 discusses a model in which there is hidden bias and $\pi_{[j]}$ is not a function of $\mathbf{x}_{[j]}$.

A significance level, such as (2.4.2), cannot be calculated using (3.2.2) because $\lambda(\mathbf{x})$ is unknown. To adjust for overt bias in a study that is free of hidden bias is to address the fact that $\lambda(\mathbf{x})$ is unknown. The simplest approach is to stratify on $\mathbf{x}$.

3.2.2. Stratifying on x

Often, units are grouped into strata on the basis of the covariate $\mathbf{x}$. From the M units, select $N \leq M$ units and group them into S nonoverlapping strata with n_s units in stratum s. In selecting the N units and assigning them to strata, use only the $\mathbf{x}$'s and possibly a table of random numbers. A stratification formed in this way is called a *stratification on* $\mathbf{x}$. Renumber the units so the ith unit in stratum s has treatment assignment Z_{si} and covariate $\mathbf{x}_{si}$. Using the same notation as Chapter 2, write $\mathbf{Z}$ for the N-tuple $(Z_{11}, \ldots, Z_{S,n_S})^{\mathrm{T}}$. Write m_s for the number of treated units in stratum s, that is, $m_s = \sum_i Z_{si}$, and $\mathbf{m} = (m_1, \ldots, m_S)^{\mathrm{T}}$.

An *exact stratification on* $\mathbf{x}$ has strata that are homogeneous in $\mathbf{x}$, so two units are in the same stratum only if they have the same value of $\mathbf{x}$, that is $\mathbf{x}_{si} = \mathbf{x}_{sj}$ for all s, i, and j. Exact stratification on $\mathbf{x}$ is practical only when $\mathbf{x}$ is of low dimension and its coordinates are discrete; otherwise, it will be difficult to locate many units with the same $\mathbf{x}$.

In an exact stratification on $\mathbf{x}$, if the study is free of hidden bias, that is, if (3.2.2) holds, then all units in the same stratum have the same chance of receiving the treatment. In this case, write λ_s in place of $\lambda(\mathbf{x}_{si})$, so (3.2.2) implies:

$$\text{prob}(\mathbf{Z} = \mathbf{z}) = \prod_{s=1}^{S} \prod_{i=1}^{n_s} \lambda_s^{z_{si}}(1 - \lambda_s)^{1-z_{si}}. \tag{3.2.3}$$

In (3.2.3), the distribution of treatment assignments, $\text{prob}(\mathbf{Z} = \mathbf{z})$ is unknown because λ_s is unknown, and $m_s = \sum_i Z_{si}$ is a random variable. Consider the conditional distribution of $\mathbf{Z}$ given $\mathbf{m}$. It is a distribution on a set Ω whose elements are N-tuples of 0's and 1's such that $\mathbf{z} \in \Omega$ if and only if $m_s = \sum_i z_{si}$ for $s = 1, \ldots, S$, so Ω has $K = \prod_{s=1}^{S} \binom{n_s}{m_s}$ elements. Every treatment assignment $\mathbf{z} \in \Omega$ has the same unconditional probability in (3.2.3), namely,

$$\text{prob}(\mathbf{Z} = \mathbf{z}) = \prod_{s=1}^{S} \lambda_s^{m_s}(1 - \lambda_s)^{n_s - m_s}. \tag{3.2.4}$$

It follows that the conditional probability given $\mathbf{m}$ is constant, $\text{prob}(\mathbf{Z} = \mathbf{z}|\mathbf{m}) = 1/K$. Of course, this is the distribution of $\mathbf{Z}$ in a uniform randomized experiment; see §2.3.2.

In short, if an observational study is free of hidden bias, and if one stratifies exactly on $\mathbf{x}$, then the conditional distribution of the treatment assignment $\mathbf{Z}$ given the numbers $\mathbf{m}$ of treated units in each stratum, namely $\text{prob}(\mathbf{Z} = \mathbf{z}|\mathbf{m})$, is the same as the distribution of treatment assignments in a uniform randomized experiment. This is true even though the treatment assignment probabilities $\lambda(\mathbf{x})$ are unknown. In this case, given $\mathbf{m}$, the statistical procedures discussed in Chapter 2 have the properties described there. In other words, if the study is free of hidden bias and one stratifies exactly on $\mathbf{x}$, then the study may be analyzed using methods for a uniform randomized experiment.

Be clear on a key point. This result does not say that there is no difference between an experiment and an observational study. The difference is that in a uniform randomized experiment, the assignment probabilities $\text{prob}(\mathbf{Z} = \mathbf{z})$ are known to equal $1/K$ because we forced this to be true by randomizing. In an observational study, the conclusion $\text{prob}(\mathbf{Z} = \mathbf{z}|\mathbf{m}) = 1/K$ is deduced from the premise that the study is free of hidden bias, a premise we have little reason to believe. In an observational study, this premise is subjected to strict scrutiny, asking whether evidence can support or refute it, asking how findings would change if the premise were in error. This scrutiny is the focus of Chapters 4 through 8.

3.2.3. Matching on $\mathbf{x}$

In §3.2.3, strata were formed using $\mathbf{x}$ alone. One way that matching differs from stratification is that there are constraints on the number m_s of treated units and the number $n_s - m_s$ of control units in a stratum. For instance, pair matching requires $n_s = 2$ and $m_s = 1$ for each s, while matching with multiple controls requires $n_s \geq 2$ and $m_s = 1$. A *matching on* $\mathbf{x}$ is a matched sample formed by:

(i) placing some restriction on S, $\mathbf{m}$ and $\mathbf{n} = (n_1, \ldots, n_S)^T$, and
(ii) picking a stratification that meets these restrictions based exclusively on the pattern of $\mathbf{x}$'s in the strata and possibly a table of random numbers.

For instance, a pair matched sample with $S = 100$ pairs would be formed by considering all possible stratifications with $n_s = 2$ and $m_s = 1$ for $s = 1, \ldots, 100$, and selecting one of these possible statifications based on the $\mathbf{x}$'s in the strata and possibly random numbers. An *exact matching on* $\mathbf{x}$ is a matching on $\mathbf{x}$ in which $\mathbf{x}$ is the same for all n_s units in each matched set, that is, $\mathbf{x}_{si} = \mathbf{x}_{sj}$ for $i, j = 1, \ldots, n_s$ for each s. As with exact stratification, exact matching is possible only when $\mathbf{x}$ is of low dimension and discrete.

The same argument as in §3.2.2 shows that, in an observational study that is free of hidden bias, if one matches exactly on $\mathbf{x}$, then the conditional distribution of the treatment assignment $\mathbf{Z}$ given $\mathbf{m}$ is the same as in a uniform randomized experiment, $\text{prob}(\mathbf{Z} = \mathbf{z}|\mathbf{m}) = 1/K$. If the study were free of hidden bias, it could be analyzed as if it were a matched randomized experiment, but of course, the comment at the end of §3.2.2 applies here as well.

3.2.4. An Example: Lead in the Blood of Children

Morton et al. (1982) studied lead in the blood of children whose parents worked in a factory where lead is used in making batteries. They were concerned that children were exposed to lead inadvertently brought home by their parents. Their study included 33 such children from different families—they are the exposed or treated children. The outcome R was the level of lead found in a child's blood in $\mu g/dl$ of whole blood. The covariate $\mathbf{x}$ was two-dimensional recording age and neighborhood of residence. They matched each exposed child to one control child of the same age and neighborhood whose parents were employed in other industries not using lead. Table 3.2.1 shows the levels of lead found in the children's blood in $\mu g/dl$ of whole blood.

If this study were free of hidden bias, which may or may not be the case, we would be justified in analyzing Table 3.2.1 using methods for a uniform randomized experiment with 33 matched pairs. If the null hypothesis of no treatment effect is tested using Wilcoxon's signed rank test, the one-sided significance level is less than 0.0001. The Hodges–Lehmann estimate of the size of an additive effect is 15 $\mu g/dl$ with 95% confidence interval (9.5, 20.5). If the study were free of hidden bias, this would strongly suggest that the parents who work with lead did raise the level of lead in their children's

Table 3.2.1. Lead in Children's Blood ($\mu g/dl$).

Pair	Exposed	Control	Difference	Rank
1	38	16	22	22
2	23	18	5	8
3	41	18	23	23.5
4	18	24	−6	9.5
5	37	19	18	21
6	36	11	25	26
7	23	10	13	14
8	62	15	47	32
9	31	16	15	17
10	34	18	16	18.5
11	24	18	6	9.5
12	14	13	1	2.5
13	21	19	2	4
14	17	10	7	11
15	16	16	0	1
16	20	16	4	7
17	15	24	−9	12.5
18	10	13	−3	5.5
19	45	9	36	30
20	39	14	25	26
21	22	21	1	2.5
22	35	19	16	18.5
23	49	7	42	31
24	48	18	30	28
25	44	19	25	26
26	35	12	23	23.5
27	43	11	32	29
28	39	22	17	20
29	34	25	9	12.5
30	13	16	−3	5.5
31	73	13	60	33
32	25	11	14	15.5
33	27	13	14	15.5

blood by about 15 $\mu g/dl$, a large increase compared to the level of lead found in controls. In later chapters, these data will be examined again without the premise that the study is free of hidden bias.

3.2.5. Stratifying and Matching on the Propensity Score

Often, exact stratification or matching on $\mathbf{x}$ is difficult or impossible. If $\mathbf{x}$ is of high dimension or contains continuous measurements, each of the N

units may have a different value of $\mathbf{x}$, so no stratum can contain a treated and control unit with the same $\mathbf{x}$. There are several questions. Does a large number of covariates—that is, a high-dimensional $\mathbf{x}$—make stratification and matching infeasible? Does close but inexact matching on $\mathbf{x}$ remove most of the bias due to $\mathbf{x}$? What algorithms produce good stratifications or matchings? The second and third question will be discussed in Chapter 9. The current section begins to answer the first question. As it turns out, there is a sense in which all matching problems are one dimensional, so the dimensionality of $\mathbf{x}$ is not critical by itself.

Suppose an observational study is free of hidden bias, so (3.2.2) holds. Instead of stratifying or matching exactly on $\mathbf{x}$, imagine forming strata or matched sets in which units in the same stratum have the same chance of receiving the treatment $\lambda(\mathbf{x})$. Then within a stratum or matched set, units may have different values of $\mathbf{x}$, but they have the same propensity score $\lambda(\mathbf{x})$. Formally, it may happen that $\mathbf{x}_{si} \neq \mathbf{x}_{sj}$ but always $\lambda(\mathbf{x}_{si}) = \lambda(\mathbf{x}_{sj})$. Call this *exact matching or stratification on the propensity score.* In this case, the arguments in §3.2.2 and §3.2.3 go through without changes. In those arguments, equal $\mathbf{x}$'s within strata were used only to ensure equal $\lambda(\mathbf{x})$'s. In short, in an observational study free of hidden bias, exact matching or stratification on the propensity score yields a conditional distribution of treatment assignments $\mathbf{Z}$ given $\mathbf{m}$ that is the same as a uniform randomized experiment, namely $\text{prob}(\mathbf{Z} = \mathbf{z}|\mathbf{m}) = 1/K$. In this case, the methods in Chapter 2 for uniform randomized experiments may be applied. The same conclusion is reached if strata or matched sets are formed based on $\lambda(\mathbf{x})$ and parts of $\mathbf{x}$, providing the strata or matched sets are homogeneous in $\lambda(\mathbf{x})$.

In practice, $\lambda(\mathbf{x})$ is unknown, so matching or stratification on $\lambda(\mathbf{x})$ is not possible. The use of estimated propensity scores in matching and stratification is discussed in Chapter 9. Also, §9.2 discusses balancing properties of the propensity score that are true whether or not the study is free of hidden bias.

3.3. Case-Referent Studies

3.3.1. Selecting Subjects Based on Their Outcomes

In Chapter 1, the study of DES and vaginal cancer was a case-referent study, and such a study has two features that distinguished it from the other observational studies in Chapter 1. First, the binary outcome, namely, vaginal cancer, is extremely rare, so a study which simply followed women until they developed the disease would need an enormous number of women to produce even a handful of cases of this rare cancer. This first feature is the motivation for conducting a case-referent study, but it is the second feature that characterizes such a study. In a case-referent study, cases are deliberately over-represented and referents are under-represented.

The DES study identified cases of vaginal cancer and compared them to a small number of matched referents in terms of the frequency of maternal exposure to DES. In other words, subjects are included or excluded from the study, in part, on the basis of their outcomes. Instead of comparing the outcomes of treated and untreated groups, the case-referent study compares the frequency of exposure to the treatment among cases and referents.

At first, it is not clear that this makes sense. If the outcome $\mathbf{R} = \mathbf{r}_{\mathbf{Z}}$ is affected by the treatment $\mathbf{Z}$, then selecting subjects using their outcomes may distort the frequency of exposure to the treatment. Indeed, this seems to have happened in the DES study. Of the 40 women in the study, seven had mothers who had used DES, which exceeds the frequency of exposure to DES in the general population. When the treatment has an effect, it is related to the outcome, so selecting subjects using their outcomes changes the frequency of exposure to the treatment. How can a case-referent study be interpreted?

3.3.2. Synthetic Case-Referent Studies

A synthetic case-referent study starts with the population of M subjects in §3.2.1, and draws a random sample of cases and a separate random sample of referents, possibly after stratification using the observed covariates $\mathbf{x}$. Synthetic case-referent studies are typically conducted when there is a computerized database describing the entire population of M subjects, but the study requires the costly collection of additional data not in the database. This sort of study does occur, but far more common are case-referent studies that do not use random sampling. Synthetic case-referent studies are easier to consider theoretically because the mechanism that selects subjects has known properties. The term "synthetic" was introduced by Mantel (1973), while the odds ratio property discussed in this section is due to Cornfield (1951).

Consider the data before cases and referents are sampled, as in §3.2.1, so the jth of the M subjects has observed covariate $\mathbf{x}_{[j]}$, treatment assignment $Z_{[j]}$, observed binary response $R_{[j]}$ which equals $r_{T[j]}$ if j is given the treatment and $r_{C[j]}$ if j is given the control. Divide the M subjects in the population into strata based $\mathbf{x}$, and abbreviate by $\sum_{\mathbf{x}}$ a sum over all subjects j with $\mathbf{x}_{[j]} = \mathbf{x}$, that is, write $\sum_{\mathbf{x}}$ for $\sum_{j:\, \mathbf{x}_{[j]}=\mathbf{x}}$. The subjects in the stratum with covariate value $\mathbf{x}$ are recorded in the contingency Table 3.3.1.

Table 3.3.1. Data Before Selecting Cases and Referents.

		Z	
		1	0
R	1	$\sum_{\mathbf{x}} Z_{[j]} r_{T[j]}$	$\sum_{\mathbf{x}} (1 - Z_{[j]}) r_{C[j]}$
	0	$\sum_{\mathbf{x}} Z_{[j]}(1 - r_{T[j]})$	$\sum_{\mathbf{x}} (1 - Z_{[j]})(1 - r_{C[j]})$

Table 3.3.2. Expectations in the Absence of Hidden Bias.

		Z	
		1	0
R	1	$\lambda(\mathbf{x})\sum_{\mathbf{x}} r_{T[j]}$	$\{1-\lambda(\mathbf{x})\}\sum_{\mathbf{x}} r_{C[j]}$
	0	$\lambda(\mathbf{x})\sum_{\mathbf{x}}(1-r_{T[j]})$	$\{1-\lambda(\mathbf{x})\}\sum_{\mathbf{x}}(1-r_{C[j]})$

If there is no hidden bias, then $E(Z_{[j]}) = \lambda(\mathbf{x}_{[j]})$, so the entries in Table 3.3.1 have as expectations the values in Table 3.3.2.

The odds ratio or cross-product ratio in Table 3.3.2 is

$$\frac{\left(\lambda(\mathbf{x})\sum_{\mathbf{x}} r_{T[j]}\right)\left(\{1-\lambda(\mathbf{x})\}\sum_{\mathbf{x}}(1-r_{C[j]})\right)}{\left(\lambda(\mathbf{x})\sum_{\mathbf{x}}(1-r_{T[j]})\right)\left(\{1-\lambda(\mathbf{x})\}\sum_{\mathbf{x}} r_{C[j]}\right)} = \frac{\left(\sum_{\mathbf{x}} r_{T[j]}\right)\left(\sum_{\mathbf{x}}(1-r_{C[j]})\right)}{\left(\sum_{\mathbf{x}}(1-r_{T[j]})\right)\left(\sum_{\mathbf{x}} r_{C[j]}\right)}, \tag{3.3.1}$$

and this odds ratio is a measure of the magnitude of a treatment effect. Notice that the odds ratio is one if there is no treatment effect in the stratum defined by j, that is if $r_{T[j]} = r_{C[j]}$ for all $[j]$ with $\mathbf{x}_{[j]} = \mathbf{x}$, and the odds ratio is greater than one if the treatment has a positive effect in the stratum defined by $\mathbf{x}$, $r_{T[j]} \geq r_{C[j]}$ for all j with $\mathbf{x}_{[j]} = \mathbf{x}$ and $r_{T[j]} \neq r_{C[j]}$ for some j with $\mathbf{x}_{[j]} = \mathbf{x}$. Hamilton (1979) discusses a wide variety of related measures under the model of a positive effect.

Tables 3.3.1 and 3.3.2 describe the initial population of M subjects. Consider a synthetic case-referent study formed from Table 3.3.1. Draw a random sample without replacement consisting of a fraction $k_{1\mathbf{x}}$ of the cases, the first row of the table, and a random sample consisting of a fraction $k_{0\mathbf{x}}$ from the referents, the second row, with $0 < k_{1\mathbf{x}} \leq 1$ and $0 < k_{0\mathbf{x}} \leq 1$. The resulting table of counts for the synthetic case-referent study has as its expectations as shown in Table 3.3.3.

The key observation is that the odds ratio computed from Table 3.3.3 equals the odds ratio (3.3.1) before case-referent sampling. In other words, in the absence of hidden bias, the data from a synthetic case-referent study provides a direct estimate of the population odds ratio (3.3.1) at each $\mathbf{x}$.

Table 3.3.3. Expected Counts in a Synthetic Case-Referent Study Absent Hidden Bias.

		Z	
		1	0
R	1	$k_{1\mathbf{x}}\lambda(\mathbf{x})\sum_{\mathbf{x}} r_{T[j]}$	$k_{1\mathbf{x}}\{1-\lambda(\mathbf{x})\}\sum_{\mathbf{x}} r_{C[j]}$
	0	$k_{0\mathbf{x}}\lambda(\mathbf{x})\sum_{\mathbf{x}}(1-r_{T[j]})$	$k_{0\mathbf{x}}\{1-\lambda(\mathbf{x})\}\sum_{\mathbf{x}}(1-r_{C[j]})$

As in §3.2, when attention shifts from the population of M subjects to the $N < M$ subjects included in the case-referent study, the notation in §3.2.3 is used; that is, the ith, subject in the sth stratum defined by $\mathbf{x}$ has unbracketed subscript s, i. As in §3.2, let $m_s = \sum_i Z_{si}$ be the number of treated or exposed subjects in stratum s of the case-referent study. If the study were free of hidden bias, so (3.2.2) holds, and if the treatment had no effect, so $r_{Tsi} = r_{Csi}$ for each s, i, then after synthetic case-referent sampling, the conditional distribution of the treatment assignments $\mathbf{Z}$ given $\mathbf{m}$ is uniform on Ω. This means, for instance, that the Mantel–Haenszel statistic may be used to test the null hypothesis of no effect in a synthetic case-referent study that is free of hidden bias.

3.3.3. Selection Bias in Case-Referent Studies

Unlike the synthetic study in §3.3.2, most case-referent studies do not use random sampling. It is common to use all cases that are made available by some process, for instance, all new cases admitted to one or more hospitals in a given time interval, together with referents selected by an ostensibly similar process, for instance, patients with some other illness admitted to the same hospitals at the same time, or neighbors or coworkers of the cases.

Nonrandom selections of cases and referents may distort the odds ratio in (3.3.1). For instance, if cases of lung cancer at a hospital were compared to referents who were selected as patients with cardiac disease in the same hospital, the odds ratio linking lung cancer with cigarette smoking would be too small, because smoking causes both lung cancer and cardiac disease. In this case, there is a *selection bias*, that is, a bias that was not the result of the manner in which subjects were assigned to treatment in the population, but rather a bias introduced by the nonrandom process of selecting cases and referents. Selection bias is discussed further in Chapter 7.

*3.4. Inexact Matching Followed by Stratification

3.4.1. An Example: Vasectomy and Myocardial Infarction

Walker et al. (1981) studied the possible effect of vasectomy on increased risk of myocardial infarction (MI), a possibility suggested by animal studies where increased risks were observed. The study contained 4830 pairs of

* This section may be skipped without loss of continuity.

men, one vasectomized and one control, matched for year of birth and calendar time of follow-up. The data were not matched for two other variables, obesity and smoking history recorded as binary traits, both of which are believed to be related to the risk of MI. This section describes a method for controlling covariates that were not controlled by the matching.

The outcome is a binary variable indicating whether a MI occurred during the follow-up period; however, the method discussed here may be used with outcomes of any kind. McNemar's test statistic will be used; see §2.4.2 for discussion of this test. In most of the 4830 pairs, no MI occurred. Pairs containing no MI or two MI's are said to be concordant, and it is not difficult to verify that these pairs do not contribute to McNemar's test. There were 36 discordant pairs, that is, pairs in which one person had an MI and the other did not—only these affect the test. Walker's (1982) data for these 36 pairs of men are given in Table 3.4.1. The score q_{si} is 1 if the ith man in pair s had an MI and is 0 otherwise.

In each pair, there are two possibilities. A pair may be exactly matched for obesity and smoking, so the two matched men are the same on these variables, or else the men may differ. If they differ, then there as six ways they may differ, that is, there are six patterns of imbalance in the covariates. For instance, one possible imbalance occurs if one man in a pair is a nonsmoker who is not obese and the other is a smoker who is not obese; this is (os, oS) in Table 3.4.1. Notice that, in counting the patterns of imbalance in covariates, we consider only the covariates and not vasectomy or MI. The 36 pairs are grouped into seven classes, one class that is perfectly matched and six classes for the six patterns of imbalance.

3.4.2. Adjusting Inexactly Matched Pairs by Stratifying Pairs

Suppose that an observational study is free of hidden bias, so (3.2.2) holds, and pairs of treated and control units are matched inexactly for $\mathbf{x}$, so $\mathbf{x}_{s1}$ may not equal $\mathbf{x}_{s2}$. How does one control imbalances in $\mathbf{x}$ that remain after matching? The method described in this section is useful when matching has failed to control a few coordinates of $\mathbf{x}$ containing discrete variables, as in §3.4.1.

Using (3.2.2), for any two distinct units j and k, $\text{prob}(Z_{[j]} = 1, Z_{[k]} = 0) = \lambda(\mathbf{x}_{[j]})\{1 - \lambda(\mathbf{x}_{[k]})\}$ and $\text{prob}(\mathbf{Z}_{[j]} = 0, Z_{[k]} = 1) = \lambda(\mathbf{x}_{[k]})\{1 - \lambda(\mathbf{x}_{[j]})\}$. Pair matching selects units so that $m_s = Z_{s1} + Z_{s2} = 1$ for each s. Therefore, conditionally given that a pair contains exactly one treated unit, the chance that the first unit is the treated unit is

$$\text{prob}(Z_{s1} = 1 | Z_{s1} + Z_{s2} = 1) = \frac{\lambda(\mathbf{x}_{s1})\{1 - \lambda(\mathbf{x}_{s2})\}}{\lambda(\mathbf{x}_{s1})\{1 - \lambda(\mathbf{x}_{s2})\} + \lambda(\mathbf{x}_{s2})\{1 - \lambda(\mathbf{x}_{s1})\}}. \tag{3.4.1}$$

Table 3.4.1. Vasectomy and Myocardial Infarction.

Class c	Covariate imbalance $(\mathbf{x}_{s1}, \mathbf{x}_{s2})$	q_{s1}	q_{s2}	Z_{s1}	Walker's id #, s
0	No imbalance	1	0	0	4
		1	0	1	8
		1	0	1	10
		0	1	1	11
		1	0	0	12
		1	0	1	16
		0	1	0	17
		0	1	0	23
		1	0	0	26
		0	1	0	27
		1	0	1	30
		0	1	0	35
		1	0	1	36
1	(os, oS)	0	1	0	3
		1	0	1	6
		0	1	1	9
		0	1	1	14
		0	1	0	15
		0	1	1	20
		0	1	1	21
		1	0	0	22
		0	1	0	24
		0	1	1	29
		0	1	0	32
2	(os, Os)	0	1	1	1
		0	1	0	28
		1	0	0	33
3	(os, OS)	0	1	0	19
		0	1	0	25
4	(oS, Os)	0	1	0	2
		0	1	0	7
		1	0	0	34
5	(oS, OS)	0	1	0	13
		1	0	0	18
		0	1	1	31
6	(Os, OS)	0	1	1	5

Key: O = obese, o = not obese, S = smoker, s = nonsmoker.

Divide the S pairs into $C + 1$ classes, where class 0 contains the exactly matched pairs with $\mathbf{x}_{s1} = \mathbf{x}_{s2}$, and the other C classes contain the C patterns of imbalance in $\mathbf{x}$. Let ι_c be the set of pairs exhibiting imbalance c, so $\iota_0 \cup \iota_1 \cup \cdots \cup \iota_C = \{1, \ldots, S\}$. In class $c = 0$, there is an exact match, that is $\mathbf{x}_{s1} = \mathbf{x}_{s2}$ for $s \in \iota_0$. In the other classes, there is an imbalance, $\mathbf{x}_{s1} \neq \mathbf{x}_{s2}$ for $s \in \iota_c$, for $1 \le c \le C$. Write $\tilde{n}_c$ for the number of pairs in ι_c.

Renumber the two units in each pair so that every pair s in class c has the same value of $\mathbf{x}_{s1}$, and every pair in class c has the same value of $\mathbf{x}_{s2}$, as in Table 3.4.1. This notational change simplifies the appearance of various quantities but it does not change their values.

With this notation, (3.4.1) takes the same value for all pairs in the same class. Write ρ_c for the common value of (3.4.1) in class c; that is, prob$(\mathbf{Z}_{s1} = 1 | Z_{s1} + Z_{s2} = 1) = \rho_c$ for all $s \in \iota_c$. Notice that $\rho_0 = \frac{1}{2}$, but the other ρ_c are unknown since $\lambda(\mathbf{x})$ is unknown.

For $c = 1, \ldots, C$, write $\tilde{m}_c$ for the number of pairs in class s in which the treated unit is the first unit,

$$\tilde{m}_c = \sum_{s \in \iota_c} Z_{s1} \qquad \text{and write} \quad \tilde{\mathbf{m}} = \begin{bmatrix} \tilde{m}_1 \\ \vdots \\ \tilde{m}_C \end{bmatrix}.$$

Then $\tilde{m}_c$, like m_s, is a random variable, since it depends on the Z's. Note the distinction between $\mathbf{m}$ and $\tilde{\mathbf{m}}$, both of which are used below. With this notation, the distribution of treatment assignments within pairs is

$$\begin{aligned} \operatorname{pr}(\mathbf{Z} = \mathbf{z} | \mathbf{m}) &= \prod_{c=0}^{C} \prod_{s \in \iota_c} \rho_c^{z_{s1}} \{1 - \rho_c\}^{z_{s2}}, \\ &= \left(\frac{1}{2}\right)^{\tilde{n}_0} \prod_{c=1}^{C} \rho_c^{\tilde{m}_c} \{1 - \rho_c\}^{\tilde{n}_c - \tilde{m}_c}. \end{aligned} \tag{3.4.2}$$

Unlike the distribution of treatment assignments in §3.2.3 for exact matching, the distribution (3.4.2) for inexact matching involves unknown parameters, the ρ_c, reflecting the remaining imbalances in $\mathbf{x}$.

Consider the conditional distribution of the treatment assignments $\mathbf{Z}$ given both $\mathbf{m}$ and $\tilde{\mathbf{m}}$. As will now be seen, conditioning on both $\mathbf{m}$ and $\tilde{\mathbf{m}}$ yields a known distribution free of the unknown ρ_c. It is a distribution on a set $\tilde{\Omega} \subseteq \Omega$ where $\mathbf{z} \in \tilde{\Omega}$ if and only if:

(i) $\mathbf{z}$ is a treatment assignment for S matched pairs, that is, $z_{s1} + z_{s2} = m_s = 1$ or, equivalently, $\mathbf{z} \in \Omega$; and
(ii) $\mathbf{z}$ exhibits the same degree of imbalance as the observed data, that is, $\tilde{m}_c = \sum_{s \in \iota_c} z_{s1}$ for $c = 1, \ldots, C$.

The set $\tilde{\Omega}$ has

$$\tilde{K} = 2^{\tilde{n}_0} \prod_{c=1}^{C} \binom{\tilde{n}_c}{\tilde{m}_c}$$

elements, and $\text{pr}(\mathbf{Z} = \mathbf{z}|\mathbf{m}, \tilde{\mathbf{m}}) = 1/\tilde{K}$ for each $\mathbf{z} \in \tilde{\Omega}$. In class $c = 0$ containing $\tilde{n}_0$ pairs, all $2^{\tilde{n}_0}$ possible assignments are equally likely. In class $c \geq 1$ containing $\tilde{n}_c$ pairs, the assignments give the treatment to the first unit in exactly $\tilde{m}_c$ pairs, and there are $\binom{\tilde{n}_c}{\tilde{m}_c}$ such assignments.

Using the known distribution $\text{pr}(\mathbf{Z} = \mathbf{z}|\mathbf{m}, \tilde{\mathbf{m}})$, significance levels are obtained in a manner similar to that in §2.4.1. Under the null hypothesis of no treatment effect, the statistic $T = t(\mathbf{Z}, \mathbf{r})$ has significance level

$$\text{prob}\{t(\mathbf{Z}, \mathbf{r}) \geq T|\mathbf{m}, \tilde{\mathbf{m}})\} = \frac{|\{\mathbf{z} \in \tilde{\Omega}: t(\mathbf{z}, \mathbf{r}) \geq T\}|}{|\tilde{\Omega}|} = \frac{|\{\mathbf{z} \in \tilde{\Omega}: t(\mathbf{z}, \mathbf{r}) \geq T\}|}{\tilde{K}}, \tag{3.4.3}$$

which parallels (2.4.4) and is simply the proportion of treatment assignments in $\tilde{\Omega}$ giving values of the test statistic at least as large as the observed value.

If the test statistic is a sum statistic in the sense of §2.4.3, that is, if $t(\mathbf{Z}, \mathbf{r}) = \mathbf{Z}^{\mathrm{T}}\mathbf{q}$ where $\mathbf{q}$ is a function of $\mathbf{r}$, then the expectation and variance of the test statistic are given in the following proposition. The proposition assumes the study is free of hidden bias in the sense that (3.2.2) holds, and it concerns the conditional expectation and variance of a sum statistic given $\mathbf{m}$, $\tilde{\mathbf{m}}$, that is, the expectation and variance of $t(\mathbf{Z}, \mathbf{r}) = \mathbf{Z}^{\mathrm{T}}\mathbf{q}$ over the distribution (3.4.3).

Proposition 3.1. *If the study is free of hidden bias, then the null expectation and variance of* $\mathbf{Z}^{\mathrm{T}}\mathbf{q}$ *are*

$$\begin{aligned} E(\mathbf{Z}^{\mathrm{T}}\mathbf{q}|\mathbf{m}, \tilde{\mathbf{m}}) &= \sum_{s \in \iota_0} \frac{q_{s1} + q_{s2}}{2} + \sum_{c=1}^{C} \tilde{m}_c\tilde{\mu}_{c1} + (\tilde{n}_c - \tilde{m}_c)\tilde{\mu}_{c2} \\ \text{var}(\mathbf{Z}^{\mathrm{T}}\mathbf{q}|\mathbf{m}, \tilde{\mathbf{m}}) &= \sum_{s \in \iota_0} \frac{(q_{s1} - q_{s2})^2}{4} + \sum_{c=1}^{C} \frac{\tilde{m}_c(\tilde{n}_c - \tilde{m}_c)\tilde{\sigma}_c^2}{\tilde{n}_c}, \end{aligned} \tag{3.4.4}$$

where

$$\tilde{\mu}_{ci} = \frac{1}{\tilde{n}_c} \sum_{s \in \iota_c} q_{si} \qquad \textit{for} \quad i = 1, 2$$

and

$$\tilde{\sigma}_c^2 = \frac{1}{\tilde{n}_c - 1} \sum_{s \in \iota_c} \{(q_{s1} - q_{s2}) - (\tilde{\mu}_{c1} - \tilde{\mu}_{c2})\}^2.$$

Proof. The proof makes use of several elementary observations. First, the conditional distribution of $\mathbf{Z}$ given $\mathbf{m}$, $\tilde{\mathbf{m}}$ is uniform on $\tilde{\Omega}$, and as a result, the Z_{si}'s in different classes are independent of each other. Second, write $\mathbf{Z}^{\mathrm{T}}\mathbf{q}$ as

$$\mathbf{Z}^{\mathrm{T}}\mathbf{q} = \sum_{c=0}^{C} \sum_{s \in \iota_c} Z_{s1}(q_{s1} - q_{s2}) + \sum_{c=0}^{C} \sum_{s \in \iota_c} q_{s2}, \tag{3.4.5}$$

where the second sum on the right is a constant since it does not involve $\mathbf{Z}$. In class $c = 0$, the Z_{s1}'s independent of each other given $\mathbf{m}$, $\tilde{\mathbf{m}}$, and each Z_{s1} equals 1 or 0 with probability 1/2. For $c \geq 1$, the sum $\sum_{s \in \iota_c} Z_{s1}(q_{s1} - q_{s2})$ in (3.4.5) is the sum of $\tilde{m}_c$ of the $(q_{s1} - q_{s2})$'s randomly selected without replacement from among the $\tilde{n}_c$ pairs $s \in \iota_c$. The proposition then follows directly from standard facts about simple random sampling without replacement. □

3.4.3. Return to the Example of Vasectomy and Mycocardial Infarction

The 36 pairs in Table 3.4.1 are divided into seven classes, numbered 0, 1, ..., 6, based on the pattern of imbalance in two unmatched covariates, obesity and smoking. In class 0, the two men are exactly matched for obesity and smoking. There are $\tilde{n}_0 = 13$ exactly matched pairs in class 0. In class 1, labeled (os, oS), each pair contains two men who are not obese, of whom exactly one is a smoker. There are $\tilde{n}_1 = 11$ pairs with this imbalance. In class 2, labeled (os, Os), each pair contains two nonsmokers, of whom exactly one is obese, and so on. Each pair contains one man with a vasectomy and one without, indicated by Z_{s1} in Table 3.4.1. The last column gives Walker's (1982) pair identification number s, so pair $s = 3$ is the first pair in class $c = 1$, labeled (os, oS). In this pair, both men are not obese, and only the second man is a smoker; he is also the man with a vasectomy, since $Z_{31} = 0$, so $Z_{32} = 1$, and he had an MI since $q_{32} = 1$.

McNemar's statistic is the number of vasectomized men who had an MI, namely, $\sum_s \sum_i Z_{si} q_{si} = 20$ of a possible 36. Proposition 3.1 gives the moments of the statistic adjusting for obesity and smoking in addition to the variables used to form matched pairs. Table 3.4.2 gives a few intermediate calculations. Using these in (3.4.4) gives an expectation of 19.015 and a variance of 6.813 for the McNemar statistic under the null hypothesis of no treatment effect. In words, the number of vasectomized men who had an MI, namely, 20, is quite close to the expectation 19.015 in the absence of a treatment effect. The standardized deviate with continuity correction is $(20 - 19.015 - \frac{1}{2})/\sqrt{6.813} = 0.186$, and this is small when compared with the standard normal distribution, so there is no indication of an effect of vasectomy on the risk of MI. This would be a correct test if the study were free of hidden bias once adjustments had been made for the two matched and the two unmatched covariates.

Had smoking and obesity been ignored, the usual expectation for McNemar's statistic would have been $36/2 = 18$ vasectomized MI's rather than 19.015, so the adjustment for smoking and obesity moved the expected count closer to the observed 20. In other words, if vasectomy had no effect on the risk of MI, we would nonetheless have expected more than half of

Table 3.4.2. Intermediate Calculations for Vasectomy and Myocardial Infarction.

c	$\tilde{n}_c$	$\tilde{m}_c$	$\tilde{\mu}_{c1}$	$\tilde{\mu}_{c2}$	$\tilde{\sigma}_c^2$
0	13	6			
1	11	6	0.182	0.818	0.655
2	3	1	0.333	0.667	1.333
3	2	0	0.000	1.000	0.000
4	3	0	0.333	0.667	1.333
5	3	1	0.333	0.667	1.333
6	1	1	0.000	1.000	0.000

the vasectomized men to exhibit MI's, because both vasectomy and MI are related to smoking and obesity in the data shown in Table 3.4.2.

*3.5. Small Sample Inference with an Unknown Propensity Score

3.5.1. Conditional Inference Under a Logit Model for the Propensity Score

In §3.2, in the absence of hidden bias, the distribution of treatment assignments prob($\mathbf{Z} = \mathbf{z}$) in (3.2.2) was unknown because the propensity score $\boldsymbol{\lambda} = (\lambda_1, \ldots, \lambda_S)^T$ was unknown. However, by conditioning on the number of treated subjects in each stratum $\mathbf{m}$, the conditional distribution of treatment assignments, prob($\mathbf{Z} = \mathbf{z} | \mathbf{m}$), was known, and, in fact, was the distribution of treatment assignments in a uniform randomized experiment. Notice that the unknown parameter $\boldsymbol{\lambda}$ was eliminated by conditioning on a sufficient statistic $\mathbf{m}$. This line of reasoning generalizes.

Suppose that $\boldsymbol{\lambda}$ satisfies a logit model,

$$\log\left(\frac{\lambda_s}{1 - \lambda_s}\right) = \boldsymbol{\beta}^T \mathbf{x}_s, \tag{3.5.1}$$

where $\boldsymbol{\beta}$ is an unknown parameter. Write

$$\bar{\mathbf{m}} = \sum_{s=1}^{S} m_s \mathbf{x}_s, \tag{3.5.2}$$

so $\bar{\mathbf{m}}$ is the sum of the $\mathbf{x}_s$ weighted by the number of treated subjects m_s in stratum s. Under the model (3.5.1), $\bar{\mathbf{m}}$ sufficient for $\boldsymbol{\beta}$, so prob($\mathbf{Z} = \mathbf{z} | \bar{\mathbf{m}}$) is a

* This section may be skipped without loss of continuity.

known distribution free of the unknown parameter $\boldsymbol{\beta}$. See Cox (1970) for a detailed discussion of logit models, including a discussion of the sufficiency of $\overline{\mathbf{m}}$.

Let $\overline{\Omega}$ be the set containing all treatment assignments $\mathbf{z}$ that give rise to the same value of $\overline{\mathbf{m}}$, that is,

$$\overline{\Omega} = \left\{\mathbf{z}\colon z_{si} \in \{0, 1\} \text{ for } i = 1, \ldots, n_s, s = 1, \ldots, S, \text{ and } \overline{\mathbf{m}} = \sum_{s=1}^{S} \mathbf{x}_s \sum_{i=1}^{n_s} z_{si}\right\}.$$

Notice that $\overline{\Omega}$ is a larger set than Ω, in the sense that $\Omega \subseteq \overline{\Omega}$, so $\overline{\Omega}$ contains at least as many treatment assignments $\mathbf{z}$ as Ω. This is true because every $\mathbf{z} \in \Omega$ gives rise to the same $\mathbf{m}$, and hence also to the same $\overline{\mathbf{m}}$.

Under the logit model (3.5.1), the conditional distribution of treatment assignments given $\overline{\mathbf{m}}$ is constant on $\overline{\Omega}$,

$$\text{prob}(\mathbf{Z} = \mathbf{z}|\overline{\mathbf{m}}) = \frac{1}{|\overline{\Omega}|} \quad \text{for each} \quad \mathbf{z} \in \overline{\Omega}. \tag{3.5.3}$$

As a result, in the absence of hidden bias, under the model (3.5.1), the known distribution (3.5.3) forms the basis for a permutation test; in particular, a test statistic $T = t(\mathbf{Z}, \mathbf{r})$ has significance level

$$\text{prob}\{t(\mathbf{Z}, \mathbf{r}) \geq T|\overline{\mathbf{m}}\} = \frac{|\{\mathbf{z} \in \overline{\Omega}\colon t(\mathbf{z}, \mathbf{r}) \geq T\}|}{|\overline{\Omega}|}, \tag{3.5.4}$$

for testing the null hypothesis of no treatment effect. The significance level (3.5.4) is the proportion of treatment assignments $\mathbf{z}$ in $\overline{\Omega}$ giving a value of the test statistic $t(\mathbf{z}, \mathbf{r})$ at least equal to the observed value of T.

The procedure just described is useful in small samples when the strata are so thin that the set Ω is too small to be useful. For instance, if Ω contains fewer than 20 treatment assignments, then no pattern of responses can be significant at the 0.05 level using (2.4.4). Such a small Ω arises when the sample size N is small and this small sample size is thinly spread over many strata; for instance, strata with $m_s = 0$ or $m_s = n_s$ contribute nothing to (2.4.4). The set $\overline{\Omega}$, being larger than Ω, is useful when Ω is too small to permit reasonable permutation inferences. An example is considered in the next section.

3.5.2. An Example: A Small Observational Study Embedded in a Clinical Trial

Table 3.5.1 describes 14 patients taken from a clinical trial of the drug treatment combination PACCO in the treatment of nonsmall cell bronchogenic carcinoma (Whitehead, Rosenbaum, and Carbone, 1984; Rosenbaum, 1984b, §3). This phase II trial contained two minor variations of what was intended to be the same treatment; however, when the responses were tabulated, all of the patients who responded to therapy had received the same variation of

Table 3.5.1. Fourteen Lung Cancer Patients from a Phase II Trial of Pacco.

Patient	Tumor response r_{si}	Treat-ment Z_{si}	Cell type	Previous treatment	Perform-ance status	Stratum s
1	0	0	Squamous	None	0	1
2	0	0	Large cell	None	1	2
3	0	0	Squamous	Radiation	1	3
4	0	0	Squamous	Radiation	1	3
5	0	0	Squamous	Radiation	2	4
6	1	1	Squamous	Radiation	1	3
7	0	1	Squamous	Radiation	1	3
8	0	1	Adenocarcinoma	Radiation	1	5
9	1	1	Squamous	None	1	6
10	0	1	Large cell	None	2	7
11	0	1	Squamous	Radiation and chemotherapy	2	8
12	0	1	Squamous	Chemotherapy	1	9
13	0	1	Squamous	None	0	1
14	2	1	Squamous	None	1	6

the treatment. The question is whether this is evidence that the treatments differ, given the characteristics of the patients involved. As presented here, the example is adapted to illustrate the method.

The outcome is tumor response, where 0 signifies no response, 1 signifies a partial response, and 2 signifies a complete response. The covariate information describes the cell type, previous treatment, and performance status. In $\mathbf{x}_s$, previous treatment was coded as three binary variables, cell type as two binary variables, and performance status was taken as a single variable with three scored categories, so $\mathbf{x}_s$ has dimension six.

Table 3.5.2 describes the strata based on the covariates. The 14 patients are divided into nine strata, with most strata containing a single patient. The set Ω has $2 \times 1 \times 6 \times 1 \times 1 \times 1 \times 1 \times 1 \times 1 = 12$ treatment assignments, so no matter what responses are observed, the smallest possible significance level is $1/12 = 0.083$. The set Ω is too small to be useful for a permutation test.

The set $\bar{\Omega}$ is somewhat larger, containing 28 treatment assignments. These are all the treatment assignments that give rise to the observed value of $\bar{\mathbf{m}}$. This means that a treatment assignment $\mathbf{z}$ must have the correct number of treated patients with each cell type, each previous treatment, and the correct average performance status. It contains all treatment assignments $\mathbf{z}$ such that:

(i) nine patients receive treatment one, and of those nine;
(ii) one has adenocarcinoma;

Table 3.5.2. Strata for the PACCO Trial.

Stratum s	Cell type	Previous treatment	Performance status	n_s	m_s	$\binom{n_s}{m_s}$
1	Squamous	None	0	2	1	2
2	Large cell	None	1	1	0	1
3	Squamous	Radiation	1	4	2	6
4	Squamous	Radiation	2	1	0	1
5	Adenocarcinoma	Radiation	1	1	1	1
6	Squamous	None	1	2	2	1
7	Large cell	None	2	1	1	1
8	Squamous	Radiation and chemotherapy	2	1	1	1
9	Squamous	Chemotherapy	1	1	1	1

(iii) one has large cell carcinoma;
(iv) seven have squamous cell carcinoma;
(v) three have had only previous radiation therapy;
(vi) one has had only previous chemotherapy;
(vii) one has had both previous chemotherapy and previous radiation therapy; and
(viii) the average performance status is $10/9 = 1.1$.

In other words, these 28 treatment assignments resemble the observed treatment assignment in the sense that similar patients received the treatment.

The test statistic is the total of the response scores for treated patients, $T = t(\mathbf{Z}, \mathbf{r}) = \mathbf{Z}^{\mathrm{T}}\mathbf{r} = 2 + 1 + 1 = 4$. Of the 28 treatment assignments $\mathbf{z}$ in $\bar{\Omega}$, eight have $\mathbf{z}^{\mathrm{T}}\mathbf{r} = 4$, thirteen have $\mathbf{z}^{\mathrm{T}}\mathbf{r} = 3$, four have $\mathbf{z}^{\mathrm{T}}\mathbf{r} = 2$, and three have $\mathbf{z}^{\mathrm{T}}\mathbf{r} = 1$. Therefore, under the null hypothesis, the distribution (3.5.4) of $\mathbf{Z}^{\mathrm{T}}\mathbf{r}$ assigns probability $8/28 = 0.29$ to 4, $13/28 = 0.46$ to 3, $4/28 = 0.14$ to 2, and $3/28 = 0.11$ to 1. In other words, it is not surprising to find that all the responses occurred in treatment group one—this would happen by chance 29% of the time if the treatments did not differ in their effects.

Algorithms for computations involving $\bar{\Omega}$ are discussed in Rosenbaum (1984b).

3.6. Bibliographic Notes

Direct adjustment is surveyed by Bishop, Fienberg, and Holland (1975, §4.3), Cochran and Rubin (1973), Fleiss (1981, §14), Kitagawa (1964), and Mosteller and Tukey (1977, §11). The analysis of matched pairs and matched sets is surveyed by Breslow and Day (1980, §5), Cochran and Rubin (1973),

Fleiss (1981, §8), Gastwirth (1988, §11), and Kleinbaum, Kupper, and Morgenstern (1982, §18). Statistical procedures that may be derived from an assumption of random assignment of treatments within strata have long been applied to observational studies; see, for instance, the influential paper by Mantel and Haenszel (1959). Cochran (1965, §3.2) viewed matching, subclassification, and model-based adjustments are different ways of doing the same thing. In an important paper, Rubin (1977) demonstrated that if treatments are randomly assigned on the basis of a covariate, then adjustments for the covariate produce appropriate estimates of a treatment effect. Rubin (1978) develops related ideas from a Bayesian view. Rosenbaum (1984b, 1987a) obtains known permutation distributions by conditioning on a sufficient statistic for unknown assignment probabilities, as in §3.2. The propensity score is discussed by Rosenbaum and Rubin (1983), and its link in §3.2.5 to permutation inference is discussed by Rosenbaum (1984b). Cole (1979, p. 16) says the first true case-referent study was conducted in 1926. Cornfield (1951) showed that case-referent sampling did not alter the odds ratio. He also argued that if the disease is rare, as it is in most case-referent studies, then the odds ratio approximates another measure, the relative risk. See also Greenhouse (1982). Mantel's (1973) paper is a careful discussion of case-referent studies; see also Hamilton (1979), Rosenbaum (1987b), and Holland and Rubin (1988). Further references concerning case-referent studies are given in Chapter 6. Inexact matching followed by stratification in §3.4 is discussed by Rosenbaum (1988). Conditional tests given a sufficient statistic for the propensity score in §3.5 are discussed in Rosenbaum (1984b).

3.7. Problems

3.7.1. The Permutation Distribution of the Covariance Adjusted Estimate of Effect

1. Problems 1 and 2 consider a test statistic motivated by linear regression, but instead of referring the statistic to a theoretical distribution, these problems consider its permutation distribution. Write $\mathbf{X}$ for the matrix with N rows, numbered (s, i), $s = 1, \ldots, S$, $i = 1, \ldots, n_s$, where row (s, i) is $\mathbf{x}_s^T$. Assume that $\mathbf{X}$ has more rows than columns and has rank equal to the number of columns. Consider the linear regression model $\mathbf{R} = \mathbf{X}\boldsymbol{\theta} + \mathbf{Z}\tau + \mathbf{e}$, where $\mathbf{e}$ is a vector of unobserved errors, but do not assume the model is correct. The least squares estimate of τ under this model is

$$t(\mathbf{Z}, \mathbf{R}) = \frac{\mathbf{R}^T(\mathbf{I} - \mathbf{H})\mathbf{Z}}{\mathbf{Z}^T(\mathbf{I} - \mathbf{H})\mathbf{Z}} \quad \text{where} \quad \mathbf{H} = \mathbf{X}(\mathbf{X}^T\mathbf{X})^{-1}\mathbf{X}^T.$$

In a uniform randomized experiment (or in an observational study free of hidden bias), consider the permutation distribution (2.5) of this $t(\mathbf{Z}, \mathbf{R})$ under the null hypothesis of no treatment effect. Show that the significance level (2.4.4) for the covariance adjusted estimate $t(\mathbf{Z}, \mathbf{R})$ equals the significance level for the total response in the treated group, $\mathbf{Z}^{T}\mathbf{r}$, which in turn equals the significance level for the difference in sample means (2.4.1). (*Hint*: Use Problem 10 in §2.12.5. How does $\mathbf{X}^{T}\mathbf{z}$ vary as $\mathbf{z}$ ranges over Ω?)

2. Continuing Problem 1, show that the same conclusion holds if the significance level is based on (3.5.4) rather than (2.4.4), that is, if the permutation distribution is the conditional distribution of $\mathbf{Z}$ given $\overline{\mathbf{m}}$ under the model (3.5.1).

Rosenbaum (1984b, §2.5)

3.7.2. Conditional Tests Given a Sufficient Statistic for the Propensity Score

3. Suppose that the observational study is free of hidden bias, that the model (3.5.1) describes the propensity score, and the treatment has no effect. Consider testing the hypothesis H_0: $\theta = 0$ against H_A: $\theta > 0$ in the logit model

$$\log\left(\frac{\lambda_s}{1 - \lambda_s}\right) = \boldsymbol{\beta}^{T}\mathbf{x}_s + \theta q_s,$$

where $\mathbf{q}$ is some function of $\mathbf{r}$, so $\mathbf{q}$ is fixed under the null hypothesis. Verify that the uniformly most powerful unbiased test uses the statistic $t(\mathbf{Z}, \mathbf{r}) = \mathbf{Z}^{T}\mathbf{q}$ and the significance level given by (3.5.4). In other words, if the test statistic $t(\mathbf{Z}, \mathbf{r})$ is a sum statistic $\mathbf{Z}^{T}\mathbf{q}$, then the test in §3.5.1 is the same as the exact test of a hypothesis about the parameter of a logit model.

Cox (1970, §4.2); Rosenbaum (1984b, §4.2)

References

Bishop, Y., Fienberg, S., and Holland, P. (1975). *Discrete Multivariate Analysis.* Cambridge, MA: MIT Press.

Breslow, N. and Day, N. (1980). *The Analysis of Case-Control Studies.* Volume 1 of *Statistical Methods in Cancer Research.* Lyon, France: International Agency for Research on Cancer of the World Health Organization.

Canner, P. (1984). How much data should be collected in a clinical trial? *Statistics in Medicine*, **3**, 423–432.

Canner, P. (1991). Covariate adjustment of treatment effects in clinical trials. *Controlled Clinical Trials*, **12**, 359–366.

Cochran, W.G. (1957). The analysis of covariance. *Biometrics*, **13**, 261–281.

Cochran, W.G. (1965). The planning of observational studies of human populations (with Discussion). *Journal of the Royal Statistical Society*, Series A, **128**, 134–155.

Cochran, W.G. (1968). The effectiveness of adjustment by subclassification in removing bias in observational studies. *Biometrics*, **24**, 205–213.

Cochran, W.G. and Rubin, D.B. (1973). Controlling bias in observational studies: A review. *Sankya*, Series A, **35**, 417–446.

Cole, P. (1979). The evolving case-control study. *Journal of Chronic Diseases*, **32**, 15–27.

Coleman, J., Hoffer, T., and Kilgore, S. (1982). Cognitive outcomes in public and private schools. *Sociology of Education*, **55**, 65–76.

Cornfield, J. (1951). A method of estimating comparative rates from clinical data: Applications to cancer of the lung, breast and cervix. *Journal of the National Cancer Institute*, **11**, 1269–1275.

Cox, D.R. (1970). *The Analysis of Binary Data*. London: Methuen.

Cox, D.R. (1958). *The Planning of Experiments*. New York: Wiley.

Fleiss, J. (1981). *Statistical Methods for Rates and Proportions*. New York: Wiley.

Gastwirth, J. (1988). *Statistical Reasoning in Law and Public Policy*. New York: Academic Press.

Goldberger, A. and Cain, G. (1982). The causal analysis of cognitive outcomes in the Coleman, Hoffer, and Kilgore report. *Sociology of Education*, **55**, 103–122.

Greenhouse, S. (1982). Cornfield's contributions to epidemiology. *Biometrics*, **38S**, 33–46.

Hamilton, M. (1979). Choosing a parameter for 2×2 table or $2 \times 2 \times 2$ table analysis. *American Journal of Epidemiology*, **109**, 362–375.

Holland, P. and Rubin, D. (1988). Causal inference in retrospective studies. *Evaluation Review*, **12**, 203–231.

Kitagawa, E. (1964). Standardized comparisons in population research. *Demography*, **1**, 296–315.

Kleinbaum, D., Kupper, L., and Morgenstern, H. (1982). *Epidemiologic Research: Principles and Quantitative Methods*. Belmont, CA: Wadsworth.

Mantel, N. (1973). Synthetic retrospective studies. *Biometrics*, **29**, 479–486.

Mantel, N. and Haenszel, W. (1959). Statistical aspects of retrospective studies of disease. *Journal of the National Cancer Institute*, **22**, 719–748.

Morton, D., Saah, A., Silberg, S., Owens, W., Roberts, M., and Saah, M. (1982). Lead absorption in children of employees in a lead related industry. *American Journal of Epidemiology*, **115**, 549–555.

Mosteller, F. and Tukey, J. (1977). *Data Analysis and Regression*. Reading, MA: Addison-Wesley.

Rosenbaum, P.R. (1984a). The consequences of adjustment for a concomitant variable that has been affected by the treatment. *Journal of the Royal Statistical Society*, Series A, **147**, 656–666.

Rosenbaum, P.R. (1984b). Conditional permutation tests and the propensity score in observational studies. *Journal of the American Statistical Association*, **79**, 565–574.

Rosenbaum, P. (1986). Dropping out of high school in the United States: An observational study. *Journal of Educational Statistics*, **11**, 207–224.

Rosenbaum, P.R. (1987a). Model-based direct adjustment. *Journal of the American Statistical Association*, **82**, 387–394.

Rosenbaum, P.R. (1987b). The role of a second control group in an observational study (with Discussion). *Statistical Science*, **2**, 292–316.

Rosenbaum, P.R. (1988). Permutation tests for matched pairs with adjustments for covariates. *Applied Statistics*, **37**, 401–411.

Rosenbaum, P. and Rubin, D. (1983). The central role of the propensity score in observational studies for causal effects. *Biometrika*, **70**, 41–55.

Rosenbaum, P. and Rubin, D. (1984). Reducing bias in observational studies using subclassification on the propensity score. *Journal of the American Statistical Association*, **79**, 516–524.

Rubin, D.B. (1977). Assignment to treatment group on the basis of a covariate. *Journal of Educational Statistics*, **2**, 1–26.

Rubin, D.B. (1978). Bayesian inference for causal effects: The role of randomization. *Annals of Statistics*, **6**, 34–58.

Ury, H. (1975). Efficiency of case-control studies with multiple controls per case: Continuous or dichotomous data. *Biometrics*, **31**, 643–649.

Walker, A. (1982). Efficient assessment of confounder effects in matched follow-up studies. *Applied Statistics*, **31**, 293–297.

Walker, A., Jick, H., Hunter, J., Danford, A., Watkins, R., Alhadeff, L., and Rothman, K. (1981). Vasectomy and non-fatal myocardial infarction. *Lancet*, 13–15.

Whitehead, R., Rosenbaum, P., and Carbone, P. (1984). Cisplatin, doxorubicin, cyclophosphamide, lomustine, and vincristine (PACCO) in the treatment of non-small cell bronchogenic carcinoma. *Cancer Treatment Reports*, **68**, 771–773.

CHAPTER 4

Sensitivity to Hidden Bias

4.1. What Is a Sensitivity Analysis?

Cornfield, Haenszel, Hammond, Lilienfeld, Shimkin, and Wynder (1959) first conducted a formal sensitivity analysis in an observational study. Their paper is a survey of the evidence available in 1959 linking smoking with lung cancer. The paper asks whether the association between smoking and lung cancer is an effect caused by smoking or whether it is instead due to a hidden bias. Can lung cancer be prevented by not smoking? Or are the higher rates of lung cancer among smokers due to some other difference between smokers and nonsmokers?

In their effort to sort through conflicting claims, Cornfield et al. (1959) derived an inequality for a risk ratio defined as the ratio of the probability of death from lung cancer for smokers divided by the probability of death from lung cancer for nonsmokers. Specifically, they write:

> If an agent, A, with no causal effect upon the risk of a disease, nevertheless, because of a positive correlation with some other causal agent, B, shows an apparent risk, r, for those exposed to A, relative to those not so exposed, then the prevalence of B, among those exposed to A, relative to the prevalence among those not so exposed, must be greater than r.
>
> Thus, if cigarette smokers have 9 times the risk of nonsmokers for developing lung cancer, and this is not because cigarette smoke is a causal agent, but only because cigarette smokers produce hormone X, then the proportion of hormone X-producers among cigarette smokers must be at least 9 times greater than that of non-smokers. If the relative prevalence of hormone X-producers is considerably less than ninefold, then hormone X cannot account for the magnitude of the apparent effect.

Their statement is an important conceptual advance. The advance consists in replacing a general qualitative statement that applies in all observational studies by a quantitative statement that is specific to what is observed in a particular study. Instead of saying that an association between treatment and outcome does not imply causation, that hidden biases can explain observed associations, they say that to explain the association seen in a particular study, one would need a hidden bias of a particular magnitude. If the association is strong, the hidden bias needed to explain it is large.

Though an important conceptual advance, the above inequality of Cornfield et al. (1959) is limited to binary outcomes, and it ignores sampling variability which is hazardous except in extremely large studies. This chapter discusses a general method of sensitivity analysis that is similar to their method in spirit and purpose.

4.2. A Model for Sensitivity Analysis

4.2.1. The Model Expressed in Terms of Assignment Probabilities

If a study is free of hidden bias, as discussed in Chapter 3, then the chance $\pi_{[j]}$ that unit j receives the treatment is a function $\lambda(\mathbf{x}_{[j]})$ of the observed covariates $\mathbf{x}_{[j]}$ describing unit j. There is *hidden bias* if two units with the same observed covariates $\mathbf{x}$ have differing chances of receiving the treatments, that is, if $\mathbf{x}_{[j]} = \mathbf{x}_{[k]}$ but $\pi_{[j]} \neq \pi_{[k]}$ for some j and k. In this case, units that appear similar have differing chances of assignment to treatment.

A sensitivity analysis asks: How would inferences about treatment effects be altered by hidden biases of various magnitudes? Suppose the π's differ at a given $\mathbf{x}$. How large would these differences have to be to alter the qualitative conclusions of a study?

Suppose that we have two units, say j and k, with the same $\mathbf{x}$ but possibly different π's, so $\mathbf{x}_{[j]} = \mathbf{x}_{[k]}$ but possibly $\pi_{[j]} \neq \pi_{[k]}$. Then units j and k might be matched together to form a matched pair or placed together in the same subclass in our attempts to control overt bias due to $\mathbf{x}$. The odds that units j and k receive the treatment are, respectively, $\pi_{[j]}/(1 - \pi_{[j]})$ and $\pi_{[k]}/(1 - \pi_{[k]})$, and the odds ratio is the ratio of these odds. Imagine that we knew that this odds ratio for units with the same $\mathbf{x}$ was at most some number $\Gamma \geq 1$, that is,

$$\frac{1}{\Gamma} \leq \frac{\pi_{[j]}(1 - \pi_{[k]})}{\pi_{[k]}(1 - \pi_{[j]})} \leq \Gamma \qquad \text{for all } j, k \text{ with} \quad \mathbf{x}_{[j]} = \mathbf{x}_{[k]}. \tag{4.2.1}$$

If Γ were 1, then $\pi_{[j]} = \pi_{[k]}$ whenever $\mathbf{x}_{[j]} = \mathbf{x}_{[k]}$, so the study would be free of hidden bias, and the methods of Chapter 3 would apply. If $\Gamma = 2$, then two units who appear similar, who have the same $\mathbf{x}$, could differ in their odds of receiving the treatment by as much as a factor of 2, so one could be twice as likely as the other to receive the treatment. In other words, Γ is a measure of the degree of departure from a study that is free of hidden bias. A sensitivity analysis will consider several possible values of Γ and show how the inferences might change. A study is sensitive if values of Γ close to 1 could lead to inferences that are very different from those obtained assuming the study is free of hidden bias. A study is insensitive if extreme values of Γ are required to alter the inference.

4.2.2. The Model Expressed in Terms of an Unobserved Covariate

When speaking of hidden biases, we commonly refer to characteristics that were not observed, that are not in $\mathbf{x}$, and therefore that were not controlled by adjustments for $\mathbf{x}$. This section will express the model of §4.2.1 in terms of an unobserved covariate, say u, that should have been controlled along with $\mathbf{x}$ but was not controlled because u was not observed. This re-expression of the model takes place without loss of generality, that is, it entails writing the same model in a different way. The models of §4.2.1 and §4.2.2 are identical in the sense that they describe exactly the same collections of probability distributions for $Z_{[1]}, \ldots, Z_{[M]}$.

Unit j has both an observed covariate $\mathbf{x}_{[j]}$ and an unobserved covariate $u_{[j]}$. The model has two parts, a logit form linking treatment assignment $Z_{[j]}$ to the covariates $(\mathbf{x}_{[j]}, u_{[j]})$ and a constraint on $u_{[j]}$, namely,

$$\log\left(\frac{\pi_{[j]}}{1-\pi_{[j]}}\right) = \kappa(\mathbf{x}_{[j]}) + \gamma u_{[j]}, \tag{4.2.2a}$$

$$0 \le u_{[j]} \le 1, \tag{4.2.2b}$$

where $\kappa(\cdot)$ is an unknown function and γ is an unknown parameter.

Suppose units j and k have the same observed covariate, so $\mathbf{x}_{[j]} = \mathbf{x}_{[k]}$, and hence $\kappa(\mathbf{x}_{[j]}) = \kappa(\mathbf{x}_{[k]})$. In adjusting for $\mathbf{x}$, these two units might be matched together or might be placed in the same stratum. Consider the ratio of the odds that units j and k receive the treatment

$$\frac{\pi_{[j]}(1-\pi_{[k]})}{\pi_{[k]}(1-\pi_{[j]})} = \exp\{\gamma(u_{[j]} - u_{[k]})\}. \tag{4.2.3}$$

In other words, two units with the same $\mathbf{x}$ differ in their odds of receiving the treatment by a factor that involves the parameter γ and the difference in their unobserved covariates u.

The following proposition says that the model (4.2.2) is the same as the inequality (4.2.1). The proposition shows that the model (4.2.2) involves no additional assumptions beyond (4.2.1)—the logit form, the constraint on u, the absence of interaction terms linking **x** and u—these are not additional assumptions but rather are implied by (4.2.1).

Proposition 4.1. *With* $e^\gamma = \Gamma \geq 1$, *there is a model of the form* (4.2.2) *which describes the* $\pi_{[1]}, \ldots, \pi_{[M]}$ *if and only if* (4.2.1) *is satisfied.*

PROOF. Assume the model (4.2.2). Then by (4.2.2b), it follows that $-1 \leq u_{[j]} - u_{[k]} \leq 1$, and so from (4.2.3),

$$\exp(-\gamma) \leq \frac{\pi_{[j]}(1-\pi_{[k]})}{\pi_{[k]}(1-\pi_{[j]})} \leq \exp(\gamma) \qquad \text{whenever} \quad \mathbf{x}_{[j]} = \mathbf{x}_{[k]}. \qquad (4.2.4)$$

In other words, if (4.2.2) holds, then (4.2.1) holds with $\Gamma = \exp(\gamma)$.

Conversely, assume the inequality (4.2.1) holds. For each value **x** of the observed covariate, find that unit k with $\mathbf{x}_{[k]} = \mathbf{x}$ having the smallest π, so

$$\pi_{[k]} = \min_{\{j:\, \mathbf{x}_{[j]} = \mathbf{x}\}} \pi_{[j]};$$

then set $\kappa(\mathbf{x}) = \log\{\pi_{[k]}/(1-\pi_{[k]})\}$ and $u_{[k]} = 0$. If $\Gamma = 1$, then $\mathbf{x}_{[j]} = \mathbf{x}_{[k]}$ implies $\pi_{[j]} = \pi_{[k]}$, so set $u_{[j]} = 0$. If $\Gamma > 1$ and there is another unit j with the same value of **x**, then set

$$u_{[j]} = \frac{1}{\gamma}\log\left(\frac{\pi_{[j]}}{1-\pi_{[j]}}\right) - \frac{\kappa(\mathbf{x})}{\gamma} = \frac{1}{\gamma}\log\left(\frac{\pi_{[j]}(1-\pi_{[k]})}{\pi_{[k]}(1-\pi_{[j]})}\right). \qquad (4.2.5)$$

Now (4.2.5) implies (4.2.2a). Since $\pi_{[j]} \geq \pi_{[k]}$, it follows that $u_{[j]} \geq 0$. Using (4.2.1) and (4.2.5), it follows that $u_{[j]} \leq 1$. So (4.2.2b) holds. □

The constraint (4.2.2b) may be viewed in various ways. First, it may be seen simply as the formal expression of the statement (4.2.1) that subjects who appear similar in terms of **x** may differ in their odds of receiving the treatment by at most Γ. This faithful translation of the model into nontechnical terms is often helpful in discussing the results of a sensitivity analysis.

A second view of the constraint (4.2.2b) relates it to the risk ratio inequality of Cornfield et al. (1959). In that inequality, the unobserved variable is a binary trait, $u = 0$ or $u = 1$. Of course, a binary trait satisfies the constraint (4.2.2b). Indeed, in later calculations, the relevant u's will take the extreme values of 0 or 1. In other words, the constraint (4.2.2b) could, for most purposes, be replaced by the assumption that u is a binary variable taking values 0 or 1.

Finally, the constraint (4.2.2b) may be seen as a restriction on the scale of the unobserved covariate, a restriction needed if the numerical value of γ is to have meaning. This last view suggests consideration of other restrictions on the scale, for instance, restrictions that do not confine u to a finite range. As is seen later in Problem 2, it is not difficult to work with alternative restrictions on u, though the attractive nontechnical translation of the parameter Γ is then lost.

Throughout this chapter, the model (4.2.1) or its equivalent model (4.2.2) is assumed to hold.

4.2.3. The Distribution of Treatment Assignments After Stratification

As in §3.2.2, group units into S strata based on the observed covariate $\mathbf{x}$, with n_s units in stratum s, of whom m_s received the treatment. Under the equivalent models (4.2.1) or (4.2.2), the conditional distribution of the treatment assignment $\mathbf{Z} = (Z_{11}, \ldots, Z_{S,n_S})$ given $\mathbf{m}$ is no longer constant, as it was in §3.2.2 for a study free of hidden bias. Instead, the distribution is

$$\mathrm{pr}(\mathbf{Z} = \mathbf{z}|\mathbf{m}) = \frac{\exp(\gamma \mathbf{z}^{\mathrm{T}} \mathbf{u})}{\sum_{\mathbf{b} \in \Omega} \exp(\gamma \mathbf{b}^{\mathrm{T}} \mathbf{u})} = \prod_{s=1}^{S} \frac{\exp(\gamma \mathbf{z}_s^{\mathrm{T}} \mathbf{u}_s)}{\sum_{\mathbf{b}_s \in \Omega_s} \exp(\gamma \mathbf{z}_s^{\mathrm{T}} \mathbf{u}_s)}, \tag{4.2.6}$$

where

$$\mathbf{z} = \begin{bmatrix} \mathbf{z}_1 \\ \vdots \\ \mathbf{z}_S \end{bmatrix}, \qquad \mathbf{u} = \begin{bmatrix} \mathbf{u}_1 \\ \vdots \\ \mathbf{u}_S \end{bmatrix},$$

and Ω_s is the set containing the $\binom{n_s}{m_s}$ different n_s-tuples with m_s ones and $n_s - m_s$ zeros. The steps leading from (4.2.2a) to (4.2.6) form Problem 3.

Notice what (4.2.6) says. Given $\mathbf{m}$, the distribution of treatment assignments $\mathbf{Z}$ no longer depends of the unknown function $\kappa(\mathbf{x})$, but it continues to depend on the unobserved covariate u. In words, stratification on $\mathbf{x}$ was useful in that it eliminated part of the uncertainty about the unknown π's, specifically, the part due to $\kappa(\mathbf{x})$, but stratification on $\mathbf{x}$ was insufficient to render all treatment assignments equally probable, as they would be in an experiment.

There are two exceptions, two cases in which (4.2.6) gives equal probabilities $1/K$ to all K treatment assignments in Ω. If $\gamma = 0$, then the unobserved covariate u is not relevant to treatment assignment; see (4.2.2). If $u_{si} = u_{sj}$ for all $1 \le i < j \le n_s$ and for $s = 1, \ldots, S$, then units in the same stratum have the same value of u. In both cases, (4.2.6) equals $1/K$ for all $\mathbf{z} \in \Omega$. In other words, there is no hidden bias if u is either unrelated to treatment assignment or if u is constant for all units in the same stratum.

For any specific $(\gamma, \mathbf{u})$, the distribution (4.2.6) is a distribution of treatment assignments $\mathbf{Z}$ on Ω. If $(\gamma, \mathbf{u})$ were known, (4.2.6) could be used as a basis for permutation inference, as in §2. Since $(\gamma, \mathbf{u})$ is not known, a sensitivity analysis will display the sensitivity of inferences to a range of assumptions about $(\gamma, \mathbf{u})$. Specifically, for several values of γ, the sensitivity analysis will determine the most extreme inferences that are possible for $\mathbf{u}$ satisfying the constraint (4.2.2b), that is, for $\mathbf{u}$ in the N-dimensional unit cube $U = [0, 1]^N$. Specific procedures are discussed in the remaining sections of this chapter.

4.3. Matched Pairs

4.3.1. Sensitivity of Significance Levels: The General Case

With matched pairs, the model and methods have a particularly simple form. For this reason, matched pairs will be discussed first and separately, though §4.3 could be formally subsumed within the more general discussion in §4.4.

With S matched pairs, each stratum s, $s = 1, \ldots, S$, has $n_s = 2$ units, one of which received the treatment, $1 = m_s = Z_{s1} + Z_{s2}$, so $Z_{s2} = 1 - Z_{s1}$. The model (4.2.6) describes S independent binary trials with unequal probabilities, namely,

$$\operatorname{pr}(\mathbf{Z} = \mathbf{z} \mid \mathbf{m}) = \prod_{s=1}^{S} \left[\frac{\exp(\gamma u_{s1})}{\exp(\gamma u_{s1}) + \exp(\gamma u_{s2})}\right]^{z_{s1}} \left[\frac{\exp(\gamma u_{s2})}{\exp(\gamma u_{s1}) + \exp(\gamma u_{s2})}\right]^{1-z_{s1}}. \tag{4.3.1}$$

In a randomized experiment or in a study free of hidden bias, $\gamma = 0$ and every unit has an equal chance of receiving each treatment, so (4.3.1) is the uniform distribution on Ω. In contrast, if $\gamma > 0$ in (4.3.1), then the first unit in pair s is more likely to receive the treatment than the second if $u_{s1} > u_{s2}$.

As seen in Chapter 2, common test statistics for matched pairs are sign-score statistics, including Wilcoxon's signed rank statistic and McNemar's statistic. They have the form

$$T = t(\mathbf{Z}, \mathbf{r}) = \sum_{s=1}^{S} d_s \sum_{i=1}^{2} c_{si} Z_{si}$$

where c_{si} is binary, $c_{si} = 1$ or $c_{si} = 0$, and both $d_s \geq 0$ and c_{si} are functions of $\mathbf{r}$, and so are fixed under the null hypothesis of no treatment effect. In a randomized experiment, $t(\mathbf{Z}, \mathbf{r})$ is compared to its randomization distribution (2.4.4) under the null hypothesis, but this is not possible here because the treatment assignments $\mathbf{Z}$ have distribution (4.3.1) in which $(\gamma, \mathbf{u})$ is unknown. Specifically, for each possible $(\gamma, \mathbf{u})$, the statistic $t(\mathbf{Z}, \mathbf{r})$ is the sum of S independent random variables, where the sth variable equals d_s with probability

$$p_s = \frac{c_{s1} \exp(\gamma u_{s1}) + c_{s2} \exp(\gamma u_{s2})}{\exp(\gamma u_{s1}) + \exp(\gamma u_{s2})}, \tag{4.3.2}$$

and equals 0 with probability $1 - p_s$. A pair is said to be concordant if $c_{s1} = c_{s2}$. If $c_{s1} = c_{s2} = 1$ then $p_s = 1$ in (4.3.2), while if $c_{s1} = c_{s2} = 0$ then $p_s = 0$, so concordant pairs contribute a fixed quantity to $t(\mathbf{Z}, \mathbf{r})$ for all possible $(\gamma, \mathbf{u})$.

Though the null distribution of $t(\mathbf{Z}, \mathbf{r})$ is unknown, for each fixed γ, the null distribution is bounded by two known distributions. With $\Gamma = \exp(\gamma)$, define p_s^+ and p_s^- in the following way:

$$p_s^+ = \begin{cases} 0 & \text{if } c_{s1} = c_{s2} = 0, \\ 1 & \text{if } c_{s1} = c_{s2} = 1, \\ \dfrac{\Gamma}{1+\Gamma} & \text{if } c_{s1} \neq c_{s2}, \end{cases} \quad \text{and} \quad p_s^- = \begin{cases} 0 & \text{if } c_{s1} = c_{s2} = 0, \\ 1 & \text{if } c_{s1} = c_{s2} = 1, \\ \dfrac{1}{1+\Gamma} & \text{if } c_{s1} \neq c_{s2}. \end{cases}$$

Then using (4.2.2b), it follows that $p_s^- \leq p_s \leq p_s^+$ for $s = 1, \ldots, S$. Define T^+ to be the sum of S independent random variables, where the sth takes the value d_s with probability p_s^+ and takes the value 0 with probability $1 - p_s^+$. Define T^- similarly with in p_s^- place of p_s^+. The following proposition says that, for all $\mathbf{u} \in U$, the unknown null distribution of the test statistic $T = t(\mathbf{Z}, \mathbf{r})$ is bounded by the distributions of T^+ and T^-.

Proposition 4.2. *If the treatment has no effect, then for each fixed* γ,

$$\text{prob}(T^+ \geq a) \geq \text{prob}\{T \geq a|\mathbf{m}\} \geq \text{prob}(T^- \geq a) \qquad \textit{for all } a \textit{ and } \mathbf{u} \in U.$$

For each γ, Proposition 4.2 places bounds on the significance level that would have been appropriate had $\mathbf{u}$ been observed. The sensitivity analysis for a significance level involves calculating these bounds for several values of γ.

The bounds in Proposition 4.2 are attained for two values of $\mathbf{u} \in U$, and this has two practical consequences. Specifically, the upper bound $\text{prob}(T^+ \geq a)$ is the distribution of $t(\mathbf{Z}, \mathbf{r})$ when $u_{si} = c_{si}$ and the lower bound $\text{prob}(T^- \geq a)$ is the distribution of $t(\mathbf{Z}, \mathbf{r})$ when $u_{si} = 1 - c_{si}$. The first consequence is that bounds in Proposition 4.2 cannot be improved unless additional information is given about the value of $\mathbf{u} \in U$. Second, the bounds are attained at values of $\mathbf{u}$ which perfectly predict the signs c_{si}. For conventional statistics like McNemar's statistic and Wilcoxon's signed rank statistic, this means that the bounds are attained for values of $\mathbf{u}$ that exhibit a strong, near perfect relationship with the response $\mathbf{r}$.

The bounding distributions of T^+ and T^- have easily calculated moments. For T^+, the expectation and variance are

$$E(T^+) = \sum_{s=1}^{S} d_s p_s^+ \quad \text{and} \quad \text{var}(T^+) = \sum_{s=1}^{S} d_s^2 p_s^+(1 - p_s^+). \tag{4.3.3}$$

For T^- the expectation and variance are given by the same formulas with p_s^- in place of p_s^+. As the number of pairs S increases, the distributions of T^+ and T^- are approximated by Normal distributions, providing the number of discordant pairs increases with S and provided the d_s are reasonably well-behaved, as they are for the McNemar and Wilcoxon statistics.

The method is illustrated in §4.3.2 for the McNemar test and in §4.3.3 for the Wilcoxon signed rank test.

4.3.2. Sensitivity Analysis for McNemar's Test: Simplified Formulas and an Example

In a study of the effects of smoking on lung cancer, Hammond (1964) paired 36,975 heavy smokers to nonsmokers on the basis of age, race, nativity, rural versus urban residence, occupational exposures to dusts and fumes, religion, education, marital status, alcohol consumption, sleep duration, exercise, severe nervous tension, use of tranquilizers, current health, history of cancer other than skin cancer, history of heart disease, stroke, or high blood pressure. Though the paired smokers and nonsmokers were similar in each of these ways, they may differ in many other ways. For instance, it has at times been suggested that there might be a genetic predisposition to smoke (Fisher, 1958).

Of the $S = 36{,}975$ pairs, there were 122 pairs in which exactly one person died of lung cancer. Of these, there were 12 pairs in which the nonsmoker died of lung cancer and 110 pairs in which the heavy smoker died of lung cancer. In other words, 122 of the pairs are discordant for death from lung cancer. If this study were a randomized experiment, which it was not, or if the study were free of hidden bias, which we have little reason to believe, then McNemar's test would compare the 110 lung cancer deaths among smokers to a binomial distribution with 122 trials and probability of success 1/2, yielding a significance level less that 0.0001. In the absence of hidden bias, there would be strong evidence that smoking causes lung cancer. How much hidden bias would need to be present to alter this conclusion?

Let $d_s = 1$ for all s, $Z_{si} = 1$ if the ith person in pair s is a smoker and $Z_{si} = 0$ if this person is a nonsmoker, and $c_{si} = 1$ if this person died of lung cancer and $c_{si} = 0$ otherwise. Then the sign-score statistic

$$T = t(\mathbf{Z}, \mathbf{r}) = \sum_{s=1}^{S} d_s \sum_{i=1}^{2} c_{si} Z_{si}$$

is the number of smokers who died of lung cancer. A pair is concordant if $c_{s1} = c_{s2}$ and is discordant if $c_{s1} \neq c_{s2}$. No matter how the treatment, smoking, is assigned within pair s, if neither person died of lung cancer, then pair s contributes 0 to $t(\mathbf{Z}, \mathbf{r})$, while if both people died of lung cancer, then pair s contributes 1 to $t(\mathbf{Z}, \mathbf{r})$, so in either case, a concordant pair contributes a fixed quantity to $t(\mathbf{Z}, \mathbf{r})$. Removing concordant pairs from consideration subtracts a constant from $t(\mathbf{Z}, \mathbf{r})$ and does not alter the significance level. Therefore, set the concordant pairs aside before computing $T = t(\mathbf{Z}, \mathbf{r})$, leaving the 122 discordant pairs, so T is the number of discordant pairs in which the smoker died of lung cancer, $T = 110$, that is, T is McNemar's statistic.

With the concordant pairs removed, T^+ and T^- have binomial distributions with 122 trials and probabilities of success $p^+ = \Gamma/(1 + \Gamma)$ and $p^- = 1/(1 + \Gamma)$, respectively. Under the null hypothesis of no effect of smoking, for each $\gamma \geq 0$, Proposition 4.2 gives an upper and lower bound on the significance level, $\text{prob}\{T \geq 110|\mathbf{m}\}$, namely, for all $\mathbf{u} \in U$

$$\sum_{a=110}^{122} \binom{122}{a} (p^+)^a (1-p^+)^{122-a} \geq \text{prob}\{T \geq 110|\mathbf{m}\}$$

$$\geq \sum_{a=110}^{122} \binom{122}{a} (p^-)^a (1-p^-)^{122-a}. \quad (4.3.4)$$

In a randomized experiment or a study free of hidden bias, the sensitivity parameter γ is zero, so $p^+ = p^- = \frac{1}{2}$, the upper and lower bounds in (4.3.4) are equal, and both bounds give the usual significance level for McNemar's statistic. For $\gamma \geq 0$, (4.3.4) gives a range of significance levels reflecting uncertainty about $\mathbf{u}$.

Table 4.3.1 gives the sensitivity analysis for Hammond's data. For six values of Γ, the table gives the upper and lower bounds (4.3.4) on the significance level. For $\Gamma = 4$, one person in a pair may be four times as likely to smoke as the other because they have different values of the unobserved covariate u. In the case $\Gamma = 4$, the significance level might be less than 0.0001 or it might be as high as 0.0036, but for all $\mathbf{u} \in U$ the null hypothesis of no effect of smoking on lung cancer is not plausible. The null hypothesis of no effect begins to become plausible for at least some $\mathbf{u} \in U$ with $\Gamma = 6$. To attribute the higher rate of death from lung cancer to an unobserved covariate $\mathbf{u}$ rather than to an effect of smoking, that unobserved covariate would need to produce a six-fold increase in the odds of smoking, and it would need to be a near perfect predictor of lung cancer. As will be seen by comparison in later examples, this is a high degree of insensitivity to hidden bias—in many other studies, biases smaller than $\Gamma = 6$ could explain the association between treatment and response.

Table 4.3.1. Sensitivity Analysis for Hammond's Study of Smoking and Lung Cancer: Range of Significance Levels for Hidden Biases of Various Magnitudes.

Γ	Minimum	Maximum
1	<0.0001	<0.0001
2	<0.0001	<0.0001
3	<0.0001	<0.0001
4	<0.0001	0.0036
5	<0.0001	0.03
6	<0.0001	0.1

4.3.3. Sensitivity Analysis for the Signed Rank Test: Simplified Formulas and an Example

Wilcoxon's signed rank statistic for S matched pairs is computed by ranking the absolute differences $|r_{s1} - r_{s2}|$ from 1 to S and summing the ranks of the pairs in which the treated unit had a higher response than the matched control. In the notation of §4.3.1, d_s is the rank of $|r_{s1} - r_{s2}|$ with average ranks used for ties, and $c_{s1} = 1$, $c_{s2} = 0$ if $r_{s1} > r_{s2}$ or $c_{s1} = 0$, $c_{s2} = 1$ if $r_{s1} < r_{s2}$, so in both cases the pairs are discordant, and $c_{s1} = 0$, $c_{s2} = 0$ if $r_{s1} = r_{s2}$ so the pairs are concordant or tied.

If there are no ties among the absolute differences $|r_{s1} - r_{s2}|$ and no zero differences, then the expectation and variance (4.3.3) of T^+ have simpler forms

$$E(T^+) = \frac{p^+ S(S+1)}{2},$$

and

$$\operatorname{var}(T^+) = p^+(1 - p^+)\frac{S(S+1)(2S+1)}{6}, \tag{4.3.5}$$

where

$$p^+ = \frac{\Gamma}{1+\Gamma};$$

see Problem 4. With $p^- = 1/(1 + \Gamma)$ in place of p^+, the same expressions give the expectation and variance of T^-. If $\gamma = 0$ so $e^\gamma = \Gamma = 1$, then $p^+ = p^- = \frac{1}{2}$, and these expressions for the moments of T^+ and T^- are the same as the usual formulas for the expectation and variance of Wilcoxon's statistic in a randomized experiment (Lehmann, 1975, p. 128).

Ties and zero differences occur in practice, and in this case the general formula (4.3.3) may be used. Notice that ties do not alter the expectation of T^+ and T^-, but they alter the variance.

Consider the example in Table 3.2.1 concerning lead in the blood of children whose parents work in a battery factory. The Wilcoxon signed rank statistic for the differences in children's lead levels is 527. As seen in §3.2.4, if the study were free of hidden bias, this would constitute strong evidence of an effect of parental exposure to lead on children's lead levels. Specifically, if the study were free of hidden bias, that is if $\Gamma = 1$, the expectation of the signed rank statistic under the null hypothesis of no effect is 280 with variance 3130.63. The standardized deviate $(527 - 280)/\sqrt{3130.63}$ is 4.41, yielding an approximate one-sided significance level of less than 0.0001.

In Table 3.2.1, notice the several tied differences for which average ranks were used, and the one zero difference which is a concordant pair. Because of the ties and zero difference, the exact formula (4.3.3) was used, though in this case (4.3.5) gives nearly identical results; for instance, (4.3.3) gives 280.5 and 3132.25 for the null expectation and variance with deviate still equal to 4.41.

Table 4.3.2. Sensitivity Analysis for Lead in the Blood of Children: Range of Significance Levels for the Signed-Rank Statistic.

	Range of significance levels	
Γ	Minimum	Maximum
1	<0.0001	<0.0001
2	<0.0001	0.0019
3	<0.0001	0.0143
4	<0.0001	0.0401
4.25	<0.0001	0.0485
5	<0.0001	0.0764

If the study were free of hidden bias, there would be strong evidence that parents' occupational exposures to lead increased the level of lead in their children's blood. The sensitivity analysis asks how this conclusion might be changed by hidden biases of various magnitudes. If Γ were 2, then matched children might differ in their odds of exposure to lead by a factor of 2 due to hidden bias. In this case, $E(T^+) = 373.33$, $E(T^-) = 186.67$, $\text{var}(T^+) = \text{var}(T^-) = 2782.78$, so the deviates are $(527 - 373.33)/\sqrt{2782.78} = 2.90$ and $(527 - 186.67)/\sqrt{2782.78} = 6.45$. The range of significance levels when $\Gamma = 2$ is therefore from less than 0.0001 to 0.0019. A hidden bias of size $\Gamma = 2$ is insufficient to explain the observed difference between exposed and control children.

Table 4.3.2 gives the sensitivity analysis for the significance levels from Wilcoxon's signed rank test, that is, the range of possible significance levels for various values of Γ. The table shows that to explain away the observed association between parental exposure to lead and child's lead level, a hidden bias or unobserved covariate would need to increase the odds of exposure by more than a factor of $\Gamma = 4.25$. The association cannot be attributed to small hidden biases, but it is somewhat more sensitive to bias than Hammond's study of heavy smokers in §4.3.3.

4.3.4. Sensitivity Analysis of the Hodges–Lehmann Point Estimate in Matched Pairs

In §3.2.4, the effect of parental exposure to lead on the level of lead in the child's blood was estimated to be 15 μg of lead per decaliter of whole blood. This estimate is based on two premises:

(i) the treatment has an additive effect τ, so a child's lead level is increased by τ as a consequence of parental exposure; and
(ii) the study is free of hidden bias.

If the effect is additive, then the adjusted responses $\mathbf{R} - \tau\mathbf{Z}$ satisfy the null hypothesis of no effect. The estimate of 15 $\mu g/dl$ is the Hodges–Lehmann estimate, obtained as the value $\hat{\tau}$ such that, if the signed rank statistic is computed after subtracting $\hat{\tau}$ from the responses of exposed children, then this statistic equals its expectation $E\{t(\mathbf{Z}, \mathbf{R} - \tau\mathbf{Z})\} = \bar{\bar{t}} = S(S+1)/4$ under the null hypothesis of no effect in the absence of hidden bias. More precisely, as discussed in §2.7.2, to allow for the discreteness of the signed rank statistic, $\hat{\tau}$ is defined as

$$\hat{\tau} = \text{SOLVE}\{\bar{\bar{t}} = t(\mathbf{Z}, \mathbf{R} - \hat{\tau}\mathbf{Z})\}$$
$$= \frac{\inf\{\tau: \bar{\bar{t}} > t(\mathbf{Z}, \mathbf{R} - \tau\mathbf{Z})\} + \sup\{\tau: \bar{\bar{t}} < t(\mathbf{Z}, \mathbf{R} - \tau\mathbf{Z})\}}{2}. \quad (4.3.6)$$

If there is hidden bias, then the signed rank statistic computed from $\mathbf{R} - \tau\mathbf{Z}$ does not generally have expectation $S(S+1)/4$. If $E\{t(\mathbf{Z}, \mathbf{R} - \tau\mathbf{Z})\} \neq S(S+1)/4$, then there is no point in trying to find a $\hat{\tau}$ so that $t(\mathbf{Z}, \mathbf{R} - \hat{\tau}\mathbf{Z})$ is nearly $S(S+1)/4$; rather, $\hat{\tau}$ should be found so $t(\mathbf{Z}, \mathbf{R} - \hat{\tau}\mathbf{Z})$ is close to $E\{t(\mathbf{Z}, \mathbf{R} - \tau\mathbf{Z})\}$. Under the model (4.2.1) or (4.2.2), the expectation of the signed rank statistic computed from $\mathbf{R} - \tau\mathbf{Z}$ is between bounded by the expectations of T^+ and T^- in Proposition 4.2, that is, bounded by

$$\bar{\bar{t}}_{\min} = \frac{p^- S(S+1)}{2} \quad \text{and} \quad \bar{\bar{t}}_{\max} = \frac{p^+ S(S+1)}{2},$$

where (4.3.7)

$$p^- = \frac{1}{1+\Gamma} \quad \text{and} \quad p^+ = \frac{\Gamma}{1+\Gamma}.$$

In other words, for a given Γ, the expectation $\bar{\bar{t}} = E\{t(\mathbf{Z}, \mathbf{R} - \tau\mathbf{Z})\}$ is not known but is bounded by two known numbers, $\bar{\bar{t}}_{\min} \leq \bar{\bar{t}} \leq \bar{\bar{t}}_{\max}$. Moreover, these bounds are sharp in that they are attained for particular values of the unobserved covariate $\mathbf{u} \in U$. Consider calculating $\hat{\tau} = \text{SOLVE}\{\bar{\bar{t}} = t(\mathbf{Z}, \mathbf{R} - \hat{\tau}\mathbf{Z})\}$ for each $\bar{\bar{t}}$ in the interval $[\bar{\bar{t}}_{\min}, \bar{\bar{t}}_{\max}]$; this would produce the set of possible Hodges–Lehmann estimates. In fact, the minimum and maximum estimates in this set are easily determined. Since the signed-rank statistic is order-preserving, the smallest $\hat{\tau} = \text{SOLVE}\{\bar{\bar{t}} = t(\mathbf{Z}, \mathbf{R} - \hat{\tau}\mathbf{Z})\}$ is found for $\bar{\bar{t}} = \bar{\bar{t}}_{\max}$ and the largest $\hat{\tau}$ is found for $\bar{\bar{t}} = \bar{\bar{t}}_{\min}$.

In short, the sensitivity analysis consists of calculating the range of possible Hodges–Lehmann estimates of an additive treatment effect for biases of various magnitudes. This is done by calculating $\hat{\tau} = \text{SOLVE}\{\bar{\bar{t}} = t(\mathbf{Z}, \mathbf{R} - \hat{\tau}\mathbf{Z})\}$ for $\bar{\bar{t}} = \bar{\bar{t}}_{\max}$ and $\bar{\bar{t}} = \bar{\bar{t}}_{\min}$ for several values of Γ.

Table 4.3.3 shows the sensitivity analysis for the data on lead in the blood of children. If there is no hidden bias, that is, if $\Gamma = 1$, then $\bar{\bar{t}}_{\max} = \bar{\bar{t}}_{\min} = S(S+1)/4 = 33(33+1)/4 = 280.5$, and the range of Hodges–Lehmann estimates of effect is the single number 15$\mu g/dl$, namely, the usual Hodges–Lehmann estimate for a randomized experiment. If $\Gamma = 2$, then

Table 4.3.3. Sensitivity Analysis for Lead in Children's Blood: Range of Hodges–Lehmann Estimates of Effect for Biases of Various Magnitudes.

Γ	Minimum	Maximum
1	15	15
2	10.25	19.5
3	8	23
4	6.5	25
5	5	26.5

two subjects with the same observed covariates may differ by a factor of 2 in their odds of receiving the treatment. In this case, $\bar{\bar{t}}_{\max} = p^+ S(S+1)/2 = 2 \times 33(33+1)/6 = 374$ and $\bar{\bar{t}}_{\min} = p^- S(S+1)/2 = 33(33+1)/6 = 187$, and solving $\hat{\tau} = \text{SOLVE}\{\bar{\bar{t}} = t(\mathbf{Z}, \mathbf{R} - \hat{\tau}\mathbf{Z})\}$ gives 10.25 and 19.5, respectively. In words, if $\Gamma = 2$, the estimated effect might be as small as 10.25 $\mu g/dl$ or as high as 19.5 $\mu g/dl$. Keep in mind that the median lead level in $\mu g/dl$ among controls is 16, so an effect of 10.25 is a 64% increase above the level found among controls. For $\Gamma = 5$, the estimated effect is between 5 $\mu g/dl$ and 26 $\mu g/dl$, though the smaller estimate does not differ significantly from zero; see Table 4.3.2.

To illustrate the computations in greater detail, consider the upper bound 19.5 for $\Gamma = 2$. As noted above, $\bar{\bar{t}}_{\min} = 187$. By direct calculation with the signed rank statistic, $t(\mathbf{Z}, \mathbf{R} - 19.5\mathbf{Z}) = 193.5$, which is a bit too high, while $t(\mathbf{Z}, \mathbf{R} - 19.5001\mathbf{Z}) = 186$, which is a bit too low. Applying (4.3.6) with $\inf\{\tau: 187 > t(\mathbf{Z}, \mathbf{R} - \tau\mathbf{Z})\} = 19.5$ and $\sup\{\tau: 187 < t(\mathbf{Z}, \mathbf{R} - \tau\mathbf{Z})\} = 19.5$ yields the estimate 19.5.

4.3.5. Sensitivity Analysis for Confidence Intervals in Matched Pairs

This section discusses sensitivity analysis for a confidence interval for an additive effect τ. A $1 - \alpha$ confidence interval is the set of all values of τ that are not rejected in an α-level test. For each $(\gamma, \mathbf{u})$, there is a two-sided α-level confidence interval derived from a test statistic $t(\mathbf{Z}, \mathbf{R} - \tau\mathbf{Z})$. For each fixed γ, the sensitivity analysis will report the union of these intervals as $\mathbf{u}$ ranges over U. Call this union the sensitivity interval. A value τ is in the interval if and only if there is some $\mathbf{u} \in U$ such that τ is not rejected in an α-level test.

Let T_τ^+ and T_τ^- be the bounding random variables in §4.3.1 computed from the adjusted responses $\mathbf{R} - \tau\mathbf{Z}$, and let $T_\tau = t(\mathbf{Z}, \mathbf{R} - \tau\mathbf{Z})$. Fix $\gamma \geq 0$

Table 4.3.4. Sensitivity Analysis for Lead in Children's Blood: Confidence Intervals for an Additive Effect for Biases of Various Magnitudes.

Γ	95% Confidence interval
1	(9.5, 20.5)
2	(4.5, 27.5)
3	(1.0, 32.0)
4	(−1.0, 36.5)
5	(−3.0, 41.5)

and suppose $\alpha/2 \geq \text{prob}(T_\tau^+ \geq a_1)$ and $\alpha/2 \geq \text{prob}(T_\tau^- \leq a_2)$. Then Proposition 4.2 implies $\alpha/2 \geq \text{prob}\{T_\tau \geq a_1|\mathbf{m}\}$ for all $\mathbf{u} \in U$, and $\alpha/2 \geq \text{prob}\{T_\tau \leq a_2|\mathbf{m}\}$ for all $\mathbf{u} \in U$. If $T_\tau \geq a_1$ or $T_\tau \leq a_2$, then τ is rejected at level α for all $\mathbf{u} \in U$, so τ is excluded from the sensitivity interval. When $\gamma = 0$, this yields the confidence interval in §2.6.2 for a randomized experiment, but as γ increases this interval becomes larger reflecting uncertainty about the impact of $\mathbf{u}$.

When using the Normal approximation to the distribution of T_τ^+ and T_τ^-, the endpoints of the 95% confidence interval are

$$\inf\left\{\tau\colon \frac{T_\tau - E(T_\tau^+)}{\sqrt{\text{var}(T_\tau^+)}} \leq 1.96\right\} \quad \text{and} \quad \sup\left\{\tau\colon \frac{T_\tau - E(T_\tau^-)}{\sqrt{\text{var}(T_\tau^-)}} \geq -1.96\right\}.$$

The procedure is illustrated in Table 4.3.4 using the signed rank test for the data on lead in children's blood. If the study were free of hidden bias, that is if $\Gamma = 1$, the 95% confidence interval for the additive effect τ in $\mu g/dl$ would be (9.5, 20.5), as in §3.2.4. If $\Gamma = 2$, matched children might differ by a factor of two in their odds of exposure to lead due to differences in the unobserved covariate. In this case, the 95% confidence interval is longer, (4.5, 27.5), though the smallest plausible effect 4.5 is still 28% of the median lead level 16 $\mu g/dl$ among controls. For $\Gamma = 4$, slightly negative effects become just plausible, though large positive effects are also plausible. In comparing the tables, keep in mind that Tables 4.3.4 describes a two-sided 95% confidence interval while Table 4.3.2 describes a one-sided significance test.

Table 4.3.5 contains some of the calculations leading to Table 4.3.4 for the case $\Gamma = 2$. The top half of Table 4.3.5 shows that 27.5 is the sup of all τ leading to a deviate greater than -1.96. The bottom half of the table shows that 4.5 is the inf of all τ leading to a deviate less than 1.96.

Table 4.3.5. Illustrative Computations for $\Gamma = 2$.

τ	$t(\mathbf{Z}, \mathbf{R} - \tau\mathbf{Z})$	$E(T_\tau^-)$	$\text{var}(T_\tau^-)$	Deviate
27.4999	89.00	187.00	2609.00	−1.92
27.5000	86.00	187.00	2607.89	−1.98
τ	$t(\mathbf{Z}, \mathbf{R} - \tau\mathbf{Z})$	$E(T_\tau^+)$	$\text{var}(T_\tau^+)$	Deviate
4.5000	476.50	374.00	2766.61	1.95
4.4999	479.00	374.00	2769.00	2.00

4.4. Sensitivity Analysis for Sign-Score Statistics

4.4.1. The General Method

As noted in §2.4.2, many common statistical tests are sign-score statistics. All sign-score statistics permit a sensitivity analysis similar to that in §4.3 for matched pairs. The purpose of §4.4 is to discuss this class of problems, first in general terms in §4.4.1, and then in specific situations in the later parts of §4.4. Section 4.5 will discuss a larger class of test statistics requiring computations that are just slightly more complex.

Recall that a sign-score statistic has the form

$$t(\mathbf{Z}, \mathbf{r}) = \sum_{s=1}^{S} d_s \sum_{i=1}^{n_s} c_{si} Z_{si} = \sum_{s=1}^{S} d_s B_s \qquad \text{where} \quad B_s = \sum_{i=1}^{n_s} c_{si} Z_{si}, \tag{4.4.1}$$

c_{si} is binary, $c_{si} = 1$ or $c_{si} = 0$, d_s is nonnegative, $d_s \geq 0$, and both d_s and c_{si} are functions of $\mathbf{r}$. For binary responses $\mathbf{r}$, the signs are the responses themselves, $c_{si} = r_{si}$, and the scores are constant, $d_s = 1$; this yields Fisher's exact test for a 2×2 table, McNemar's statistic for matched pairs, and the Mantel–Haenszel statistic for a $2 \times 2 \times S$ table. In Wilcoxon's signed rank statistic for matched pairs, the signs, c_{si} identify the unit in matched set i with the largest response and the score d_s is the rank of the difference between the larger and the smaller responses in a matched pair. In the median test, the signs c_{si} identify units with responses above the median response of the combined treated and control groups.

The main fact about sign-score statistics is similar to Proposition 4.2; it bounds the unknown distribution of $t(\mathbf{Z}, \mathbf{r})$ by two known distributions, namely, the distributions at the most extreme $\mathbf{u}$'s $\in U$. Sign-score statistics are special in the following sense: The $\mathbf{u}$'s $\in U$ that provide the bounds may be determined immediately from the structure of the statistic itself.

Under the null hypothesis of no treatment effect, $\mathbf{r}$ is fixed, so d_s and c_{si} are fixed because they are functions of $\mathbf{r}$. Write $\mathbf{u}^+$ and $\mathbf{u}^-$ for the N-tuples with $u_{si}^+ = c_{si}$ and $u_{si}^- = 1 - c_{si}$, respectively. Fix γ, and let T^+ be the random variable $\sum_s d_s \sum_i c_{si} Z_{si}$ when $\mathbf{Z}$ has the distribution (4.2.6) with $\mathbf{u} = \mathbf{u}^+$, and define T^- similarly with $\mathbf{u} = \mathbf{u}^-$.

Proposition 4.3. *If the treatment has no effect and $T = t(\mathbf{Z}, \mathbf{r})$ is a sign-score statistic, then for each fixed $\gamma \geq 0$,*

$$\text{prob}(T^+ \geq a) \geq \text{prob}\{T \geq a|\mathbf{m}\} \geq \text{prob}(T^- \geq a) \quad \textit{for all } a \textit{ and } \quad \mathbf{u} \in U.$$

The proof is given in the Appendix to this chapter.

As in §4.3, the sensitivity analysis consists of calculating the bounds, $\text{prob}(T^+ \geq a)$ and $\text{prob}(T^- \geq a)$, for a range of values of γ. For $\gamma = 0$, the bounds are equal, $\text{prob}(T^+ \geq a) = \text{prob}(T^- \geq a)$, and their common value is the usual significance level for a randomized experiment. As γ increases, the bounds move apart reflecting uncertainty about hidden biases. The general method is applied to particular cases in subsequent sections.

4.4.2. Matching with Multiple Controls

When controls are plentiful, it is common to match several controls to each treated unit, so each stratum s contains a single treated unit and one or more controls, $1 = \sum_i Z_{si}$ for $s = 1, \ldots, S$. In this case, the model (4.2.6) may be written

$$\text{pr}(\mathbf{Z} = \mathbf{z}|\mathbf{m}) = \prod_{s=1}^{S} \frac{\exp\left(\gamma \sum_{i=1}^{n_s} \gamma z_{si} u_{si}\right)}{\sum_{i=1}^{n_s} \exp(\gamma u_{si})} \quad \text{for} \quad \mathbf{z} \in \Omega, \tag{4.4.2}$$

since Ω_s contains just the n_s vectors $\mathbf{z}_s$ whose coordinates include a single one and $n_s - 1$ zeros. Compare (4.4.2) and (4.3.1). Also, B_s is a binary random variable. From (4.4.2), the B_s, $s = 1, \ldots, S$, are mutually independent and

$$p_s = \text{pr}(B_s = 1|\mathbf{m}) = \frac{\sum_{i=1}^{n_s} c_{si} \exp(\gamma u_{si})}{\sum_{i=1}^{n_s} \exp(\gamma u_{si})}. \tag{4.4.3}$$

This probability is bounded by its values at $\mathbf{u}^+$ and $\mathbf{u}^-$ in §4.4.1, namely,

$$p_s^- = \frac{c_{s+}}{c_{s+} + (n_s - c_{s+})\Gamma} \leq p_s \leq \frac{\Gamma c_{s+}}{\Gamma c_{s+} + n_s - c_{s+}} = p_s^+, \tag{4.4.4}$$

where $c_{s+} = \sum_i c_{si}$ and $\Gamma = e^\gamma$. Moreover, T^+ and T^- are the random variables that equal $\sum d_s B_s$ when, respectively, $p_s = p_s^+$ and $p_s = p_s^-$. It follows that the moments of T^+ are again given by formula (4.3.3), and the moments of T^- are given by the same formula with p_s^- in place of p_s^+.

If the outcome is binary, say 1 for survived and 0 for died, then the Mantel–Haenszel (1959) statistic is the usual test of the null hypothesis of no treatment effect; see §2.4.2. This statistic $T = t(\mathbf{Z}, \mathbf{r})$ is a sign-score statistic with $d_s = 1$ and $c_{si} = r_{si}$ in (4.4.1), so T is the number treated subjects who survived. In (4.4.1), $B_s = 1$ if the treated subject in matched set s survived and $B_s = 0$ otherwise. The bounds on the approximate significance level are obtaining by referring the standardized deviates,

$$\frac{|T - \sum p_s^+| - \frac{1}{2}}{\sqrt{\sum p_s^+(1 - p_s^+)}} \quad \text{and} \quad \frac{|T - \sum p_s^-| - \frac{1}{2}}{\sqrt{\sum p_s^-(1 - p_s^-)}}, \tag{4.4.5}$$

to tables of the standard normal distribution. When $\gamma = 0$, these two deviates are equal, and the square of either deviate equals the usual Mantel–Haenszel statistic.

In general, for any sign-score statistic used in matching with one or more controls, the deviates are

$$\frac{T - \sum d_s p_s^+}{\sqrt{\sum d_s^2 p_s^+(1 - p_s^+)}} \quad \text{and} \quad \frac{T - \sum d_s p_s^-}{\sqrt{\sum d_s^2 p_s^-(1 - p_s^-)}}.$$

4.4.3. Sensitivity Analysis with Multiple Controls: The Example of Vitamin C and Cancer

Recall from Chapter 1 the study by Cameron and Pauling (1976) of vitamin C and advanced cancer. In this study, patients treated with vitamin C were each matched with ten controls on the basis of gender, age, primary cancer, and histological tumor type. In this section, the Cameron and Pauling study will help to clarify what sensitivity analyses can and cannot do.

Focus on the 18 treated patients and their 180 matched controls having primary cancers of the colon and rectum. The subsequent randomized trial by Moertel et al. (1985) concerned patients with cancers of the colon and rectum. Cameron and Pauling (1976) used as a test statistic the number of matched sets in which the treated patient survived longer than the average survival among the 10 controls. This is the same as counting the number of treated patients living longer than the mean for all 11 patients in the set, and this count is a sign-score statistic in which $c_{si} = 1$ if subject (s, i) lived longer than the mean survival in set s and $c_{si} = 0$ otherwise. In fact, 16 of the 18 treated patients survived longer than the mean in their matched set.

Given the presence of some extreme responses and certain other features of the data, it is more appropriate to count the number of treated responses that exceed the median response in their matched set, though this too is 16 of the 18 matched sets. This statistic is also a sign-score statistic; in fact, it is a version of the standard median test, though here it is applied within matched sets and then combined across matched sets. The median test is a good test in large samples if the data are from a double exponential distri-

Table 4.4.1. Sensitivity Analysis for Vitamin C and Advanced Cancer.

Γ	p_s^+	p_s^-	Deviate at $\mathbf{u}^+$	Deviate at $\mathbf{u}^-$	Range of significance levels
1	0.45	0.45	3.46	3.46	[0.0003, 0.0003]
2	0.63	0.29	2.07	5.28	[0.019, <0.0001]
3	0.71	0.22	1.38	6.62	[0.082, <0.0001]

bution and the treatment has a small effect; see Hájek and Sidák (1967, §3.1.1, p. 88) or Hettmansperger (1984, §3). Here, $c_{si} = 1$ if subject (s, i) lived longer than the median survival in set s and $c_{si} = 0$ otherwise. Then for each s, $5 = \sum_i c_{si}$, because each set contains $n_s = 11$ subjects, and the sixth largest response is the median. Then in (4.4.4), $p_s^- = 5/(5 + 6\Gamma)$ and $p_s^+ = 5\Gamma/(5\Gamma + 6)$, and $p_s^- = p_s^+ = 5/11$ if there is no hidden bias in the sense that $\Gamma = 1$. Now expression (4.4.5) yields Table 4.4.1. The table suggests that the longer survival among patients receiving vitamin C would be highly significant in a randomized experiment, and is insensitive to small biases, but it becomes sensitive at about $\Gamma = 3$.

The sensitivity analysis just performed was based on the 16/18 treated subjects who lived longer than the typical subject in the matched set, and this closely parallels the analysis in the original study. One might wonder how the results might change with a different test statistic. In fact, with the same 18 matched sets but a different choice of test statistic, the results are far less sensitive, the upper bound on the significance level being 0.045 for $\Gamma = 10$; see Rosenbaum (1988) for detailed discussion. The test described in Table 4.4.1 made no use of the fact that most treated subjects had survival times much higher than the typical survival time in the subject's matched set, and this is the reason for the diverging findings from different test statistics. The test used for illustration in Rosenbaum (1988) yielded less sensitivity to bias in this example, but it does not have properties that would form a basis for recommending it for general use.

The first observation to make is that, with a suitable test, this study is insensitive to extremely large biases, $\Gamma = 10$, and yet the findings were contradicted by the randomized experiment by Moertel et al. (1985); see §1.2. This observation contains an important lesson about what sensitivity analyses can and cannot do. A sensitivity analysis can indicate the magnitude of the bias required to alter the qualitative conclusions of an observational study, but it cannot indicate what biases are present. Large biases have occurred in some studies. Chapters 5 through 8 are concerned with the collection of data that may indicate the presence or magnitude of hidden biases.

A second observation is of a technical nature, and it concerns the choice of test statistic. In this example, the choice of test statistic had a substantial impact on the sensitivity analysis. At present, little firm advice is available about the choice of test statistic for use in sensitivity analyses. Aside from a few special cases, statistical theory has emphasized the creation of test statistics that perform well in large samples against alternative hypotheses that are close to the null hypothesis, so-called local alternatives. The idea is that, in large samples, any reasonable test can detect a large departure from the null hypothesis, so tests should be designed to perform well against local alternatives. The situation is different in a sensitivity analysis. Large departures from the null hypothesis are of interest even in large samples, because only large departures are insensitive to hidden bias. An open area for research is the development of best tests for use in sensitivity analysis. For the most part, this chapter discusses traditional test statistics which, at least, are known to perform well in the absence of bias, that is, when $\Gamma = 1$.

4.4.4. Case-Referent Studies with Multiple Matched Referents

In the study of DES and vaginal cancer in Chapter 1, eight women with vaginal cancer—that is, eight cases—were each matched to four women without the disease, that is, four referents, and the cases and referents were compared in terms of their frequency of prenatal exposure to the treatment, DES. This study design—a case-referent study with several referents matched to each case—is quite common. If a disease or other outcome is rare, it may be difficult to locate additional cases, but referents may be plentiful; hence the tendency to match several referents to each case.

Note the difference between the study designs in §4.4.2 and §4.4.4. In matching with multiple controls in §4.4.2, each matched set contained one treated subject and several untreated controls, so $m_s = 1 = \sum_{i=1}^{n_s} Z_{si}$ for each s, but the design imposed no restrictions on the outcomes in a matched set. In a study in which cases are matched to several referents, matched set s may contain any number of treated subjects, that is, m_s can be 0, 1, ..., or n_s; however, each matched set contains exactly one case, so $1 = \sum_{i=1}^{n_s} R_{si}$. In §4.4.2, Ω_s contained n_s possible treatment assignments so the distribution of treatment assignments (4.2.6) had the simple form (4.4.2), but in a case-referent study, Ω_s contains $\binom{n_s}{m_s}$ possible treatment assignments, and (4.2.6) does not simplify further.

The null hypothesis of no treatment effect says that the binary indicator of a case of disease, R_{si}, is unaffected by the treatment Z_{si}, so R_{si} is a constant r_{si} that does not change with Z_{si}. The common test of no treatment effect is the Mantel–Haenszel statistic, which is a sign-score statistic with $d_s = 1$ and $c_{si} = r_{si}$, so the statistic (4.4.1) is the number of cases exposed to the treatment, namely, $T = \sum_{s=1}^{S} \sum_{i=1}^{n_s} r_{si} Z_{si} = \sum_{s=1}^{S} B_s$ where $B_s = \sum_{i=1}^{n_s} r_{si} Z_{si}$. Since $\sum_{i=1}^{n_s} r_{si}$ is zero or one in a case-referent study, B_s is also zero or one, as in §4.4.2.

Recall that in Proposition 4.3, the distribution of the test statistic T is bounded by the distributions of T^+ and T^- obtained by assuming $\mathbf{u} = \mathbf{u}^+$ or $\mathbf{u} = \mathbf{u}^-$. Since $T = \sum_{s=1}^{S} B_s$ in matched case-referent studies, these bounds may be obtained using the following bounds on $p_s = \text{prob}(B_s = 1|\mathbf{M})$ at $\mathbf{u} = \mathbf{u}^-$ or $\mathbf{u} = \mathbf{u}^+$:

$$P_s^- = \frac{m_s}{m_s + (n_s - m_s)\Gamma} \le p_s \le \frac{m_s\Gamma}{m_s\Gamma + (n_s - m_s)} = P_s^+ \qquad \text{for all} \quad \mathbf{u} \in U. \tag{4.4.6}$$

For the derivation of (4.4.6), see Problem 5. If there is no treatment effect and no hidden bias, so $\Gamma = 1$, then $p_s = P_s^+ = P_s^- = m_s/n_s$; this says the chance the case was exposed to the treatment equals the proportion of subjects in the matched set who were exposed to the treatment. Note that (4.4.4) and (4.4.6) are similar in form but different in detail. The large sample approximations to the bounds on the significance levels for the Mantel–Haenszel statistic are obtained by referring the deviates,

$$\frac{|T - \sum P_s^+| - \frac{1}{2}}{\sqrt{\sum P_s^+(1 - P_s^+)}} \quad \text{and} \quad \frac{|T - \sum P_s^-| - \frac{1}{2}}{\sqrt{\sum P_s^-(1 - P_s^-)}}, \tag{4.4.7}$$

to tables of the standard normal distribution. As with (4.4.5), these two deviates are equal if hidden bias is absent, in the sense that $\Gamma = 1$, and in this case their square is the usual Mantel–Haenszel statistic. These computations will now be performed for five studies. This will both illustrate the computations and give an indication of the varied results that a sensitivity analysis may produce.

4.4.5. Matched Case-Control Studies: Five Examples

Recall from Chapter 1 that in the case-referent study by Herbst, Ulfelder, and Poskanzer (1971) of DES and vaginal cancer, each of eight cases of vaginal cancer was matched to four referents using day of birth and type of service, ward or private. Table 4.4.2 contains the data together with some of the calculations for the sensitivity analysis.

In Table 4.4.2, there are $S = 8$ matched sets, $s = 1, \ldots, 8$, each containing one case and four referents. In seven matched sets, exactly one patient had in utero exposure to DES, that is, $m_s = 1$, but in set $s = 5$ there were no exposures to DES, $m_s = 0$. In all of the seven matched sets, it is the case who was exposed to DES, that is, $B_s = 1$ for $s \ne 5$. The relationship between DES and vaginal cancer appears to be extremely strong, though of course there are only eight cases. The one matched set with no exposed patients is concordant, and it could be removed without changing the value of the statistics in (4.4.7); see Problem 6.

To test the null hypothesis of no effect, suppose for the moment that DES does not cause vaginal cancer. Consider one of the seven matched sets

Table 4.4.2. DES and Vaginal Cancer: Data and Computations.

s	B_s	m_s	$P_s^- = P_s^+$ for $\Gamma = 1$	P_s^- for $\Gamma = 2$	P_s^+ for $\Gamma = 2$
1	1	1	0.20	0.11	0.33
2	1	1	0.20	0.11	0.33
3	1	1	0.20	0.11	0.33
4	1	1	0.20	0.11	0.33
5	0	0	0.00	0.00	0.00
6	1	1	0.20	0.11	0.33
7	1	1	0.20	0.11	0.33
8	1	1	0.20	0.11	0.33
Total	7	7	1.40	0.78	2.33

with $n_s = 5$ patients of whom $m_s = 1$ was exposed to DES. If the study were free of hidden bias, so $\Gamma = 1$, then each of the five patients has the same chance of being exposed, so the chance that the one exposed patient is the case is 1/5 or 0.20, as in Table 4.4.1. This would lead us to expect $1.40 = 7 \times 0.20$ cases to be exposed to DES, though, in fact, $T = 7$ cases were exposed, with a variance of $7 \times 0.20 \times 0.80 = 1.12$, and a deviate of $(|7 - 1.4| - \frac{1}{2})/\sqrt{1.12} = 4.82$, whose square $4.82^2 = 23.2$ is the usual Mantel–Haenszel statistic. If the study were free of hidden bias, there would be strong evidence that DES causes vaginal cancer. If hidden bias were present to the extent that matched subjects might differ in their odds of exposure to DES by a factor of two, so $\Gamma = 2$, then the chance that the case was exposed to DES might be as low as $m_s/\{m_s + \Gamma(n_s - m_s)\} = 1/(1 + 2 \times 4) = 0.11$ or as high as $\Gamma m_s/\{\Gamma m_s + (n_s - m_s)\} = 2/(2 + 4) = 0.33$, so the expected number of exposed cases might be as low as $7 \times 0.11 = 0.78$ with variance $7 \times 0.11 \times 0.89 = 0.69$, or as high as $7 \times 0.33 = 2.33$ with variance $7 \times 0.33 \times 0.67 = 1.56$, whereas seven exposed cases were observed. For $\Gamma = 2$, the deviates in (4.4.7) are

$$\frac{(|7 - 2.33| - \frac{1}{2})}{\sqrt{1.56}} = 3.34 \quad \text{and} \quad \frac{(|7 - 0.78| - \frac{1}{2})}{\sqrt{0.69}} = 6.88,$$

yielding a range of significance levels from less than 0.0001 to at most 0.0004. A hidden bias of magnitude $\Gamma = 2$ cannot reasonably explain the strong association seen between DES and vaginal cancer. Table 4.4.3 gives results for other values of Γ. Only beyond $\Gamma = 7$ is hidden bias a plausible explanation of the association between DES and vaginal cancer. The hidden bias would need to create a seven-fold increase in the odds of exposure to DES.

Table 4.4.4 compares the sensitivity of five matched case-referent studies. The purpose of the table is to gain some insight into the types of results a

Table 4.4.3. Sensitivity Analysis for DES and Vaginal Cancer.

Γ	Deviate$^-$	Deviate$^+$	Range of signifance levels
1	4.82	4.82	<0.0001
2	6.88	3.34	<0.0001 to 0.0004
4	9.78	2.27	<0.0001 to 0.012
6	12.00	1.77	<0.0001 to 0.038
7	12.96	1.61	<0.0001 to 0.054

sensitivity analysis may produce. The study of DES and vaginal cancer by Herbst et al. (1976) has just been discussed. The study by Mack et al. (1976) compared each of $S = 63$ cases of endometrial cancer to four referents matched for age and marital status. Cases and referents, who had been drawn from a large retirement community, were compared with respect to use of estrogens. The data are given in the detailed form needed for computations in Breslow and Day (1980, §5.3).

Jick et al. (1973) studied coffee consumption and myocardial infarction in white patients aged 40 to 69 years who were not alcoholic. Matching on age, gender, history of myocardial infarction, smoking, time of entry into the study, and hospital, they formed 27 matched pairs and 88 matched sets

Table 4.4.4. Comparison of the Sensitivity of Five Matched Case Control Studies: Upper Bounds on the Significance Level for Several Values of Γ.

Γ	1	2	3	4
DES and vaginal cancer Herbst et al. (1971), $n_s = 5$ for $s = 1, \ldots, 8 = S$	<0.0001	0.0004	0.0038	0.012
Estrogen and endometrial cancer Mack et al. (1976), $n_s = 5$ for $s = 1, \ldots, 63 = S$	<0.0001	0.0004	0.013	0.068
Coffee and myocardial infarction Jick et al. (1973), $n_s = 2$ for $s = 1, \ldots, 27$, $n_s = 3$ for $s = 28, \ldots, 115 = S$	0.0038	>0.2		
Prior induced abortion and ectopic pregnancy Trichopoulos, Miettinen, and Polychronopoulou, as reported by Miettinen (1969), $n_s = 5$ for $s = 1, \ldots, 18 = S$	0.0001	0.0066	0.0301	0.0668
Truck driving and herniated lumbar disc Kelsey and Hardy (1975), $n_s = 2$ for $s = 1, \ldots, 128$	0.0057	0.093	0.24	

with two referents. This portion of the study included only subjects who either drank zero cups of coffee per day or at least six cups of coffee per day.

The study by Kelsey and Hardy (1975) examined the relationship between acute herniated lumbar disc—a back injury—and driving a truck as an occupation. They matched cases of herniated lumbar disc in New Haven to referents of the same gender, age, and hospital service or radiologist's office. Each case was matched to a single referent, and there were $S = 128$ matched pairs.

The study by Trichopoulos et al. does not appear to have been published. It is described only briefly by Miettinen (1969). It concerned prior induced abortion as a risk factor for subsequent ectopic pregnancy. A total of 18 cases were each matched to five referents.

Table 4.4.4 gives the upper bound on the significance level, the lower bound being highly significant in all cases in the table. The table shows that the five studies are quite different in their sensitivity to hidden bias. Least sensitive is the study of DES and vaginal cancer, despite its small size, because of the strong association between treatment and outcome seen in the data in Table 4.4.2. To attribute this association to an unobserved covariate, one must postulate a covariate with a dramatic relationship with both exposure to DES and vaginal cancer.

Most sensitive is the study of coffee and myocardial infarction, where the upper bound on the significance level exceeds the conventional 0.05 for all $\Gamma \geq 1.3$. An unobserved covariate strongly related to myocardial infarction but only weakly related to coffee consumption could explain the observed association, despite the small significance level of 0.0038 for the usual Mantel–Haenszel statistic. Coffee may, in fact, cause myocardial infarctions, and this study may be free of hidden biases; however, fairly small biases, too small to easily detect, could readily explain the observed association.

The studies of endometrial cancer and of ectopic pregnancy are about equally sensitive to bias. The former study is more than three times larger, but the latter exhibits a stronger relationship between treatment and outcome.

Sensitivity to hidden bias is an aspect of the conclusions of an observational study, an aspect relevant to causal interpretation. However, sensitivity to hidden bias is not a fault of a study or its authors, as insensitivity to bias is not an accomplishment.

The study of herniated lumbar discs is different from the others in its structure, and a slightly different sensitivity analysis was performed. That study distinguished three types of cases:

(i) surgical cases, in which records indicate the herniated disc was seen by a surgeon during surgery;
(ii) probable cases, having the same symptoms as surgical cases but without an explicit record that a surgeon saw the herniated disc; and
(iii) possible cases, having some but not all of the symptoms of a herniated disc.

To give greatest weight to the surgical cases, numerical scores were assigned to each matched pair, specifically score $d_s = 3$ was assigned to pairs with a surgical case, score $d_s = 2$ was assigned to probable cases, and score $d_s = 1$ was assigned to possible cases. Set $c_{si} = 1$ for cases and $c_{si} = 0$ for referents. Then the sign-score statistic $T = \sum_s d_s \sum_i c_{si} Z_{si}$ is the total of the case scores for matched sets in which the case was exposed to the treatment, specifically, the total of the case scores for matched pairs in which the case was a truck diver. The upper bounds on the significance level are obtained using the deviates

$$\frac{|T - \sum d_s P_s^+| - \frac{1}{2}}{\sqrt{\sum d_s^2 P_s^+(1 - P_s^+)}} \quad \text{and} \quad \frac{|T - \sum d_s P_s^-| - \frac{1}{2}}{\sqrt{\sum d_s^2 P_s^-(1 - P_s^-)}}, \tag{4.4.8}$$

which generalize the Mantel–Haenszel deviates in (4.4.7) to the case where $d_s \neq 1$ for some s. (The continuity correction should only be used if the scores d_s are integers, as is true in this example.)

4.4.6. The $2 \times 2 \times S$ Contingency Table

Table 4.4.5 concerns a case-referent study of the possible effects of the drug allopurinol as a cause of rash (Boston Collaborative Drug Surveillance Program, 1972). The table has two strata, males and females. Here, $r_{si} = 1$ for a case of rash, $r_{si} = 0$ for a referent, $Z_{si} = 1$ for use of allopurinol, $Z_{si} = 0$ otherwise, and $r_{s+} = \sum_i r_{si}$ is the number of cases in stratum s. The Mantel–Haenszel statistic, as discussed in §2.4.2, is based on the number of cases who used allopurinol, namely $T = \sum_s \sum_i Z_{si} r_{si} = 5 + 10 = 15$. If allopurinol had no effect, then in a randomized experiment or in the absence of hidden bias, T would have expectation 5.43 and variance 4.70, yielding a deviate of $(|15 - 5.43| - \frac{1}{2})/\sqrt{4.70} = 4.19$, with an approximate one-sided significance level less than 0.0001. In a randomized experiment or in the absence of hidden bias, Table 4.4.5 would constitute strong evidence that allopurinol causes rash.

Table 4.4.5. Allopurinol and Rash.

			Z_{1i}		$n_1 = 719$
Males, $s = 1$			Allopurinol	Other	$m_1 = 38$
	r_{1i}	Rash cases	5	36	$r_{1+} = 41$
		Noncases	33	645	
			Z_{2i}		$n_2 = 605$
Females, $s = 2$			Allopurinol	Other	$m_2 = 29$
	r_{2i}	Rash cases	10	58	$r_{2+} = 68$
		Noncases	19	518	

The procedure for conducting a sensitivity analysis for the Mantel–Haenszel statistic is as follows. The basis for the procedure is discussed in the Appendix, §4.6.3; it uses a normal approximation to a standard distribution, namely, the extended hypergeometric distribution. Determine the unique root $\tilde{E}_s$ of the quadratic equation

$$\tilde{E}_s^2(\Gamma - 1) - \tilde{E}_s\{(\Gamma - 1)(m_s + r_{s+}) + n_s\} + \Gamma r_{s+} m_s = 0. \tag{4.4.9}$$

with $\max(0, r_{s+} + m_s - n_s) \le \tilde{E}_s \le \min(r_{s+}, m_s)$, calculate

$$\tilde{V}_s = \frac{1}{\dfrac{1}{\tilde{E}_s} + \dfrac{1}{r_{s+} - \tilde{E}_s} + \dfrac{1}{m_s - \tilde{E}_s} + \dfrac{1}{n_s - r_{s+} - m_s + \tilde{E}_s}}, \tag{4.4.10}$$

and refer the deviate $(|T - \sum \tilde{E}_s| - \frac{1}{2})/\sqrt{\sum \tilde{V}_s}$ to tables of the standard normal distribution. Repeat with Γ replaced by $1/\Gamma$ to determine the other bound on the significance level.

To illustrate, for $\Gamma = 2$, $s = 1$, equation (4.4.9) is

$$\tilde{E}_1^2(2 - 1) - \tilde{E}_1\{(2 - 1)(38 + 41) + 719\} + 2 \times 41 \times 38 = 0$$

or

$$\tilde{E}_1^2 - 789\tilde{E}_1 + 3{,}116 = 0,$$

with roots 794.08 and 3.92, of which the correct root is $\tilde{E}_1 = 3.92$ since $\max(0.41 + 38 - 719) = 0 \le 3.92 \le 38 = \min(41, 38)$. Substituting $\tilde{E}_1 = 3.92$ into (4.4.10) yields $V_1 = 3.20$. In the same way, $\tilde{E}_2 = 5.66$, $V_2 = 4.21$, and $(|T - \sum \tilde{E}_s| - \frac{1}{2})/\sqrt{\sum \tilde{V}_s} = (|15 - 9.59| - \frac{1}{2})/\sqrt{7.41} = 1.80$, giving an upper bound on the one-sided significance level of 0.036.

Computed in this way, the upper bounds on the significance levels from Table 4.4.5 for $\Gamma = 1, 2$, and 3 are 0.0001, 0.036, and 0.30. The study is insensitive to a bias that would double the odds of exposure to allopurinol but sensitive to a bias that would triple the odds. It is more sensitive to bias than three of the five studies in §4.4.5 and less sensitive than the other two.

When $\mathbf{u} = \mathbf{u}^+$ and the treatment has no effect, expressions (4.4.9) and (4.4.10) yield, respectively, large sample approximations to the expectation and variance of the upper left corner cell count in the sth 2×2 table, namely $\sum_i Z_{si} r_{si}$. Consider the following 2×2 table, which has the same marginal totals as the observed table.

Table 4.4.6. Table of Approximate Expectations.

	$Z_{si} = 1$	$Z_{si} = 0$	Total
$r_{si} = 1$	$\tilde{E}_s$	$r_{s+} - \tilde{E}_s$	r_{s+}
$r_{si} = 0$	$m_s - \tilde{E}_s$	$n_s - r_{s+} - m_s + \tilde{E}_s$	$n_s - r_{s+}$
Total	m_s	$n_s - m_s$	n_s

Expression (4.4.9) may be motivated by setting the odds ratio in Table 4.4.6 equal to Γ,

$$\Gamma = \frac{\tilde{E}_s(n_s - r_{s+} - m_s + \tilde{E}_s)}{(m_s - \tilde{E}_s)(r_{s+} - \tilde{E}_s)}, \tag{4.4.11}$$

and noticing that (4.4.9) is obtained by rearranging (4.4.11). Roughly speaking, if there is no treatment effect, so the association between Z and r is solely due to failure to control the unobserved covariate u, then the odds ratio linking Z and r is at most Γ. A detailed derivation of (4.4.9) and (4.4.10) is given in the Appendix to this chapter.

The approximate expectation (4.4.9) and variance (4.4.10) are appropriate when each marginal total of table s is large, that is, when m_s, $n_s - m_s$, r_{s+}, and $n_s - r_{s+}$ are each large. If table s has a small marginal total, the exact moments of $\sum_i Z_{si} r_{si}$ may be used in place of the approximations. The exact expectation and variance are

$$\bar{E}_s = \frac{\displaystyle\sum_{t=\max(0,m_s+r_{s+}-n_s)}^{\min(m_s,r_{s+})} t \binom{r_{s+}}{t}\binom{n_s - r_{s+}}{m_s - t}\Gamma^t}{\displaystyle\sum_{t=\max(0,m_s+r_{s+}-n_s)}^{\min(m_s,r_{s+})} \binom{r_{s+}}{t}\binom{n_s - r_{s+}}{m_s - t}\Gamma^t} \tag{4.4.12}$$

and

$$\bar{V}_s = \frac{\Gamma m_s r_{s+} - \{n_s - (m_s + r_{s+})(1 - \Gamma)\}\bar{E}_s}{(1 - \Gamma)} - \bar{E}_s^2. \tag{4.4.13}$$

As an example of a table with small counts, suppose $n_s = 8$, $m_s = 4$, and $r_{s+} = 3$. For $\Gamma = 2$, the exact moments in this hypothetical table are $\bar{E}_s = 1.87$ and $\bar{V}_s = 0.52$, while the approximations are $\tilde{E}_s = 1.82$ and $V_s = 0.45$. In this case, the exact expectation in (4.4.12) is the sum has four terms, namely, from $t = 0 = \max(0, 4 + 3 - 8)$ to $t = \min(4, 3) = 3$. (Expression (4.4.13) is derived from an observation of Johnson, Kotz, and Kemp (1992, p. 280, expression (6.160)).)

Returning to the case-referent study of allopurinol and rash, Table 4.4.7

Table 4.4.7. Comparison of Results Using Exact and Approximate Moments of T.

Γ		Expectation	Variance	Deviate
1	Exact	5.43	4.70	4.19
	Approximate	5.43	4.69	4.19
2	Exact	9.60	7.44	1.80
	Approximate	9.59	7.41	1.80
3	Exact	12.97	9.17	0.51
	Approximate	12.94	9.13	0.52

compares the calculations based on the exact and approximate moments of T. In this example, the approximation is quite satisfactory.

The sensitivity analysis for binary responses has an intimate connection with setting confidence limits for an odds ratio in a $2 \times 2 \times S$ contingency table, so a program such as Cytel Corporation's StatXact can do certain sensitivity calculations. This link has not been stressed in this chapter because it does not apply with outcomes that are not binary.

4.4.7. Comparing Rates in a Treated Group to a Population of Controls

Some observational studies compare a treated group of finite size to control group that is, in effect, infinite in size, that is, a control group in which sampling variability is negligible. This is particularly common in studies in occupational health, where the mortality in an industry is often compared to mortality rates for the nation as a whole. For instance, in a study of the hazards of asbestos, Nicholson, Selikoff, Seidman, Lilis, and Formby (1979) studied a treated group consisting of 544 male miners of chrysotile asbestos at the Thetford Mines in Quebec. All had been miners for at least 20 years as of 1961, and their mortality between 1961 and 1977 is under study. These 544 minors were observed for a total of 7408 years, that is, there were 7408 person-years of observation. The control group was constructed from the mortality experience for all males in Canada. By reweighting national mortality rates that were specific to age, gender and cause of death, the investigators computed mortality rates for a constructed population of Canadians having the same distribution of ages and the same number of years of observation as the 544 miners. For instance, they found that the constructed population would have a mortality rate of 0.01232 from noninfectious pulmonary diseases other than cancer, so 544 individuals selected at random from this constructed population would be expected to have $0.01232 \times 544 = 6.70$ deaths from this disease. Because of the enormous size of the population of Canada, this mortality rate is effectively free of sampling variability; however, since miners may differ from the typical Canadian in terms of covariates other than age, the rate may be affected by hidden biases. In fact, the 544 miners experienced 30 deaths from noninfectious pulmonary diseases other than cancer, or $30/6.70 = 4.48$ times more than expected. The ratio $30/6.70 = 4.48$ is called the standardized mortality ratio. In the absence of hidden bias, the usual test of no treatment effect views the 30 deaths as either a binomial count with probability 0.01232 and sample size 544 or as a Poisson count with expectation 6.70; see, for instance, Armitage (1977, p. 389), Gastwirth (1988, §10), Mosteller and Tukey (1977, §11E), or Hakulinen (1981). The binomial test may be viewed as asking whether the 544 miners look like a random sample of size 544 from an infinite population with rate 0.01232.

A small change in the method in §4.4.6 for 2 × 2 tables addresses the infinite sample size in the control group. The large sample argument in §4.4.6 assumed that the margins, m_s, $n_s - m_s$, r_{s+}, $n_s - r_{s+}$, of each 2 × 2 table were large, with all margins increasing at about the same rate. When comparing 544 miners to a reweighting of all of Canada, a more realistic asymptotic argument assumes that $m_s = 544$ is fixed, $n_s \to \infty$, with $r_{s+}/n_s \to \zeta_s = 0.01232$. At $\mathbf{u} = \mathbf{u}^+$ under the null hypothesis of no effect, $\sum_i Z_{si} r_{si}$ is asymptotically binomial with sample size m_s and probability

$$v_s = \frac{\zeta_s \Gamma}{(1 - \zeta_s) + \zeta_s \Gamma};$$

see the Appendix, §4.6.3, for proof. If there is no hidden bias, then $\Gamma = 1$ and $v_s = \zeta_s$, leading to the usual binomial distribution for $\sum_i Z_{si} r_{si}$. With $T = \sum_s \sum_i Z_{si} r_{si}$, the deviate is

$$\frac{|T - \sum m_s v_s| - \frac{1}{2}}{\sqrt{\sum m_s v_s (1 - v_s)}}.$$

For the bound at $\mathbf{u} = \mathbf{u}^-$, the same calculation is performed with $1/\Gamma$ in place of Γ; see the Appendix, §4.6.4. Gastwirth and Greenhouse (1987) perform a similar calculation starting from the inequality of Cornfield et al. (1959).

In the example, there is one stratum, $S = 1$, and $m_1 = 544$, $\zeta = 0.01232$. Table 4.4.8 gives the sensitivity analysis for the upper bound on the significance level. In the absence of hidden bias, $\Gamma = 1$, 6.70 deaths were expected, the deviate is 8.86 with significance level less than 0.0001, and the standardized mortality ratio is $30/6.70 = 4.5$. A bias of magnitude $\Gamma = 2$ could raise the expected number of deaths to 13.24, though the observed 30 deaths are significantly more than 13.24, yielding a deviate of at least 4.52, a significance level less than 0.0001, and a standardized mortality ratio of at least $30/13.24 = 2.3$. This study is less sensitive to bias than the studies of allopurinol in §4.4.6 or coffee in §4.4.5, but it is more sensitive than the studies of smoking in §4.3.3 or lead exposure in §4.3.3.

Table 4.4.8. Sensitivity Analysis for the Asbestos Miners' Standardized Mortality Ratio.

Γ	v_1	$v_1 m_1$	Deviate
1	0.01232	6.70	8.86
2	0.02434	13.24	4.52
3	0.03607	19.62	2.27
4	0.04752	25.85	0.73

4.4.8. Other Sign-Score Statistics: Censored Data, Median Tests

All sign-score statistics permit a sensitivity analysis based on Proposition 4.3. Sections 4.3 and 4.4 have presented details and examples for methods that are widely used. This final part of §4.4 briefly mentions several other sign-score statistics that are useful in particular situations.

(i) *Censored Survival Data in Matched Pairs.* When the outcome is the time until a particular event occurs, it may happen that for some individuals the event has not yet occurred at the time the data are analyzed, in which case the outcome is censored. This is often true in studies of human survival, in which some individuals are alive at the end of the study. It is also often true in studies that compare the effectiveness of various punishments for criminal acts, in which the outcome is the time until the act is repeated. O'Brien and Fleming (1987) propose and evaluate their Prentice–Wilcoxon statistic for censored survival times in matched pairs. This statistic may be written as a sign-score statistic, so the sensitivity analysis follows directly from the general procedure in §4.4.2 with each $n_s = 2$.

(ii) *Median, Quantile, and Sign Tests.* In comparing treated and control groups, the median test combines the groups, finds the median of the combination, and asks how many treated subjects have responses above the combined median. Though inefficient for responses obtained from a Normal distribution, the median test has good properties in large samples for data from a distribution more prone to extreme observations, specifically the double exponential distribution; see Hájek and Sidák (1967, §3.1.1, p. 88) or Hettmansperger (1984, §3) for detailed discussion. The median test is usually attributed to G. Brown and A. Mood.

Other quantiles are sometimes used in place of the median. A version of the median test was used in §4.4.3 for matching with multiple controls. In pair matching, the median test becomes the sign test. The median test, the stratified or matched median test, the sign test, and other quantile tests are all sign-score tests.

4.5. Sensitivity Analysis for Comparing Two Unmatched Groups

4.5.1. Why Is There a Difference Between Sum Statistics and Sign-Score Statistics?

This section discusses the sensitivity analysis for statistics T such as Wilcoxon's rank sum statistic used to compare unmatched treated and control groups. Here, there will be a single stratum, $S = 1$, so the subscript s will be omitted. As defined in §2.43, a sum statistic has the form

$$T = \sum_{i=1}^{n} Z_i q_i, \tag{4.5.1}$$

where the q_i are functions of the responses and so are fixed under the null hypothesis of no treatment effect. For instance, in Wilcoxon's rank sum statistic, the q_i are the ranks of the responses r_i, and T is the sum of the ranks in the treated group.

The sensitivity analysis for sign-score statistics in §4.2 and §4.3 and for sum statistics in the current section are similar in certain ways and different in others. They are identical in interpretation and similar in their broad outlines, but they differ in an important detail of implementation. In Proposition 4.3, the distribution of the test statistic T is bounded at all points "a" by the distributions of two other random variables, T^+ and T^-, which are obtained from the distribution (4.2.6) at two points $\mathbf{u}^+$ and $\mathbf{u}^-$. In contrast, with sum statistics, there is a point $\mathbf{u}^+ \in U$ that provides an upper on the probability $\text{prob}(T \geq a|\mathbf{m})$; however, the point $\mathbf{u}^+$ changes as "a" changes. This distinction has a noticeable effect on calculations, but no effect on the interpretation of the sensitivity analysis.

Section 4.5 is organized as follows. The technical aspects of the general method are developed in §4.5.2. Then §4.5.3 shows the calculations needed for the significance level for Wilcoxon's rank sum statistic. The Hodges–Lehmann point estimate is discussed in §4.5.5 and the confidence interval is discussed in §4.5.6. Censored data is discussed in §4.5.7.

4.5.2. Sensitivity Analysis for Sum Statistics: The General Method

This section obtains bounds on the behavior of a sum statistic $T = \sum Z_i q_i$ under the null hypothesis of no treatment effect. Specifically, bounds will be obtained for the tail area or significance level $\text{prob}(T \geq a|\mathbf{m})$ and for the expectation $E(T)$. These are the basis for the sensitivity analysis for significance levels, confidence intervals, and point estimates, as developed in greater detail later in §4.5. For this purpose, in §4.5.2, assume that the null hypothesis of no treatment effect holds, and that $\Gamma \geq 1$. Renumber the n subjects so that $q_1 \geq q_2 \geq \cdots \geq q_n$. Since no quantity depends on the numbering of the n subjects, this is just a notational convenience.

The maximum value of $\text{prob}(T \geq a|\mathbf{m})$ from model (4.2.6) with $\mathbf{u} \in U$ is found at point $\mathbf{u}$ of the form $u_1 = 1$, $u_2 = 1, \ldots, u_k = 1$, $u_{k+1} = 0$, $u_{k+2} = 0, \ldots, u_n = 0$, that is, at a point whose coordinates consist of k ones followed by $n - k$ zeros for some number k. Let U^+ be the set of all such $\mathbf{u}$, for $k = 0, 1, \ldots, n$. Similarly, the minimum value of $\text{prob}(T \geq a|\mathbf{m})$ is found at a point whose coordinates are k zeros followed by $n - k$ ones for some k.

Write U^- for the set of all such points. The value of k may change as "a" is changed. The maximum and minimum values of $E(T)$ are also found at points of, respectively, U^+ and U^-. These facts are proved in the Appendix to this chapter, specifically in §4.6.3.

For a given $\mathbf{u}$, let $\mu_{\mathbf{u}}$ and $\sigma_{\mathbf{u}}$ be the expectation and standard deviation of T. Easily calculated approximations to $\mu_{\mathbf{u}}$ and $\sigma_{\mathbf{u}}$ will be discussed shortly. The bounds on $E(T)$ are then

$$\min_{\mathbf{u} \in U^-} \mu_{\mathbf{u}} \leq E(T) \leq \max_{\mathbf{u} \in U^+} \mu_{\mathbf{u}}. \tag{4.5.2}$$

Large sample approximations to the bounds on $\text{prob}(T \geq a | \mathbf{m})$ are obtained by referring the standardized deviates,

$$\max_{\mathbf{u} \in U^-} \frac{a - \mu_{\mathbf{u}}}{\sigma_{\mathbf{u}}} \quad \text{and} \quad \min_{\mathbf{u} \in U^+} \frac{a - \mu_{\mathbf{u}}}{\sigma_{\mathbf{u}}}, \tag{4.5.3}$$

to tables of the standard normal distribution. The effort required to compute the maximum and minimum in (4.5.3) is of the same order of magnitude as computing the rank sum statistic $2n$ times, an easy task with a computer.

Large sample approximations to $\mu_{\mathbf{u}}$ and $\sigma_{\mathbf{u}}$ will now be given when $\mathbf{u}$ has k coordinates equal to 1 and $n - k$ coordinates equal to zero. The approximations are developed in the Appendix, §4.6.3; they involve approximations $\tilde{E}$ and $\tilde{V}$ to the expectation and variance of $\mathbf{Z}^{\mathrm{T}}\mathbf{u}$, where $\mathbf{Z}^{\mathrm{T}}\mathbf{u}$ is the number of treated subjects with $u = 1$. As in §4.4.6, the calculation involves solving the quadratic equation

$$\tilde{E}^2(\Gamma - 1) - \tilde{E}\{(\Gamma - 1)(m + k) + n\} + \Gamma k m = 0 \tag{4.5.4}$$

for the root $\tilde{E}$ with $\max(0, k + m - n) \leq \tilde{E} \leq \min(k, m)$. If $k = 0$, then $\tilde{E} = 0$, but if $k = n$, then $\tilde{E} = m$, and in either case let $\tilde{V} = 0$; otherwise, let

$$\tilde{V} = \frac{1}{\dfrac{1}{\tilde{E}} + \dfrac{1}{k - \tilde{E}} + \dfrac{1}{m - \tilde{E}} + \dfrac{1}{n - k - m + \tilde{E}}}. \tag{4.5.5}$$

Then calculate the mean and variance of the scores q_i separately for subjects with $u_i = 1$ and $u_i = 0$,

$$\begin{aligned} \bar{q}_1 &= \frac{1}{k} \sum_{i: u_i = 1} q_i, & \bar{q}_0 &= \frac{1}{n - k} \sum_{i: u_i = 0} q_i, \\ w_1 &= \frac{1}{k - 1} \sum_{i: u_i = 1} (q_i - \bar{q}_1)^2, & w_0 &= \frac{1}{n - k - 1} \sum_{i: u_i = 0} (q_i - \bar{q}_0)^2, \end{aligned} \tag{4.5.6}$$

where we define $\bar{q}_1 = 0$ if $k = 0$, $\bar{q}_0 = 0$ if $k = n$, $w_1 = 0$ if $k \leq 1$, $w_0 = 0$ if $k \geq n - 1$.

Finally, if $k \neq 0$ and $k \neq n$, then approximate $\mu_{\mathbf{u}}$ by

$$\tilde{E}\bar{q}_1 + (m - \tilde{E})\bar{q}_0 \tag{4.5.7}$$

and $\sigma_{\mathbf{u}}$ by

$$\sqrt{(w_1 - w_0)\tilde{E} - (\tilde{E}^2 + \tilde{V})\left(\frac{w_1}{k} + \frac{w_0}{n-k}\right) + \frac{m(n-k-m+2\tilde{E})w_0}{n-k} + \tilde{V}(\bar{q}_1 - \bar{q}_0)^2}; \tag{4.5.8}$$

otherwise, if $k = 0$ or $k = n$,

$$\mu_{\mathbf{u}} = \frac{m}{n}\sum q_i \quad \text{and} \quad \sigma_{\mathbf{u}} = \sqrt{\frac{m(n-m)}{n(n-1)}\sum (q_i - \bar{q})^2} \quad \text{where} \quad \bar{q} = \frac{1}{n}\sum q_i. \tag{4.5.9}$$

In these calculations, $\tilde{E}$ and $\tilde{V}$ approximate the expectation and variance of $\mathbf{Z}^{\mathrm{T}}\mathbf{u}$, that is, of the number of treated subjects with $u_i = 1$. If the exact expectation $\bar{E}$ and variance $\bar{V}$ of $\mathbf{Z}^{\mathrm{T}}\mathbf{u}$ were used in place of $\tilde{E}$ and $\tilde{V}$, then (4.5.7) and (4.5.8) would give the exact moments of T. While the exact moments $\bar{E}$ and $\bar{V}$ can be used, their computation involves somewhat more effort; moreover, the approximations $\tilde{E}$ and $\tilde{V}$ perform well and are widely used in other contexts. Exact and approximate moments will be compared in an example in §4.5.3 and they are discussed in detail in §4.6.3.

4.5.3. Simplified Formulas for the Rank Sum Test When There Are No Ties

In the absence of ties among the responses, Wilcoxon's (1945) rank sum statistic T involves ranking all n responses from 1 to n and summing the ranks in the treated group. With the n subjects arranged as in §4.5.2, the rank sum is $T = \sum Z_i q_i$ with $q_1 = n$, $q_2 = n-1, \ldots, q_n = 1$. This leads to simplifications of the formulas in §4.5.3. In particular, using familiar arguments about sums of integers and sums of squares of integers (Lehmann, 1975, p. 329), it follows that for $\mathbf{u} \in U^+$ having k ones followed by $n-k$ zeros:

$$\bar{q}_0 = \frac{n-k+1}{2}, \qquad \bar{q}_1 = \frac{k+1}{2} + n - k = \frac{2n-k+1}{2},$$
$$w_0 = \frac{(n-k)(n-k+1)}{12}, \qquad w_1 = \frac{k(k+1)}{12}. \tag{4.5.10}$$

Substituting (4.5.10) in (4.5.7) gives

$$\frac{\tilde{E}n + m(n-k+1)}{2}, \tag{4.5.11}$$

as the approximation to $\mu_{\mathbf{u}}$, and similarly substituting (4.5.10) in (4.5.8) gives

$$\sqrt{\frac{\{(n+1)(2k-n) + 2m(n-k+1)\}\tilde{E} - (n+2)\tilde{E}^2}{12} + \frac{m(n-k-m)(n-k+1)}{12} + \frac{\tilde{V}(n-1)(n+\frac{2}{3})}{4}} \tag{4.5.12}$$

as the approximation to $\sigma_{\mathbf{u}}$.

The same formulas may by used for an element, say $\mathbf{u}^*$, of U^-, by proceeding as follows. To approximate $\mu_{\mathbf{u}*}$ and $\sigma_{\mathbf{u}*}$ for $\mathbf{u}^* \in U^-$, determine $\mathbf{u} = \mathbf{1} - \mathbf{u}^*$, so $\mathbf{u} \in U^+$, replace Γ by $1/\Gamma$, and apply formulas (4.5.11) and (4.5.12). This works because the distribution (4.2.6) with $(\Gamma, \mathbf{u}^*)$ is the same as with $(1/\Gamma, \mathbf{u})$; see Appendix, §4.6.4, for detailed discussion.

If ties are present, average ranks are given to tied responses, and the general moment formulas in §4.5.2 may be used.

4.5.4. An Example: Chromosome Damage from Contaminated Fish

Skerfving, Hansson, Mangs, Lindsten, and Ryman (1974) studied 23 subjects who had eaten large quantities of fish contaminated with methylmercury. Each of the 23 exposed subjects had eaten at least three meals a week of contaminated fish for more than 3 years. The control group consisted of 16 subjects who did not regularly consume contaminated fish and who ate far less fish of all kinds. Table 4.5.1 gives their data on two outcome measures, the amount of mercury found in the subject's blood, recorded in ng/g, and the percent of cells exhibiting a particular chromosome abnormality called C_u cells, specifically, asymmetrical or incomplete symmetrical chromosome aberrations as recorded in cells cultured between 48 and 120 hours. Although the original study examined several types of chromosome abnormalities, the discussion here will consider only this specific type of abnormality.

The table shows that subjects who ate contaminated fish had much higher levels of mercury in their blood and somewhat higher frequencies of chromosome aberrations. The data contain some extreme observations, for instance, subject 26 with a mercury level of 1100 ng/g and subject 24 with 9.5% of cells exhibiting chromosome aberrations. There are ties among the responses, particularly the subjects with 0% chromosome aberrations.

Consider first the level of mercury found in the blood. There are four pairs of tied observations. For instance, subjects #18 and #29 both had mercury levels of 70 ng/g, so instead of giving ranks 23 and 24 to these observations, they are both given the average rank of 23.5. The sum of the ranks in the exposed group for level of mercury is $T = 642$. In the absence of hidden bias, that is with $\Gamma = 1$, the calculations in §4.5.2 yield an approximate expectation and variance for T of 460 and 1202.92 under the null hypothesis of no treatment effect, so the deviate is $(642 - 460)/\sqrt{1202.92} = 5.25$, with significance level <0.0001. If there were no hidden bias, there would be strong evidence that eating fish contaminated with mythelmercury causes an increase in mercury levels in the blood.

Now, suppose $\Gamma = 2$, so that one subject may be twice as likely as another to eat contaminated fish because they differ with respect to an unobserved covariate for which adjustments are required. The largest signif-

Table 4.5.1. Methymercury in Fish and Human Chromosome Damage.

Id.		% Cu cells	Mercury in blood
1	Control	2.7	5.3
2	Control	0.5	15
3	Control	0	11
4	Control	0	5.8
5	Control	5	17
6	Control	0	7
7	Control	0	8.5
8	Control	1.3	9.4
9	Control	0	7.8
10	Control	1.8	12
11	Control	0	8.7
12	Control	0	4
13	Control	1	3
14	Control	1.8	12.2
15	Control	0	6.1
16	Control	3.1	10.2
17	Exposed	0.7	100
18	Exposed	4.6	70
19	Exposed	0	196
20	Exposed	1.7	69
21	Exposed	5.2	370
22	Exposed	0	270
23	Exposed	5	150
24	Exposed	9.5	60
25	Exposed	2	330
26	Exposed	3	1,100
27	Exposed	1	40
28	Exposed	3.5	100
29	Exposed	2	70
30	Exposed	5	150
31	Exposed	5.5	200
32	Exposed	2	304
33	Exposed	3	236
34	Exposed	4	178
35	Exposed	0	41
36	Exposed	2	120
37	Exposed	2.2	330
38	Exposed	0	62
39	Exposed	2	12.8

icance level for all $\mathbf{u} \in U$ is approximated using the deviate

$$\min_{\mathbf{u} \in U^+} \frac{T - \mu_{\mathbf{u}}}{\sigma_{\mathbf{u}}} \tag{4.5.13}$$

which equals 4.40 if the approximate expectation and variance in §4.5.2 are used. This says that a hidden bias of magnitude $\Gamma = 2$ could not begin to explain the higher levels of mercury found in the blood of subjects who ate contaminated fish—no covariate that doubled the odds of eating contaminated fish could reasonably be expected to produce the observed results.

The minimum in (4.5.13) is calculated directly, evaluating $(T - \mu_{\mathbf{u}})/\sigma_{\mathbf{u}}$ for $k = 0, 1, \ldots, 39$ and taking the minimum. In this case, the minimum occurs at $k = 21$, that is, at a $\mathbf{u} = (1, 1, \ldots, 1, 0, 0, \ldots, 0)$ containing $k = 21$ ones followed by $n - k = 39 - 21 = 18$ zeros. The neighborhood of the minimum is quite flat. For $k = 20$, 21, and 22, the deviates $(T - \mu_{\mathbf{u}})/\sigma_{\mathbf{u}}$ are, respectively, 4.4035, 4.3985, and 4.3987. If exact expectations and variances are used in place of the approximations, the minimum still occurs at $k = 21$ and the deviate is 4.33. There is no need to calculate the maximum deviate in (4.5.3), since we know from the case of $\Gamma = 1$ that the minimum significance level will be less than 0.0001.

For $\Gamma = 5, 20$, and 35, the minimum deviates are 3.48, 2.43, and 2.10, respectively, with approximate significance levels 0.0003, 0.0075, and 0.0179. Even if one subject was thirty-five times more likely than another to eat contaminated fish—an extreme departure from randomization—this would not explain the high levels of mercury in the exposed group. This is a high level of insensitivity to hidden bias, a higher level than encountered in previous examples.

More important than the level of mercury is the possible damage to chromosomes. The rank sum for the percentage of abnormal cells is 551.5, with deviate 2.65 and significance level 0.004 in the absence of hidden bias, that is, with $\Gamma = 1$. A difference of this magnitude would provide strong evidence of an effect in a randomized experiment. For $\Gamma = 2$ and 3, the deviates are 1.78 and 1.28, respectively, with significance levels 0.0375 and 0.10. An unobserved covariate that tripled the odds of eating contaminated fish could explain the difference in the frequency of chromosome damage in exposed and control groups. For chromosome aberrations, the study is insensitive to small biases, but sensitive to biases of moderate size.

The percentages of abnormal cells have many ties. Twelve subjects have no abnormal cells, so all twelve are given the average rank of 6.5. There are many other ties as well. If average ranks are used in computing the rank sum statistic T but are ignored in computing the expectation and variance, that is, if the simpler formulas in §2.5.3 are used, the minimum deviates for $\Gamma = 2$ and 3 are 1.75 and 1.26 instead of the 1.78 and 1.28 given above. Even with many ties, the moment formulas that ignored ties gave similar results from the view of practical interpretation. If instead the exact expectation and variance of T in §2.5.2 are used, then the deviates for $\Gamma = 2$ and 3 are 1.74 and 1.23, a negligible change.

4.5.5. Hodges–Lehmann Point Estimates of an Additive Effect

This section conducts a sensitivity analysis for a Hodges–Lehmann (1963) estimate of an additive effect obtained from the rank sum test. The approach is similar to the sensitivity analysis in §4.3.5 for the Hodges–Lehmann estimate for matched pairs derived from the signed rank test. The model of an additive treatment effect makes little sense for coarse data with many ties, so in building point and interval estimates for an additive effect, it is assumed that there are no ties, and the moment formulas in §4.5.3 are used.

If the treatment has an additive effect τ, then the adjusted responses $\mathbf{R} - \tau\mathbf{Z}$ satisfy the null hypothesis of no effect. The Hodges–Lehmann estimate of τ is the value $\hat{\tau}$ such that $\mathbf{R} - \hat{\tau}\mathbf{Z}$ appears to precisely satisfy the null hypothesis of no effect, in the sense that the rank sum statistic $T = t(\mathbf{Z}, \mathbf{R} - \hat{\tau}\mathbf{Z})$ computed from $\mathbf{R} - \hat{\tau}\mathbf{Z}$ equals its null expectation, say $\bar{\bar{t}}$. As with the signed rank statistic in §4.3.5, the rank sum statistic $t(\mathbf{Z}, \mathbf{R} - \tau\mathbf{Z})$ declines in discrete steps as τ increases, so an equation of the form $t(\mathbf{Z}, \mathbf{R} - \hat{\tau}\mathbf{Z}) = \bar{\bar{t}}$ may not have an exact solution. As a result, $\hat{\tau}$ is defined as the average of the smallest value that is too large and the largest value that is too small, that is, $\hat{\tau}$ is defined by $\hat{\tau} = \text{SOLVE}\{\bar{\bar{t}} = t(\mathbf{Z}, \mathbf{R} - \hat{\tau}\mathbf{Z})\}$ in (4.3.6).

When $\Gamma \neq 1$, the null expectation $\bar{\bar{t}}$ of the rank sum statistic is $\mu_{\mathbf{u}}$ which is not known because $\mathbf{u}$ is not known. However, bounds on $\bar{\bar{t}}$ are available from (4.5.2),

$$\bar{\bar{t}}_{\min} = \min_{\mathbf{u} \in U^-} \mu_{\mathbf{u}} \le \bar{\bar{t}} \le \max_{\mathbf{u} \in U^+} \mu_{\mathbf{u}} = \bar{\bar{t}}_{\max}, \qquad \text{say, for all} \quad \mathbf{u} \in U. \quad (4.5.14)$$

Since $t(\mathbf{Z}, \mathbf{R} - \tau\mathbf{Z})$ declines as τ increases, the maximum $\hat{\tau}$ is found as the solution $\text{SOLVE}\{\bar{\bar{t}}_{\min} = t(\mathbf{Z}, \mathbf{R} - \hat{\tau}\mathbf{Z})\}$ while the minimum $\hat{\tau}$ is found as $\text{SOLVE}\{\bar{\bar{t}}_{\max} = t(\mathbf{Z}, \mathbf{R} - \hat{\tau}\mathbf{Z})\}$.

The procedure is as follows: For each fixed Γ, compute $\bar{\bar{t}}_{\max}$ and $\bar{\bar{t}}_{\min}$ using (4.5.11) in §4.5.3. With these fixed values, compute $\text{SOLVE}\{\bar{\bar{t}}_{\max} = t(\mathbf{Z}, \mathbf{R} - \hat{\tau}\mathbf{Z})\}$ and $\text{SOLVE}\{\bar{\bar{t}}_{\min} = t(\mathbf{Z}, \mathbf{R} - \hat{\tau}\mathbf{Z})\}$ which give the range of possible Hodges–Lehmann estimates given uncertainty about the value of the unobserved covariate $\mathbf{u}$.

4.5.6. An Example: Point Estimates of the Increase in Chromosome Abnormalities

Continuing the example from §4.5.4, this section examines point estimates of the increase in chromosome abnormalities. If there is no hidden bias, that is, if $\Gamma = 1$, then the null expectation of the rank sum statistic is $\bar{\bar{t}} = 460 = m(n + 1)/2 = 23(39 + 1)/2$. In fact, the rank sum statistic T for the proportion of cells with abnormalities is far higher, namely, $T = 551.5$. The Hodges–Lehmann estimate is the value $\hat{\tau}$ which, when subtracted from the responses for the 23 exposed subjects, gives a rank sum as close as possible to 460. Now, if 1.7 is subtracted from the 23 exposed proportions in Table 4.5.1 before the rank sum is computed, then $T = 464.5$ which is just slightly too high, but if 1.70001 is subtracted, then $T = 458$ which is just slightly too low; so the Hodges–Lehmann estimate is $\hat{\tau} = 1.7$. Were this a randomized experiment, eating contaminated fish would be estimated to increase the percent of abnormalities by the additive effect 1.7%.

If there were a hidden bias of magnitude $\Gamma = 2$, then one subject might be twice as likely another to eat contaminated fish because they have differing values of the unobserved covariate. In this case, the expectation of the rank sum statistic in the absence of a treatment effect, namely $\bar{\bar{t}}$, is not known because it depends on $\mathbf{u}$. Still, for $\mathbf{u} \in U$, the expectation $\bar{\bar{t}}$ is at most $\bar{\bar{t}}_{\max} = 491.57$ and at least $\bar{\bar{t}}_{\min} = 428.43$. The maximum $\bar{\bar{t}}_{\max}$ is found at $\mathbf{u} = (1, 1, \ldots, 1, 0, 0, \ldots, 0)$ with $k = 20$ ones followed by $n - k = 39 - 20 = 19$ zeros, while $\bar{\bar{t}}_{\min}$ found at $\mathbf{u}^* = (0, 0, \ldots, 0, 1, 1, \ldots, 1)$ with $k = 19$ zeros followed by $n - k = 39 - 19 = 20$ ones. These bounds on $\bar{\bar{t}}$ are calculated using (4.5.14) and (4.5.11), together with the observation in §4.5.3 about the relationship between the max over U^+ and the min over U^-.

The next step is to calculate $\text{SOLVE}\{\bar{\bar{t}}_{\max} = t(\mathbf{Z}, \mathbf{R} - \hat{\tau}\mathbf{Z})\}$ with $\bar{\bar{t}}_{\max} = 491.57$. Now $t(\mathbf{Z}, \mathbf{R} - 0.9999\mathbf{Z}) = 497$ which is a little too large, but $t(\mathbf{Z}, \mathbf{R} - 1\mathbf{Z}) = 490.5$ which is a little too small; hence $1.0 = \text{SOLVE}\{491.57 = t(\mathbf{Z}, \mathbf{R} - \hat{\tau}\mathbf{Z})\}$. In words, if 1% is subtracted from the percentages of abnormalities in the exposed group, then the rank sum just about equals the maximum possible expectation for $\Gamma = 2$. In the same way, $\text{SOLVE}\{\bar{\bar{t}}_{\min} = t(\mathbf{Z}, \mathbf{R} - \hat{\tau}\mathbf{Z})\} = \text{SOLVE}\{428.43 = t(\mathbf{Z}, \mathbf{R} - \hat{\tau}\mathbf{Z})\} = 2.0$. Because we do not know $\mathbf{u}$, we cannot calculate a single Hodges–Lehmann estimate; however, if $\Gamma = 2$, the largest possible estimate is 2% and the smallest is 1%, all of these values being significantly different from 0% as seen in §4.5.4.

Repeating this for $\Gamma = 3$ gives $\bar{\bar{t}}_{\max} = 509.28$, $\text{SOLVE}\{509.28 = t(\mathbf{Z}, \mathbf{R} - \hat{\tau}\mathbf{Z})\} = 0.7$, $\bar{\bar{t}}_{\min} = 410.72$, and $\text{SOLVE}\{410.72 = t(\mathbf{Z}, \mathbf{R} - \hat{\tau}\mathbf{Z})\} = 2.2$. So the range of Hodges–Lehmann estimates for $\Gamma = 3$ is from 0.7% to. 2.2%. However, from §4.5.4, not all of these estimates differ significantly from 0%. Table 4.5.2 summarizes these calculations.

Table 4.5.2. Sensitivity Analysis for Methylmercury Data.

Γ	Range of estimates of effect	Range of significance levels
1	1.7	0.004
2	1.0 to 2.0	0.0001 to 0.0375
3	0.7 to 2.2	<0.0001 to 0.1

Although the null hypothesis of no treatment effect begins to be plausible at about $\Gamma = 3$, the estimated effect is still at least 0.7%, which is not a small number given the levels of abnormalities found in the control group. Going beyond the table, the smallest estimated effect equals zero when $\Gamma = 8.9$.

The model of an additive effect is not fully consistent with the data in Table 4.5.1, because the exposed group has percentages of abnormalities that are not only higher but also more variable. An alternative to an analysis of the percentages R_{si} is an analysis of transformed percentages, say $\log(1 + R_{si})$ or $\sqrt{R_{si}}$, since these transformations tend to reduce the dispersion of larger values.

4.5.7. Confidence Intervals for an Additive Effect

Confidence intervals for τ are obtained by inverting the rank sum test. Fix $\Gamma \geq 1$. With Γ fixed, for each $\mathbf{u} \in U$, the distribution (4.2.6) yields a confidence interval for τ. Since $\mathbf{u}$ is unknown, the sensitivity interval is defined to be the union of these confidence intervals, the union taken over all $\mathbf{u} \in U$. A value τ is in the sensitivity interval if τ is plausible for some $\mathbf{u} \in U$, and τ is not in the sensitivity interval if it is rejected for every $\mathbf{u} \in U$. If $\Gamma = 1$, then the sensitivity interval is the single confidence interval commonly obtained from the randomization distribution. As Γ increases, the confidence set becomes longer reflecting the possible impact of the unknown $\mathbf{u}$.

Let T be the rank sum statistic under the null hypothesis of no treatment effect. From Proposition 4.6 in the Appendix, §4.6.3, $\text{prob}\{T \geq a_1 | \mathbf{m}\}$ is maximized at some $\mathbf{u} \in U^+$ and $\text{prob}\{T \leq a_0 | \mathbf{m}\}$ is maximized at some $\mathbf{u} \in U^-$. Suppose that there are numbers a_0 and a_1 so that $\text{prob}\{T \geq a_1 | \mathbf{m}\} \leq \alpha/2$ for all $\mathbf{u} \in U^+$ with equality for some $\mathbf{u} \in U^+$ and $\text{prob}\{T \leq a_0 | \mathbf{m}\} \leq \alpha/2$ for all $\mathbf{u} \in U^-$ with equality for some $\mathbf{u} \in U^-$. In other words, if $T \leq a_0$ or $T \geq a_1$, then the null hypothesis of no effect is rejected at level α for all $\mathbf{u} \in U$. An α-level sensitivity interval is then from $\sup\{\tau: t(\mathbf{Z}, \mathbf{R} - \tau\mathbf{Z}) \geq a_0\}$ to $\inf\{\tau: t(\mathbf{Z}, \mathbf{R} - \tau\mathbf{Z}) \leq a_1\}$.

In large samples, in a one-sided 0.025 level test, a value τ is rejected as too small for all for all $\mathbf{u} \in U$ if

$$\min_{\mathbf{u} \in U^+} \frac{t(\mathbf{Z}, \mathbf{R} - \mathbf{Z}\tau) - \mu_{\mathbf{u}}}{\sigma_{\mathbf{u}}} \geq 1.96, \tag{4.5.15}$$

and τ is as too large for $\mathbf{u} \in U$ if

$$\max_{\mathbf{u} \in U^-} \frac{t(\mathbf{Z}, \mathbf{R} - \mathbf{Z}\tau) - \mu_{\mathbf{u}}}{\sigma_{\mathbf{u}}} \leq -1.96, \tag{4.5.16}$$

where $\mu_{\mathbf{u}}$ and $\sigma_{\mathbf{u}}$ are approximated by (4.5.11) and (4.5.12). Since $t(\cdot, \cdot)$ is order preserving, $t(\mathbf{Z}, \mathbf{R} - \tau\mathbf{R})$ declines as τ increases, and the set of points τ satisfying neither (4.5.15) nor (4.5.16) is therefore, an interval.

4.5.8. An Example: Sensitivity Interval for the Increase in Chromosome Aberrations

In the example of §4.5.4 the Hodges–Lehmann point estimate of the increase in chromosome aberrations was $\hat{\tau} = 1.7\%$ in the absence of hidden bias, $\Gamma = 1$, and the point estimates ranged from 0.7% to 2.2% for $\Gamma = 3$. Consider now the 95% confidence interval for τ.

In the absence of hidden bias, $\Gamma = 1$, the largest value of τ that is rejected as too small is 0.2% and the smallest value that is rejected as too large is 2.8%, so the 95% confidence interval is [0.2, 2.8]. In other words, if 0.2 were subtracted from the responses of each subject who ate contaminated fish, and the rank sum test were applied to the resulting adjusted responses, the standardized deviate would just equal 1.96, whereas if 2.8 were subtracted instead, the deviate would just pass -1.96. Notice that, even in the absence of hidden bias, the confidence interval [0.2, 2.8] is fairly long.

If there were hidden bias of magnitude $\Gamma = 2$, the largest value τ barely rejected as too small in (4.5.15) is 0.0 and the smallest value τ rejected as too large in (4.5.16) is 3.5%, so the 95% sensitivity interval is [0.0, 3.5]. Keep in mind that this is a two-sided interval, while the test in §4.5.4 was a one-sided test. The sensitivity interval is the union of all the confidence intervals that might have been calculated had the unobserved covariate $\mathbf{u}$ been observed. For $\Gamma = 3$, the 95% sensitivity interval is [-0.4, 4.0].

*4.6. Appendix: Technical Results and Proofs

4.6.1. Outline and Summary

This appendix contains proofs and general results. Section 4.6.2 proves Proposition 4.3 concerning the values $\mathbf{u}^+$ and $\mathbf{u}^-$ providing the bounds for sign-score statistics. Section 4.6.3 discusses sensitivity analysis for arrangement increasing statistics. Exact and approximate moment formulas for sum statistics are derived in §4.6.4. Finally, §4.6.5 discusses the effect of certain transformations of $\mathbf{u}$, most importantly, the equivalence of $(\Gamma, \mathbf{u})$ and $(1/\Gamma, \mathbf{1} - \mathbf{u})$.

4.6.2. Bounds for Sign-Score Statistics

The proof of Proposition 4.3 will now be given. The general proof given below is based on Holley's inequality in §2.10.4; indeed, the proof is little more than an identification of the pieces in this inequality. Special cases may be proved in other ways; see, for instance, Problem 7.

Proposition 4.3. *If the treatment has no effect and* $T = t(\mathbf{Z}, \mathbf{r})$ *is a sign-score statistic, then for each fixed* $\gamma \geq 0$,

$$\text{prob}(T^+ \geq a) \geq \text{prob}\{T \geq a|\mathbf{m}\} \geq \text{prob}(T^- \geq a) \qquad \textit{for all a and} \quad \mathbf{u} \in U.$$

PROOF OF PROPOSITION 4.3. The proof uses Holley's inequality, Theorem 2.8 in the Appendix to Chapter 2. Fix a $\mathbf{u} \in U$. Set $\mathbf{v} = \mathbf{u}^+ - \mathbf{u}$. Renumber units in each subclass so $c_{s1} \geq c_{s2} \geq \cdots \geq c_{s,n_s}$ and $v_{s1} \geq v_{s2} \geq \cdots \geq v_{s,n_s}$, which is possible by the definition of $\mathbf{u}^+$ and the fact that $\mathbf{u} \in U$. With this ordering of the units, $t(\mathbf{z}, \mathbf{r})$ is an isotonic function of $\mathbf{z} \in \Omega$ in the sense of §2.10.4, since interchanging $z_{si} = 0$ and $z_{s,i+1} = 1$ leaves $\sum_i c_{si} z_{si}$ unchanged if $c_{si} = c_{s,i+1}$ or increases $\sum_i c_{si} z_{si}$ by one if $1 = c_{si} > c_{s,i+1} = 0$. To prove the first inequality in Proposition 4.3, it therefore suffices to show that the premise of Holley's inequality holds, that is, to show

$$\frac{\exp(\gamma(\mathbf{z} \vee \mathbf{z}^*)^T\mathbf{u}^+)}{\sum_{\mathbf{b}\in\Omega} \exp(\gamma\mathbf{b}^T\mathbf{u}^+)} \cdot \frac{\exp(\gamma(\mathbf{z} \wedge \mathbf{z}^*)^T\mathbf{u})}{\sum_{\mathbf{b}\in\Omega} \exp(\gamma\mathbf{b}^T\mathbf{u})} \geq \frac{\exp(\gamma\mathbf{z}^T\mathbf{u})}{\sum_{\mathbf{b}\in\Omega} \exp(\gamma\mathbf{b}^T\mathbf{u})} \cdot \frac{\exp(\gamma(\mathbf{z}^*)^T\mathbf{u}^+)}{\sum_{\mathbf{b}\in\Omega} \exp(\gamma\mathbf{b}^T\mathbf{u}^+)}$$

for all $\mathbf{z}, \mathbf{z}^* \in \Omega$. To show this, it suffices to show that

$$(\mathbf{z} \vee \mathbf{z}^*)^T\mathbf{u}^+ + (\mathbf{z} \wedge \mathbf{z}^*)^T\mathbf{u} \geq \mathbf{z}^T\mathbf{u} + (\mathbf{z}^*)^T\mathbf{u}^+ \qquad \text{for all} \quad \mathbf{z}, \mathbf{z}^* \in \Omega. \quad (4.6.1)$$

Notice how convenient Holley's inequality is: To prove that one random variable is stochastically larger than another, all that needs to be shown is that a certain linear inequality (4.6.1) holds.

First add and subtract $(\mathbf{z} \vee \mathbf{z}^*)^T\mathbf{u}$ to the left-hand side of (4.6.1), and then apply Lemma 2.9 to get

$$\begin{aligned}(\mathbf{z} \vee \mathbf{z}^*)^T\mathbf{u}^+ + (\mathbf{z} \wedge \mathbf{z}^*)^T\mathbf{u} &= \{(\mathbf{z} \vee \mathbf{z}^*) + (\mathbf{z} \wedge \mathbf{z}^*)\}^T\mathbf{u} + (\mathbf{z} \vee \mathbf{z}^*)^T(\mathbf{u}^+ - \mathbf{u}) \\ &= (\mathbf{z} + \mathbf{z}^*)^T\mathbf{u} + (\mathbf{z} \vee \mathbf{z}^*)^T(\mathbf{u}^+ - \mathbf{u}). \quad (4.6.2)\end{aligned}$$

Write the right-hand side of (4.6.1) as $(\mathbf{z} + \mathbf{z}^*)^T\mathbf{u} + (\mathbf{z}^*)^T(\mathbf{u}^+ - \mathbf{u})$ and compare this with (4.6.2). From this comparison, to prove (4.6.1) it suffices to show that $(\mathbf{z} \vee \mathbf{z}^*)^T(\mathbf{u}^+ - \mathbf{u}) \geq (\mathbf{z}^*)^T(\mathbf{u}^+ - \mathbf{u})$, or equivalently that $\mathbf{v}^T(\mathbf{z} \vee \mathbf{z}^*) \geq \mathbf{v}^T\mathbf{z}^*$; however, this follows from $v_{s1} \geq v_{s2} \geq \cdots \geq v_{s,n_s}$, since the 1's in $\mathbf{z} \vee \mathbf{z}^*$ are to the left of the 1's in $\mathbf{z}^*$. This proves the first inequality in the proposition. The second inequality is proved in the same way after replacing $\mathbf{u}^+$ by $\mathbf{u}$ and $\mathbf{u}$ by $\mathbf{u}^-$, so $\mathbf{v}$ becomes $\mathbf{u} - \mathbf{u}^-$. □

4.6.3. Some Properties of Arrangement-Increasing Statistics

Let $\mathbf{v}$ be a fixed N-tuple, and let $h(\cdot, \cdot)$ be an arrangement-increasing function as defined in §2.4.3. This section discusses properties of the expectation of $h(\mathbf{Z}, \mathbf{v})$ under the model (4.2.6), that is, properties of

$$\omega(\mathbf{v}, \mathbf{u}) = E\{h(\mathbf{Z}, \mathbf{v})\} = \frac{\sum_{\mathbf{z}\in\Omega} h(\mathbf{z}, \mathbf{v}) \exp(\gamma\mathbf{z}^T\mathbf{u})}{\sum_{\mathbf{b}\in\Omega} \exp(\gamma\mathbf{b}^T\mathbf{u})}. \quad (4.6.3)$$

These properties justify the sensitivity analysis for sum statistics in §4.6, but they will also have other uses later on.

Before discussing the properties of $\omega(\mathbf{v}, \mathbf{u})$, consider a few cases. Let $T = \mathbf{Z}^T\mathbf{q}$ be a sum statistic and set $\mathbf{v} = \mathbf{q}$. If $h(\mathbf{Z}, \mathbf{q}) = \mathbf{Z}^T\mathbf{q}$, then $h(\cdot, \cdot)$ is arrangement-increasing, and $\omega(\mathbf{q}, \mathbf{u})$ is the expectation of $\mathbf{Z}^T\mathbf{q}$. Alternatively, if $h(\mathbf{Z}, \mathbf{q}) = 1$ if $\mathbf{Z}^T\mathbf{q} \geq a$ and $h(\mathbf{Z}, \mathbf{q}) = 0$ otherwise, then $\omega(\mathbf{q}, \mathbf{u})$ is the probability that $\mathbf{Z}^T\mathbf{q} \geq a$, that is, the tail probability used to determine a significance level. Similar considerations apply with $\mathbf{v} = \mathbf{r}$ if $T = t(\mathbf{Z}, \mathbf{r})$ is any arrangement-increasing statistic.

The following proposition is an immediate consequence of the composition theorem for arrangement-increasing functions. The composition theorem is due Hollander, Proschan, and Sethuraman (1977); see also Marshall and Olkin (1979, §6.F.12) or Eaton (1987, §3.4). The composition theorem concerns expressions such as $\sum_{\mathbf{z} \in \Omega} h(\mathbf{z}, \mathbf{v}) \exp(\gamma \mathbf{z}^T\mathbf{u})$ in (4.6 3); it asserts that this expression is an arrangement-increasing function of $\mathbf{v}$ and $\mathbf{u}$ because $h(\mathbf{z}, \mathbf{v})$ and $\exp(\gamma \mathbf{z}^T\mathbf{u})$ are each arrangement-increasing functions and Ω is a symmetrical set. The following proposition says that $\omega(\mathbf{v}, \mathbf{u})$ increases—or more precisely, $\omega(\mathbf{v}, \mathbf{u})$ does not decrease—as the coordinates of $\mathbf{u}$ are permuted within strata into the same order as the coordinates of $\mathbf{v}$. In particular, the expectation $\omega(\mathbf{v}, \mathbf{u})$ is largest when $\mathbf{u}$ and $\mathbf{v}$ are ordered in the same way within each stratum and $\omega(\mathbf{v}, \mathbf{u})$ is smallest when $\mathbf{u}$ and $\mathbf{v}$ are ordered in opposite ways.

Proposition 4.4. $\omega(\mathbf{v}, \mathbf{u})$ *is arrangement-increasing.*

Pick any coordinate (s, i) and let $\boldsymbol{\varepsilon}_{si}$ be the N-tuple with a one in coordinate (s, i) and zeros in the other $N - 1$ coordinates. The following proposition from Rosenbaum and Krieger (1990) says that $\omega(\mathbf{v}, \mathbf{u} + \delta\boldsymbol{\varepsilon}_{si})$ is monotone in δ for each fixed $\mathbf{u}$ and $\mathbf{v}$. Note carefully that while this is true for each fixed $\mathbf{u}$ and $\mathbf{v}$ and for each (s, i), the direction of the monotonicity—increasing or decreasing—may change as $\mathbf{u}$, $\mathbf{v}$, and (s, i) are varied.

Proposition 4.5. *For each fixed* $\mathbf{u}$, $\mathbf{v}$, *and* (s, i), *the expectation* $\omega(\mathbf{v}, \mathbf{u} + \delta\boldsymbol{\varepsilon}_{si})$ *is monotone in* δ.

PROOF. Let $\Omega_0 = \{\mathbf{z} \in \Omega\colon z_{si} = 0\}$ and $\Omega_1 = \{\mathbf{z} \in \Omega\colon z_{si} = 1\}$, so $\Omega = \Omega_0 \cup \Omega_1$, and a sum over Ω may be broken up into a sum over Ω_0 plus a sum over Ω_1. Write

$$\omega(\mathbf{v}, \mathbf{u} + \delta\boldsymbol{\varepsilon}_{si}) = \frac{\sum_{\mathbf{z} \in \Omega_0} h(\mathbf{z}, \mathbf{v}) \exp(\gamma \mathbf{z}^T\mathbf{u}) + \exp(\gamma\delta) \sum_{\mathbf{z} \in \Omega_1} h(\mathbf{z}, \mathbf{v}) \exp(\gamma \mathbf{z}^T\mathbf{u})}{\sum_{\mathbf{b} \in \Omega_0} \exp(\gamma \mathbf{b}^T\mathbf{u}) + \exp(\gamma\delta) \sum_{\mathbf{b} \in \Omega_1} \exp(\gamma \mathbf{b}^T\mathbf{u})}$$

$$= \frac{A_0 + \exp(\gamma\delta)A_1}{D_0 + \exp(\gamma\delta)D_1}, \quad \text{say,} \tag{4.6.4}$$

where A_0, A_1, D_0, and D_1 are constants not varying with δ; moreover, D_0

and D_1 are strictly positive. Differentiating (4.6.4) with respect to δ gives

$$\frac{\partial\omega(\mathbf{v}, \mathbf{u} + \delta\boldsymbol{\varepsilon}_{si})}{\partial\delta} = \frac{(D_0 A_1 - A_0 D_1)\gamma \exp(\gamma\delta)}{\{D_0 + D_1 \exp(\gamma\delta)\}^2}, \tag{4.6.5}$$

so the sign of this derivative does not change as δ changes; hence $\omega(\mathbf{v}, \mathbf{u} + \delta\boldsymbol{\varepsilon}_{si})$ is monotone in δ. □

Let $\vec{\mathbf{v}}$ be the N-tuple $\mathbf{v}$ after its coordinates have been arranged in decreasing order within each stratum, $v_{si} \geq v_{s,i+1}$ for $i = 1, \ldots, n_s - 1$, and $s = 1, \ldots, S$. Let U^+ be the set of all N-tuples $\mathbf{b}$ of zeros and ones such that $b_{si} \geq b_{s,i+1}$, $i = 1, \ldots, n_s - 1$, and $s = 1, \ldots, S$, and let U^- be the all N-tuples $\mathbf{b}$ of zeros and ones such that $b_{si} \leq b_{s,i+1}$, $i = 1, \ldots, n_s - 1$, and $s = 1, \ldots, S$.

The following proposition bounds the unknown $\omega(\mathbf{v}, \mathbf{u})$ by two quantities that can be directly calculated. The bounds are sharp in that they are attained for some $\mathbf{u} \in U$. For instance, if $h(\mathbf{Z}, \mathbf{q}) = 1$ if $\mathbf{Z}^T\mathbf{q} \geq a$ and $h(\mathbf{Z}, \mathbf{q}) = 0$ otherwise, then Proposition 4.6 gives bounds on the one-sided significance level obtained using the sum statistic $\mathbf{Z}^T\mathbf{q}$.

Proposition 4.6. *For all* $\mathbf{u} \in U$,

$$\min_{\mathbf{b} \in U^-} \omega(\vec{\mathbf{v}}, \mathbf{b}) \leq \omega(\mathbf{v}, \mathbf{u}) \leq \max_{\mathbf{b} \in U^+} \omega(\vec{\mathbf{v}}, \mathbf{b}). \tag{4.6.6}$$

Proof. Suppose that, contrary to the upper bound in (4.6.6), there is a $\mathbf{u} \in U$ such that

$$\omega(\mathbf{v}, \mathbf{u}) > \max_{\mathbf{b} \in U^+} \omega(\vec{\mathbf{v}}, \mathbf{b}).$$

A sequence $\mathbf{u}_0, \mathbf{u}_1, \ldots, \mathbf{u}_J$ will be constructed such that $\mathbf{u} = \mathbf{u}_0$, $\mathbf{u}_J \in U^+$, and $\omega(\mathbf{v}, \mathbf{u}_j) \leq \omega(\mathbf{v}, \mathbf{u}_{j+1})$, thereby establishing a contradiction. The construction is as follows. Suppose $\mathbf{u}_j$ has at least one coordinate that is not equal to zero or one. Then form $\mathbf{u}_{j+1}$ from $\mathbf{u}_j$ by picking any one such coordinate of $\mathbf{u}_j$, and setting it to either zero or one so that $\omega(\mathbf{v}, \mathbf{u}_j) \leq \omega(\mathbf{v}, \mathbf{u}_{j+1})$; this is possible by Proposition 4.5. If every coordinate of $\mathbf{u}_j$ is either zero or one, then set $J = j + 1$ and let $\mathbf{u}_J$ be obtained by sorting the coordinates of $\mathbf{u}_j$ into decreasing order within each stratum. By Proposition 4.4, $\omega(\mathbf{v}, \mathbf{u}_j) \leq \omega(\vec{\mathbf{v}}, \mathbf{u}_J)$. Also, $\mathbf{u}_J \in U^+$, so there is a contradiction, and the upper bound in (4.6.6) is proved. The lower bound is proved similarly. □

4.6.4. Moments of Sum Statistics at Extreme Values of u

This section determines the expectation and variance of a sum statistic $\mathbf{Z}^T\mathbf{q}$ under the model (4.2.6) when $\mathbf{u}$ has binary coordinates, $u_{si} = 0$ or $u_{si} = 1$ for each (s, i). Also, large sample approximations to the expectation and variance are given. These results were used in §4.4.6 and §4.4.7 for the $2 \times 2 \times S$ contingency table and in §4.5 for the rank sum statistic.

Under model (4.2.6), a sum statistic is the sum of S independent contributions, one from each stratum. The expectation and variance of a sum statistic is, therefore, the sum of the S separate expectations and variances of the contributions from each of the S strata. As a result, it suffices to consider the case of a single stratum, $S = 1$, so in this section the subscript s is omitted. Let $u_i = 1$ or $u_i = 0$ for $i = 1, \ldots, n$, let $k = \sum u_i$, so $\mathbf{u}$ has k coordinates equal to 1 and $n - k$ equal to zero.

An important role is played by the random variable $C = \mathbf{Z}^T\mathbf{u}$, which is the number of treated units for which $u_i = 1$. The variable C has the extended hypergeometric distribution, which arises in other contexts as the conditional distribution of one binomial random variable given the sum of this variable and a second independent binomial variable. For detailed discussion of the extended hypergeometric distribution, see Johnson, Kotz, and Kemp (1992, §6.11) or Plackett (1981, §4.2). Let $\overline{E} = E(C)$ and $\overline{V} = \text{var}(C)$. The expectation and variance of $\mathbf{Z}^T\mathbf{q}$ have easily computed formulas defined in terms of $\overline{E}$ and $\overline{V}$. (For reasons that will become apparent shortly, the symbols $\overline{E}$ and $\overline{V}$ used here are the same as the symbols used in §4.4.6.)

Find the mean and variance of the q_i separately for units with $u_i = 1$ and $u_i = 0$, as follows:

$$\bar{q}_1 = \frac{1}{k}\sum_{i:u_i=1} q_i, \qquad \bar{q}_0 = \frac{1}{n-k}\sum_{i:u_i=0} q_i,$$
$$w_1 = \frac{1}{k-1}\sum_{i:u_i=1} (q_i - \bar{q}_1)^2, \qquad w_0 = \frac{1}{n-k-1}\sum_{i:u_i=0} (q_i - \bar{q}_0)^2. \tag{4.6.7}$$

Proposition 4.7. *Under model* (4.2.6) *in which* $\mathbf{u}$ *has binary coordinates, the expectation and variance of* $\mathbf{Z}^T\mathbf{q}$ *are*

$$E(\mathbf{Z}^T\mathbf{q}) = \overline{E}\bar{q}_1 + (m - \overline{E})\bar{q}_0,$$
$$\text{var}(\mathbf{Z}^T\mathbf{q}) = (w_1 - w_0)\overline{E} - (\overline{E}^2 + \overline{V})\left(\frac{w_1}{k} + \frac{w_0}{n-k}\right) + \frac{m(n-k-m+2\overline{E})w_0}{n-k} + \overline{V}(\bar{q}_1 - \bar{q}_0)^2. \tag{4.6.8}$$

Proof. Notice that in (4.2.6), $C = \mathbf{Z}^T\mathbf{u}$ is the sufficient statistic, so the conditional distribution of $\mathbf{Z}$ given $C = \mathbf{Z}^T\mathbf{u}$ does not depend on γ. More than this, $\text{prob}(\mathbf{Z} = \mathbf{z} | C = c)$ is uniform on the subset of Ω such that $\mathbf{z}^T\mathbf{u} = c$, that is, uniform on the set containing the $\binom{k}{c}\binom{n-k}{m-c}$ vectors with c coordinates equal to 1 among the k coordinates with $u_i = 1$, $m - c$ coordinates equal to 1 among the $n - k$ coordinates with $u_i = 0$, and all other coordinates equal to 0. In other words, the distribution $\text{prob}(\mathbf{Z} = \mathbf{z} | C = c)$ picks at random c of the k coordinates with $u_i = 1$ and independently picks at random $m - c$ of the $n - k$ coordinates with $u_i = 0$. It follows that $\text{prob}(\mathbf{Z}^T\mathbf{q} | C = c)$ is the distribution of the sum of c scores selected at random from among the k scores q_i such that $u_i = 1$ plus the sum of $m - c$ scores independently selected at random from among the $n - k$ scores q_i

such that $u_i = 0$. Therefore,

$$E(\mathbf{Z}^{\mathrm{T}}\mathbf{q}) = E\{E(\mathbf{Z}^{\mathrm{T}}\mathbf{q}|C)\} = E\{C\bar{q}_1 + (m - C)\bar{q}_0\} = \bar{E}\bar{q}_1 + (m - \bar{E})\bar{q}_0,$$

proving the first part of (4.6.8). Similarly,

$$\begin{aligned}
\mathrm{var}(\mathbf{Z}^{\mathrm{T}}\mathbf{q}) &= E\{\mathrm{var}(\mathbf{Z}^{\mathrm{T}}\mathbf{q}|C)\} + \mathrm{var}\{E(\mathbf{Z}^{\mathrm{T}}\mathbf{q}|C)\} \\
&= E\left\{\frac{C(k - C)w_1}{k} + \frac{(m - C)(n - k - m + C)w_0}{n - k}\right\} \\
&\quad + \mathrm{var}\{C\bar{q}_1 + (m - C)\bar{q}_0\} \\
&= E\left\{(w_1 - w_0)C - C^2\left(\frac{w_1}{k} + \frac{w_0}{n - k}\right) + \frac{m(n - k - m + 2C)w_0}{n - k}\right\} \\
&\quad + \mathrm{var}\{C(\bar{q}_1 - \bar{q}_0)\} \\
&= (w_1 - w_0)\bar{E} - (\bar{E}^2 + \bar{V})\left(\frac{w_1}{k} + \frac{w_0}{n - k}\right) + \frac{m(n - k - m + 2\bar{E})w_0}{n - k} \\
&\quad + \bar{V}(\bar{q}_1 - \bar{q}_0)^2
\end{aligned}$$

proving the second part of (4.6.8). □

There are $\binom{k}{c}\binom{n-k}{m-c}$ elements of Ω such that $\mathbf{z}^{\mathrm{T}}\mathbf{u} = c$ for each integer c, $\max(0, m + k - n) \le c \le \min(m, k)$. Using (4.2.6) gives:

$$\bar{E} = \frac{\sum_{c=\max(0,m+k-n)}^{\min(m,k)} c\binom{k}{c}\binom{n-k}{m-c}\exp(\gamma c)}{\sum_{c=\max(0,m+k-n)}^{\min(m,k)} \binom{k}{c}\binom{n-k}{m-c}\exp(\gamma c)} \tag{4.6.9}$$

and

$$\bar{V} = \frac{\sum_{c=\max(0,m+k-n)}^{\min(m,k)} c^2\binom{k}{c}\binom{n-k}{m-c}\exp(\gamma c)}{\sum_{c=\max(0,m+k-n)}^{\min(m,k)} \binom{k}{c}\binom{n-k}{m-c}\exp(\gamma c)} - \bar{E}^2. \tag{4.6.10}$$

An equivalent but simpler expression for $\bar{V}$ is given by Johnson, Kotz, and Kemp (1992, p. 280, expression (6.160)), namely,

$$\bar{V} = \frac{\Gamma mk - \{n - (m + k)(1 - \Gamma)\}\bar{E}}{(1 - \Gamma)} - \bar{E}^2, \tag{4.6.11}$$

where $\Gamma = \exp(\gamma)$.

When m, k, $n - m$, $n - k$ are large, $\bar{E}$ and $\bar{V}$ can be cumbersome to compute, but in this case a large sample approximation is available. The approximations to $\bar{E}$ and $\bar{V}$ are $\tilde{E}$ and $\tilde{V}$ where $\tilde{E}$ is the root of the quadratic equation

$$\tilde{E}^2(\Gamma - 1) - \tilde{E}\{(\Gamma - 1)(m + k) + n\} + \Gamma km = 0 \tag{4.6.12}$$

with $\max(0, k + m - n) \le \tilde{E} \le \min(k, m)$, and

$$\tilde{V} = \frac{1}{\dfrac{1}{\tilde{E}} + \dfrac{1}{k - \tilde{E}} + \dfrac{1}{m - \tilde{E}} + \dfrac{1}{n - k - m + \tilde{E}}}. \tag{4.6.13}$$

These expressions are due to Stevens (1951). Hannan and Harkness (1963) give mild conditions such that C has a limiting normal distribution with expectation $\tilde{E}$ and variance $\tilde{V}$. If instead, m is fixed as $n \to \infty$ and $k \to \infty$ with $k/n \to \zeta$, then Harkness (1965) shows that C has a limiting binomial distribution with sample size m and probability $\zeta\Gamma/\{(1 - \zeta) + \zeta\Gamma\}$.

These facts about the moments of $\mathbf{Z}^{\mathrm{T}}\mathbf{q}$ were used several times in this chapter. Consider the upper bound for the significance level in a 2×2 contingency table in §4.4.6 and §4.4.7. Here, r_i is binary, $q_i = r_i$ and $u_i = r_i$, so $k = r_+$, $\bar{q}_1 = 1$, $\bar{q}_0 = 0$, $w_1 = w_0 = 0$; therefore, from (4.6.8), $E(\mathbf{Z}^{\mathrm{T}}\mathbf{q}) = \bar{E}$ and $\operatorname{var}(\mathbf{Z}^{\mathrm{T}}\mathbf{q}) = \bar{V}$. The Normal approximation was used in §4.4.6 and the binomial approximation in §4.4.7.

Now consider the upper bound for the sum statistic in §4.5. Here, the q_i are sorted into decreasing order, and $u_1 = 1$, $u_2 = 1, \ldots, u_k = 1$, $u_{k+1} = 0, \ldots, u_n = 0$ for some k and (4.6.8) is used with the approximations in place of the exact moments.

Under mild conditions on the behavior of the scores q_{si}, the limiting Normal distribution for $\mathbf{Z}^{\mathrm{T}}\mathbf{q}$ under model (4.26) as $\min n_s \to \infty$ follows from, for example, Theorem 2.1 of Bickel and van Zwet (1978). They also provide an asymptotic expansion of the tail area and a uniform bound on the error.

4.6.5. The Effects of Certain Transformations of u

It is useful to observe that, in the model (4.2.6), certain transformations of $\mathbf{u}$ have simple consequences. For instance, consider the linear transformation $\mathbf{u}^* = (1/\beta)(\mathbf{u} - \alpha\mathbf{1})$ for $\beta > 0$. Write $\gamma^* = \beta\gamma$. Then

$$\operatorname{pr}(\mathbf{Z} = \mathbf{z} \mid \mathbf{m}) = \frac{\exp(\gamma\mathbf{z}^{\mathrm{T}}\mathbf{u})}{\sum_{\mathbf{b}\in\Omega}\exp(\gamma\mathbf{b}^{\mathrm{T}}\mathbf{u})} = \frac{\exp\{\gamma\mathbf{z}^{\mathrm{T}}(\alpha\mathbf{1} + \beta\mathbf{u}^*)\}}{\sum_{\mathbf{b}\in\Omega}\exp\{\gamma\mathbf{b}^{\mathrm{T}}(\alpha\mathbf{1} + \beta\mathbf{u}^*)\}}$$

$$= \frac{\exp(\gamma^*\mathbf{z}^{\mathrm{T}}\mathbf{u}^*)}{\sum_{\mathbf{b}\in\Omega}\exp(\gamma^*\mathbf{b}^{\mathrm{T}}\mathbf{u}^*)},$$

where the last equality follows from the observation that $\mathbf{b}^{\mathrm{T}}\mathbf{1} = \sum m_s$ for all $\mathbf{b} \in \Omega$. In other words, a change in the location of $\mathbf{u}$ does not change the model, while a change in the scale of $\mathbf{u}$ changes the sensitivity parameter γ

by a corresponding multiple. The practical consequence is that there is no need to consider other locations and scales for $\mathbf{u}$, since they are implicitly covered by (4.2.6).

A more important transformation is $\mathbf{u}^{**} = \mathbf{1} - \mathbf{u}$. Now, $\mathbf{u} \in [0, 1]^N$ if and only if $\mathbf{u}^{**} \in [0, 1]^N$. Then

$$\operatorname{pr}(\mathbf{Z} = \mathbf{z}|\mathbf{m}) = \frac{\exp(\gamma \mathbf{z}^T \mathbf{u})}{\sum_{\mathbf{b} \in \Omega} \exp(\gamma \mathbf{b}^T \mathbf{u})} = \frac{\exp\{\gamma \mathbf{z}^T(\mathbf{1} - \mathbf{u}^{**})\}}{\sum_{\mathbf{b} \in \Omega} \exp\{\gamma \mathbf{b}^T(\mathbf{1} - \mathbf{u}^{**})\}}$$

$$= \frac{\exp(-\gamma \mathbf{z}^T \mathbf{u}^{**})}{\sum_{\mathbf{b} \in \Omega} \exp(-\gamma \mathbf{b}^T \mathbf{u}^{**})},$$

where the last equality again follows from $\mathbf{b}^T\mathbf{1} = \sum m_s$ for all $\mathbf{b} \in \Omega$. In other words, replacing $\mathbf{u}$ by $\mathbf{1} - \mathbf{u}$ has the effect of changing the sign of γ, or equivalently of replacing Γ by $1/\Gamma$. This has several practical consequences. First, it suffices to consider $\gamma \geq 0$, since any distribution with $\gamma < 0$ at a point $\mathbf{u}$ is the same distribution as one with $\gamma > 0$ at $\mathbf{1} - \mathbf{u}$. Second, in sensitivity analyses for sum statistics, the sets U^+ and U^- appear; see §4.5.2, and Proposition 4.6 in §4.6.3. Now, $\mathbf{u} \in U^-$ if and only if $\mathbf{1} - \mathbf{u}$ is in U^+. It follows that the minimum of a quantity, say $\mu_{\mathbf{u}}$ in (4.5.2), as $\mathbf{u}$ ranges over U^- with $\gamma > 0$ equals a minimum of the same quantity as $\mathbf{u}$ ranges over U^+ with γ replaced by $-\gamma$.

4.7. Bibliographic Notes

Cornfield, Haenszel, Hammond, Lilienfeld, Shimkin, and Wynder (1959) proposed a sensitivity analysis for risk ratios. Independently, Bross (1966, 1967) proposed a sensitivity analysis for 2 × 2 tables. A related approach is discussed by Schlesselmann (1978). Gastwirth (1988, 1992a,b) extended the method of Cornfield et al. (1959) in several directions. Rosenbaum and Rubin (1983) discussed sensitivity analysis for a point estimate of a difference in proportions in a $2 \times 2 \times S$ table using $3S$ sensitivity parameters. Rosenbaum (1986) discussed sensitivity analysis in multiple linear regression. The methods discussed in this chapter are largely taken from Rosenbaum (1987, 1988, 1991a, 1993) and Rosenbaum and Krieger (1990).

4.8. Problems

4.8.1. Graphing SOLVE

1. View the signed rank statistic $t(\mathbf{Z}, \mathbf{R} - \tau\mathbf{Z})$ in §4.3 as a function of τ with $\mathbf{Z}$ and $\mathbf{R}$ fixed. How would this function look if it were graphed with $t(\mathbf{Z}, \mathbf{R} - \tau\mathbf{Z})$ on the vertical axis and τ on the horizontal axis? What happens as τ moves from $\tau < R_{si}$ to $\tau = R_{si}$ to $\tau > R_{si}$? Consider the estimate (4.3.6) and its relationship to this graph. How does SOLVE operate?

4.8.2. Unbounded Covariates

2. Consider S matched pairs in §4.3, and replace the assumption that $0 \le u_{si} \le 1$ by the weaker assumption that $|u_{s1} - u_{s2}| \le 1$ for at least S' pairs, where S' is a given number, $S' \le S$. This says that at least S'/S of the pairs have odds of receiving the treatment that differ by at most Γ, but the remaining $(1 - S')/S$ of the pairs may differ by far more. In this case, obtain sharp bounds analogous to those in Proposition 4.2 with two new random variables replacing T^+ and T^-.
Rosenbaum (1987)

4.8.3. Derivations

3. By conditioning on $\mathbf{m}$, derive (4.2.6) from (4.2.2a) and the independence of the $Z_{[j]}$.
4. In the absence of ties and zero differences, derive the moments (4.3.5) of Wilcoxon's signed rank statistic from the moments (4.3.3) for a sign-score statistic. Show that the expectation is unaffected by ties if average ranks are used.
5. Derive the bounds (4.4.6) by evaluating the model (4.2.6) at $\mathbf{u} = \mathbf{u}^+$ and $\mathbf{u} = \mathbf{u}^-$.
Rosenbaum (1991a, §4)
6. Show that concordant matched sets with $m_s = 0$ or $m_s = n_s$ could be removed without altering the value of the statistics (4.4.7).
7. Give a short proof of Proposition 4.3 when $1 = \sum_i Z_{si}$ for each s, as is true in §4.3 and §4.4.2.
Rosenbaum (1988)

References

Armitage, P. (1977). *Statistical Methods in Medical Research* (fourth printing), Oxford, UK: Blackwell.

Bickel, P. and van Zwet, (1978). Asymptotic expansions for the power of distribution free tests in the two-sample problem. *Annals of Statistics*, **6**, 937–1004.

Breslow, N. and Day, N. (1980). *Statistical Methods in Cancer Research, I: The Analysis of Case-Control Studies.* Lyon, France: International Agency for Research on Cancer.

Boston Collaborative Drug Project (1972). Excess of ampicillin rashes associated with allopurinol or hyperuricemia. *New England Journal of Medicine*, **286**, 505–507.

Bross, I.D.J. (1966). Spurious effects from an extraneous variable. *Journal of Chronic Diseases*, **19**, 637–647.

Bross, I.D.J. (1967). Pertinency of an extraneous variable. *Journal of Chronic Diseases*, **20**, 487–495.

Cameron, E. and Pauling, L. (1976). Supplemental ascorbate in the supportive treatment of cancer: Prolongation of survival times in terminal human cancer. *Proceedings of the National Academy of Sciences* (*USA*), **73**, 3685–3689.

Cornfield, J., Haenszel, W., Hammond, E., Lilienfeld, A., Shimkin, M., and Wynder, E. (1959). Smoking and lung cancer: Recent evidence and a discussion of some questions. *Journal of the National Cancer Institute*, **22**, 173–203.

Eaton, M. (1987). *Lectures on Topics in Probability Inequalities.* Amsterdam: Centrum voor Wiskunde en Informatica.

Fisher, R.A. (1958). Lung cancer and cigarettes? *Nature*, **182**, July 12, 108.

Gastwirth, J. (1988). *Statistical Reasoning in Law and Public Policy*. New York: Academic Press.

Gastwirth, J. (1992a). Employment discrimination: A statistician's look at analysis of disparate impact claims. *Law and Inequality*, **11**, 151–179.

Gastwirth, J. (1992b). Methods for assessing the sensitivity of statistical comparisons used in Title VII cases to omitted variables. *Jurimetrics Journal*, 19–34.

Gastwirth, J. and Greenhouse, S. (1987). Estimating a common relative risk: Application in equal employment. *Journal of the American Statistical Association*, **82**, 38–45.

Gibbons, J.D. (1982). Brown–Mood median test. In: *Encyclopedia of Statistical Sciences* (eds., S. Kotz and N. Johnson), New York: Wiley, Volume 1, pp. 322–324.

Greenhouse, S. (1982). Jerome Cornfield's contributions to epidemiology. *Biometrics*, **38S**, 33–46.

Hakulinen, T. (1981). A Mantel–Haenszel statistic for testing the association between a polychotomous exposure and a rare outcome. *American Journal of Epidemiology*, **113**, 192–197.

Hammond, E.C. (1964). Smoking in relation to mortality and morbidity: Findings in first thirty-four months of follow-up in a prospective study started in 1959. *Journal of the National Cancer Institute*, **32**, 1161–1188.

Hannan, J. and Harkness, W. (1963). Normal approximation to the distribution of two independent binomials, conditional on a fixed sum. *Annals of Mathematical Statistics*, **34**, 1593–1595.

Harkness, W. (1965). Properties of the extended hypergeometric distribution. *Annals of Mathematical Statistics*, **36**, 938–945.

Háyek, J. and Sidák, Z. (1967). *Theory of Rank Tests*. New York: Academic Press.

Herbst, A., Ulfelder, H., and Poskanzer, D. (1971). Adenocarcinoma of the vagina: Association of maternal stilbestrol therapy with tumor appearance in young women. *New England Journal of Medicine*, **284**, 878–881.

Hettmansperger, T. (1984). *Statistical Inference Based on Ranks*. New York: Wiley.

Hodges, J. and Lehmann, E. (1963). Estimates of location based on rank tests. *Annals of Mathematical Statistics*, **34**, 598–611.

Hollander, M., Proschan, F., and Sethuraman, J. (1977). Functions decreasing in transposition and their applications in ranking problems. *Annals of Statistics*, **5**, 722–733.

Hollander, M. and Wolfe, D. (1973). *Nonparametric Statistical Methods*. New York: Wiley.

Holley, R. (1974). Remarks on the FKG inequalities. *Communications in Mathematical Physics*, **36**, 227–231.

Jick, H., Miettinen, O., Neff, R., et al. (1973). Coffee and myocardial infarction. *New England Journal of Medicine*, **289**, 63–77.

Johnson, N., Kotz, S., and Kemp, A. (1992). *Univariate Discrete Distributions*. New York: Wiley.

Kelsey, J. and Hardy, R. (1975). Driving of motor vehicles as a risk factor for acute herniated lumbar intervertebral disc. *American Journal of Epidemiology*, **102**, 63–73.

Lehmann, E. (1975). *Nonparametrics: Statistical Methods Based on Ranks*. San Francisco: Holden-Day.

Mack, T., Pike, M., Henderson, B., Pfeffer, R., Gerkins, V., Arthur, B., and Brown, S. (1976). Estrogens and endometrial cancer in a retirement community. *New England Journal of Medicine*, **294**, 1262–1267.

Mann, H. and Whitney, D. (1947). On a test of whether one of two random variables is stochastically larger than the other. *Annals of Mathematical Statistics*, **18**, 50–60.

Mantel, N. and Haenszel, W. (1959). Statistical aspects of retrospective studies of disease. *Journal of the National Cancer Institute*, **22**, 719–748.

Marshall, A. and Olkin, I. (1979). *Inequalities: Theory of Majorization and Its Applications.* New York: Academic Press.

McNemar, Q. (1947). Note on the sampling error of the differences between correlated proportions or percentages. *Psychometrika*, **12**, 153–157.

Miettinen, O. (1969). Individual matching with multiple controls in the case of all or none responses. *Biometrics*, **22**, 339–355.

Moertel, C., Fleming, T., Creagan, E., Rubin, J., O'Connell, M., and Ames, M. (1985). High-dose vitamin C versus placebo in the treatment of patients with advanced cancer who have had no prior chemotherapy: A randomized double-blind comparison. *New England Journal of Medicine*, **312**, 137–141.

Morton, D., Saah, A., Silberg, S., Owens, W., Roberts, M., and Saah, M. (1982). Lead absorption in children of employees in a lead related industry. *American Journal of Epidemiology*, **115**, 549–555.

Mosteller, F. and Tukey, J. (1977). *Data Analysis and Regression.* Reading, MA: Addison-Wesley.

Nicholson, W., Selikoff, I., Seidman, H., Lilis, R., and Formby, P. (1979). Long-term mortality experience of chrysotile miners and millers in Thetford Mines, Quebec. *Annals of the New York Academy of Sciences*, **330**, 11–21.

O'Brien, P.C. and Fleming, T.R. (1987). A paired Prentice–Wilcoxon test for censored paired data. *Biometrics*, **43**, 169–180.

Plackett, R.L. (1981). *The Analysis of Categorical Data* (second edition). New York: Macmillan.

Rosenbaum, P. (1986). Dropping out of high school in the United States: An observational study. *Journal of Educational Statistics*, **11**, 207–224.

Rosenbaum, P.R. (1987). Sensitivity analysis for certain permutation inferences in matched observational studies. *Biometrika*, **74**, 13–26.

Rosenbaum, P.R. (1988). Sensitivity analysis for matching with multiple controls. *Biometrika*, **75**, 577–581.

Rosenbaum, P.R. (1989). On permutation tests for hidden biases in observational studies: An application of Holley's inequality to the Savage lattice. *Annals of Statistics*, **17**, 643–653.

Rosenbaum, P.R. (1991a). Sensitivity analysis for matched case-control studies. *Biometrics*, **47**, 87–100.

Rosenbaum, P.R. (1991b). Discussing hidden bias in observational studies. *Annals of Internal Medicine*, **115**, 901–905.

Rosenbaum, P.R. (1993). Hodges–Lehmann point estimates of treatment effect in observational studies. *Journal of the American Statistical Association*, **88**, 1250–1253.

Rosenbaum, P. and Krieger, A. (1990). Sensitivity analysis for two-sample permutation inferences in observational studies. *Journal of the American Statistical Association*, **85**, 493–498.

Rosenbaum, P. and Rubin, D. (1983). Assessing sensitivity to an unobserved binary covariate in an observational study with binary outcome. *Journal of the Royal Statistical Society*, Series B, **45**, 212–218.

Rubin, D.B. (1978). Bayesian inference for causal effects: The role of randomization. *Annals of Statistics*, **6**, 34–58.

Savage, I.R. (1964). Contributions to the theory of rank order statistics: Applications of lattice theory. *Review of the International Statistical Institute*, **32**, 52–63.

Schlesselmann, J.J. (1978). Assessing the effects of confounding variables. *American Journal of Epidemiology*, **108**, 3–8.

Skerfving, S., Hansson, K., Mangs, C., Lindsten, J., and Ryman, N. (1974). Methylmercury-induced chromosome damage in man. *Environmental Research*, **7**, 83–98.

Stevens, W.L. (1951). Mean and variance of an entry in a contingency table. *Biometrika*, **38**, 468–470.

Wilcoxon, F. (1945). Individual comparisons by ranking methods. *Biometrics*, **1**, 80–83.

CHAPTER 5
Known Effects

5.1. Detecting Hidden Bias Using Known Effects

5.1.1. Sensitivity Analysis and Detecting Hidden Bias: Their Complementary Roles

As seen in Chapter 4, observational studies vary considerably in their sensitivity to hidden bias. The study of the study of coffee and myocardial infarction in §4.4.5 is sensitive to small biases while the studies of smoking and lung cancer in §4.3.3 or DES and vaginal cancer in §4.4.5 are sensitive only to biases that are many times larger. If sensitive to small biases, a study is especially open to the criticism that a particular unrecorded covariate was not controlled because, in this case, small differences in an important covariate can readily explain the difference in outcomes in treated and control groups. Still, all observational studies are sensitive to sufficiently large biases, and large biases have occurred on occasion; see §4.4.3. A sensitivity analysis shows how biases of various magnitudes might alter conclusions, but it does not indicate whether biases are present or what magnitudes are plausible.

Efforts to detect hidden bias involve collecting data that have a reasonable prospect of revealing a hidden bias if it is present. If treatments were not assigned to subjects as if at random, if the distribution of $\mathbf{Z}$ is not uniform on Ω, if instead the distribution is tilted, favoring certain assignments $\mathbf{Z}$ over others, then this tilting or bias might leave visible traces in data we have or could obtain. Efforts to detect hidden bias use additional information beyond the treated and control groups, their covariates and outcomes of primary interest. Recall from §1.2 Sir Ronald Fisher's advice, "Make your theories elaborate." This additional information may consist of several control groups, or several referent groups in a case-referent study, or it may consist of additional outcomes for which the magnitude or direction of the treatment effect are known. A systematic difference between the outcomes in several control groups cannot be an effect of the treatment and must be a hidden bias. A systematic difference between treated and control groups in an outcome the treatment does not affect must be a hidden bias. Detecting hidden bias entails checking that treatment effects appear where they should, and not elsewhere.

Efforts to detect bias and sensitivity analysis complement one another in their strengths and limitations. It is difficult if not impossible to detect small hidden biases, but a sensitivity analysis might indicate that small biases would not materially alter the study's conclusions. All studies are sensitive to sufficiently dramatic biases, but dramatic biases are the ones most likely to be detected.

Statistical theory contributes in two ways to efforts to detect bias. First, it provides qualitative advice about what these efforts can and cannot do, and about the types of data and study designs that have a reasonable prospect of detecting hidden bias. Second, theory provides a quantitative link between detection and sensitivity analysis. When does failure to detect hidden bias provide evidence that biases are absent or small? Does failure to detect hidden bias make a study less sensitive to bias?

Chapters 5, 6, and 7 discuss three methods of detecting hidden biases, namely known effects, multiple referent groups in a case-referent study, and multiple control groups. The three methods are ordered in this way because the technical development in Chapter 5 is most similar to Chapter 4, Chapter 6 is a step away, and Chapter 7 is a larger step away. The remainder of §5.1 discusses examples of known effects and concludes by asking when an effect is known.

5.1.2. An Example: Nuclear Weapons Testing and Leukemia

Can hidden biases be detected? Is there any real hope of detecting a departure from a random assignment of treatments within strata that are homogeneous in observed covariates? Is the task ahead an impossible task? The first example is of interest because it affords several opportunities to detect hidden bias and, in each case, bias is detected. Also, there is a bit of late news produced 4 years after the publication of the original study.

Lyon, Klauber, Gardner, and Udall (1979) studied the possible effects of fallout from nuclear weapons testing in Nevada on the risk of childhood leukemia and other cancers in Utah. The study grouped counties in Utah into high or low exposure based on proximity to nuclear testing in Nevada. In addition, the counties were studied in three time periods, 1944–1950 prior to nuclear testing in Nevada, 1951–1958 during above-ground testing, and 1959–1975 during underground testing. As will be seen, this is an attractive and careful study design in that it provides several checks for hidden bias. Frequencies of childhood leukemia and other cancer were obtained from the Utah State Bureau of Vital Statistics, and person-years at risk were determined from the US Census and the Utah State Bureau of Vital Statistics; see Lyon et al. (1979) for details. This study was discussed by Land (1979), Beck and Krey (1983), Rosenbaum (1984a), Gastwirth (1988, pp. 870–878). In particular, Gastwirth discusses the outcome of related litigation.

If above ground nuclear testing has an effect, then how should that effect appear? Radiation could reasonably cause leukemias and other childhood cancers, but it should not prevent them, and more radiation should yield more cancers. If above ground testing has an effect, and if there is no hidden bias, one expects the highest levels of leukemia and other cancers in the high-exposure counties during the period of above ground testing, followed by lower but possibly still elevated levels in the low-exposure counties during the same time period, with still lower and similar levels in the high- and low-exposure counties before and after nuclear testing. This is what one would expect to see if, contrary to fact, children had been assigned at random to one of the six cohorts of children. It is an elaborate theory in Fisher's sense, one that includes several effects of known direction and several control groups.

Figure 5.1.1 shows the leukemia mortality rates from Lyon et al. (1979) for children under the age of 15. These are mortality rates per 100,000 person-years.

Figure 5.1.1 conforms to the anticipated effect in one respect but not in another. As anticipated, in the high-exposure counties, the leukemia mortal-

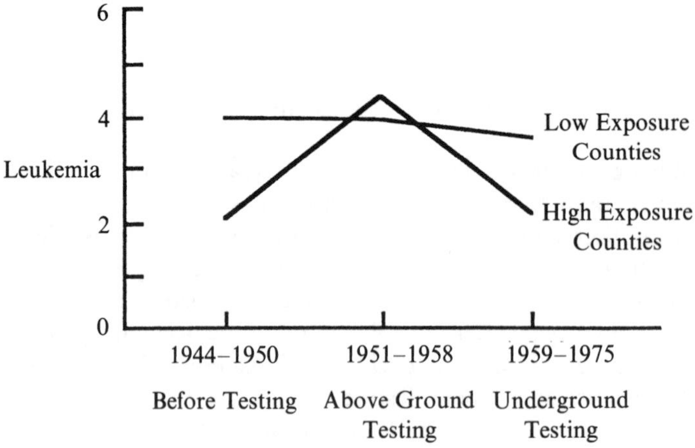

Figure 5.1.1. Leukemia morality per 100,000 person-years for children in Utah before, during, and after above-ground nuclear testing.

ity rises from 2.1 leukemias per 100,000 before testing to 4.4 during above-ground testing, and then returns to 2.2 during the period of underground testing. The low-exposure counties show a slight decline. On the other hand, the low-exposure counties had almost twice the leukemia of the high-exposure counties before nuclear testing (4.0 versus 2.1) and more than 50% more after nuclear testing went underground (3.6 versus 2.2). Indeed, the rise in leukemias in the high-exposure counties during above ground testing brought the level to just slightly more than that found in the low-exposure counties at the same time (4.4 versus 3.9). It appears that the high- and low-exposure counties are not comparable in ways that matter for the leukemia rate.

Figure 5.1.2 shows the corresponding rates for other childhood cancers. Here, the anticipated effect is not seen. For the high-exposure counties, the rate of other cancers declined during the period of above-ground testing, just as the leukemia rate was rising. This decline is not plausibly an effect of nuclear testing, and it strongly suggests that the two periods were not comparable in ways that matter for cancers other than leukemias. The low-exposure counties show a slight decline. For cancers other than leukemia, before nuclear testing, the high-exposure counties had higher rates than the low-exposure counties, but that reversed during above-ground testing. It appears that the high- and low-exposure counties are not comparable in ways that matter for cancers other than leukemias. If the differences that should not be present are subjected to a formal hypothesis test, then the differences are found to be statistically significant and not readily attributed to chance; see Rosenbaum (1984a).

In short, the mortality rates vary in ways not readily attributed to chance or to effects of above ground nuclear testing. More than this, the variation

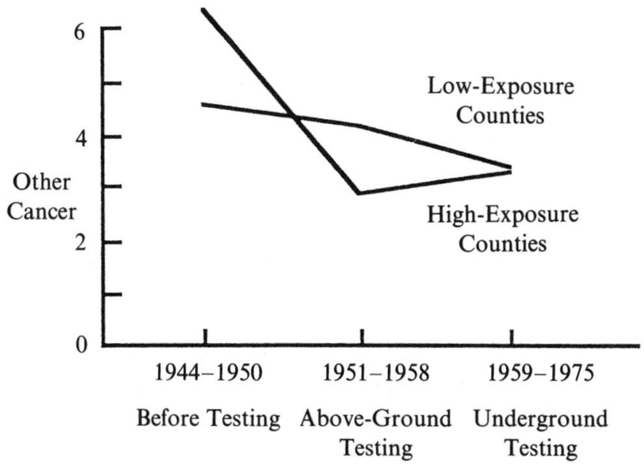

Figure 5.1.2. Other cancer morality per 100,000 person-years for children in Utah before, during, and after above-ground nuclear testing.

in mortality rates that is inconsistent with an effect of nuclear testing is as large as the variation that is consistent with an effect. This does not necessarily mean that the interesting rise in leukemia in high-exposure counties is not an effect of nuclear testing, but it is reason for extreme caution about any such claim.

The study by Lyon et al. (1979) is an excellent study, even though there is strong evidence of hidden bias. Indeed, it is an excellent study *because* there is strong evidence of hidden bias. A poor study might have looked only at leukemia and only in high-exposure counties, and it would have found no evidence of hidden bias—because it did not look—and therefore it would have left its audience with an incorrect impression. The editorial policies of major journals shape science. The *New England Journal of Medicine* wisely published the study by Lyon et al. (1979) along with a cautious editorial by Land (1979), despite the evidence of hidden bias, and one hopes it would have wisely declined to publish the imagined study which considered only leukemia in high-exposure counties, despite its absence of evidence of hidden bias.

Four years later, Beck and Krey (1983) published a study that attempted to estimate the doses of radiation received between 1951 and 1958 in different parts of Utah. They took soil samples in the high- and low-exposure counties described by Lyon et al. (1979) and measured cesium-137 and plutonium. They concluded: "Although the highest exposures were found in the extreme southwest part of Utah, as expected, the residents of the populous northern valleys around Provo, Salt Lake City, and Ogden received higher mean dose and a significantly greater population dose (person-rads) than did the residents of most counties closer to the test site [p. 18]." In other words, their sampling suggests that the "low"-exposure counties may have received higher exposures than the "high"-exposure counties. Beck and Krey also compared the radiation they found with background radiation and global fallout and expressed their view that "... bone doses to the population of southern Utah were far too low to account for the excess childhood leukemia mortality reported by Lyon et al. (1979)."

For us, the important point is that careful study design can help to detect hidden biases.

5.1.3. Specificity of Associations and Known Effects

In epidemiology, it is often said that a "specific" association between an exposure and a disease provides greater evidence that the exposure causes the disease than does a "nonspecific" association. For instance, Yerushalmy and Palmer (1959) write:

> The demonstration of high relative frequencies in the study group is thus only a first step in the process of searching for etiologic factors. The investigation must proceed to the second and more crucial consideration (which, for

> want of a better term, is denoted here as that of specificity of effect), i.e., to the demonstration the difference in relative frequencies reflects a specific and meaningful relationship between the characteristic under suspicion and the disease under consideration ... [p. 36].
>
> ... If the characteristic can be shown to be related only or mostly to the disease under study and not to many other disease entities, then our confidence that it is a cause-carrying vector for that disease is greatly increased.
>
> If, on the other hand, it is found that the characteristic is also related to numerous other diseases, including those without obvious physiologic or pathologic connection with the characteristic in question, the relationship must be assumed—until further proof—to be nonspecific [p. 37].

In a comment, Lilienfeld (1959) adds several points. He suggests that specificity does not refer to the presence or absence of an association, but to its strength or degree. He argues that the association between smoking and lung diseases is specific because these associations are strong, though there are weak associations with many diseases. Moreover, he writes:

> It is needless to point out that our interpretation of any relationship is limited by our biologic knowledge, and it may well be that an association which at present does not appear to be biologically plausible will turn out to be so when our knowledge has been extended. In fact, the finding of a biologically implausible association may be the first lead to this extension of knowledge.

A similar point is made by Sartwell (1960), though with a more critical tone. Still more critical is Rothman (1986, p. 18), who says that "... everyday experience teaches us repeatedly that single events may have many effects" and that the criterion of specificity "seems useless and misleading." Sir Austin Bradford Hill (1965, p. 297) regards specificity more favorably, but lists these and other difficulties.

If one agrees that the *number* of associations is not a particularly useful number, and instead emphasizes whether there are associations that are not plausibly effects of the treatment or exposure under study, then there is a strong link between specificity in this narrow sense and detecting hidden bias through known effects. If an observational study includes several outcomes, some of which may plausibly be affected by the treatment and others for which an actual effect is implausible, then the association between treatment and outcomes is specific in the narrow sense if strong associations are confined to outcomes for which effects are plausible. In this instance, the treatment is known not to affect certain outcomes.

5.1.4. When Is an Effect *Known*?

When is an effect known? Might we be in error in asserting that an effect is known? In discussing "specificity," Sartwell (1960) recalls the ridicule that was once attached to the idea that tiny, invisible living organisms were a cause of disease. Would anyone claim that our current knowledge is free of substantial error? There are two points.

First, as suggested at the end of §5.1.2, known effects and other attempts to detect hidden biases are not the basis for suppressing data or studies—quite the contrary. The study by Lyon et al. (1979) is a superior study because its report included information about internal inconsistencies. If we misinterpret those inconsistencies because of some error in our current knowledge, then we are far more likely to correct the error if the report includes the inconsistencies than if they are excluded. If we attempt to detect hidden bias, we report more data, not less.

Second, there is an entirely practical, albeit circular, definition of a known effect. An effect is known if, were the study to contradict it, we would have grave doubts about the study and few doubts about the effect. It is conceivable, in principle, that our understanding of time and radiation is substantially in error, as our understanding of bacterial disease was once substantially in error, so that radiation released in the 1950s had biological effects in the 1940s; however, before entertaining this possibility, we would think that perhaps the records of the Census Bureau or the Utah registry contained errors, or that some other environmental hazard was the cause.

5.2. An Outcome Known to Be Unaffected by the Treatment

5.2.1. An Additional Outcome Intended to Detect Hidden Bias

In addition to the response **R** of primary interest, the study includes another outcome, say **Y**, in an effort to detect hidden bias. In other words, **Y** was recorded in the hope of testing whether the treatment assignments **Z** given **m** are uniformly or randomly distributed over Ω, or alternatively whether there is a hidden bias.

This chapter considers two senses in which the effect of the treatment on **Y** might be known. In §5.2, **Y** is known to be unaffected by the treatment. For instance, in §5.1.3 the concern was with diseases that the treatment could not plausibly cause; see also Problem 1. In §5.3, **Y** will be known either to be unaffected or to be positively affected, that is, **Y** may be affected by the treatment, but the direction of the effect is known. In §5.2, because **Y** is unaffected by the treatment, write **y** in place of **Y**.

In an observational study, suppose a test statistic $T = t(\mathbf{Z}, \mathbf{R})$ is computed from the responses **R** of primary interest, and then T is compared to the uniform or random distribution of **Z** on Ω under the null hypothesis that $\mathbf{R} = \mathbf{r}$ is unaffected by the treatment, that is, the significance level (2.4.4) is calculated. Recall from §3.2.2 that this significance level is the basis for testing the null hypothesis of no treatment effect *if the study is free of* *hidden bias*, but the test is not generally correct if hidden bias is present. If this significance level is small, it may be evidence that the null hypothesis is untrue and the treatment does affect **R**, or it may be evidence that there is a hidden bias and the distribution of treatment assignments **Z** is not random.

The situation with **y** is different, because **y** is known to be unaffected by the treatment. From the unaffected outcome **y**, calculate a statistic $T^* = t^*(\mathbf{Z}, \mathbf{y})$ and its significance level $|\{\mathbf{z} \in \Omega: t^*(\mathbf{z}, \mathbf{y}) \geq T^*\}|/K$. If this significance level is small, then there is evidence of hidden bias, that is, evidence that **Z** is not uniformly or randomly distributed on Ω.

5.2.2. An Example: Unrelated Health Conditions in the Study of Methylmercury in Fish

Recall from §4.5.4 the study by Skerfving, Hansson, Mangs, Lindsten, and Ryman (1974) of possible chromosome damage caused by eating fish contaminated with methylmercury. The outcomes of interest were the level of mercury in the blood and the percent of cells exhibiting chromosome damage. In addition to these outcomes, Table 1 of their study described other health conditions of these 39 subjects, including other diseases such as hypertension and asthma, drugs taken regularly, diagnostic X-rays over the previous 3 years, and viral diseases such as influenza. These are outcomes since they describe the period when the exposed subjects were consuming contaminated fish. However, it is difficult to imagine that eating fish contaminated with methylmercury causes influenza or asthma, or prompts X-rays of the hip or lumbar spine. For illustration, Table 5.2.1 records the number of unrelated health conditions for each of the 39 subjects or **y** which will be assumed to be unaffected by eating contaminated fish.

If the rank sum statistic is calculated from **y**, the sum of the ranks for exposed subjects is 483.5, with average ranks used for the many ties. The null hypothesis states that the study is free of hidden bias, so the distribution of treatment assignments **Z** given **m** is random or uniform on Ω. Under the null hypothesis, the rank sum has expectation 460 and variance 932.4 allowing for the many ties. The deviate is $(483.5 - 460)/\sqrt{932.4} = 0.77$ which is far from significance at the 0.05 level. In words, in terms of health conditions methylmercury could not reasonably cause, the treated and control subjects differ no more than we would have expected had treatments been assigned at random. The hypothesis that the study is free of hidden bias is not rejected.

Several questions arise. When does such a test have a reasonable prospect of detecting hidden bias? If no evidence of hidden bias is found, does this imply reduced sensitivity to bias in the comparisons involving the outcomes of primary interest? If evidence of bias is found, what can be said about its magnitude and its impact on the primary comparisons?

Table 5.2.1. Unrelated Health Conditions and Exposure to Contaminated Fish.

Id.		Unrelated conditions
1	Control	0
2	Control	0
3	Control	0
4	Control	0
5	Control	0
6	Control	0
7	Control	0
8	Control	0
9	Control	2
10	Control	0
11	Control	0
12	Control	0
13	Control	2
14	Control	1
15	Control	4
16	Control	1
17	Exposed	0
18	Exposed	0
19	Exposed	2
20	Exposed	0
21	Exposed	0
22	Exposed	2
23	Exposed	0
24	Exposed	0
25	Exposed	1
26	Exposed	9
27	Exposed	0
28	Exposed	0
29	Exposed	1
30	Exposed	0
31	Exposed	0
32	Exposed	2
33	Exposed	0
34	Exposed	0
35	Exposed	2
36	Exposed	0
37	Exposed	6
38	Exposed	1
39	Exposed	1

5.2.3. The Power of the Test for Hidden Bias

For a particular unobserved covariate **u**, what unaffected outcome **y** would be useful in detecting hidden bias from **u**? Put another way, for a given unaffected outcome **y**, what unobserved covariate **u** can **y** hope to detect?

The power of the test for hidden bias based on $T^* = t^*(\mathbf{Z}, \mathbf{y})$ will now be studied in qualitative terms that indicate the relationship between **u** and **y** that leads to high power. Let "a" be a number such that

$$\alpha = \frac{1}{K} \sum_{\mathbf{z} \in \Omega} [t^*(\mathbf{z}, \mathbf{y}) \geq a] = \frac{|\{\mathbf{z} \in \Omega: t^*(\mathbf{z}, \mathbf{y}) \geq a\}|}{K},$$

where [event] = 1 if event occurs and [event] = 0 otherwise, so a test for hidden bias which rejects when $t^*(\mathbf{Z}, \mathbf{y}) \geq a$ has level α. Under the model (4.2.6), the power of this test is

$$\beta(\mathbf{y}, \mathbf{u}) = \sum_{\mathbf{z} \in \Omega} [t^*(\mathbf{z}, \mathbf{y}) \geq a] \frac{\exp(\gamma \mathbf{z}^T \mathbf{u})}{\sum_{\mathbf{b} \in \Omega} \exp(\gamma \mathbf{b}^T \mathbf{u})}. \tag{5.2.1}$$

All of the statistics $t^*(\mathbf{z}, \mathbf{y})$ in §2.4.2 are arrangement-increasing, meaning that the statistics are monotone-increasing as the coordinates of **z** and **y** are gradually permuted into the same order within each stratum or matched set. In other words, if in stratum s, subject i received the treatment, $z_{si} = 1$, subject j received the control $z_{sj} = 0$, but subject j had the higher value of y, that is, $y_{sj} \geq y_{si}$, then interchanging these two subjects, placing j in the treated group and i in the control group would increase—or at least not decrease—the value of the statistic $t^*(\mathbf{z}, \mathbf{y})$.

The following proposition says that the power of test $\beta(\mathbf{y}, \mathbf{u})$ increases steadily as the coordinates of **y** and **u** are permuted into the same order within each stratum or matched set. The power is greater when **y** and **u** are strongly related.

Proposition 5.1. *If $t^*(\mathbf{z}, \mathbf{y})$ is arrangement-increasing, then the power $\beta(\mathbf{y}, \mathbf{u})$ is arrangement-increasing.*

The proof follows from Proposition 4.4 in §4.6.3 which in turn is a consequence of the composition theorem for arrangement-increasing functions (Hollander, Proschan, and Sethuraman, 1977). D'Abadie and Proschan (1984, §6) describe a test with an arrangement-increasing power function as having *isotonic power*, meaning the power increases as the departure from the null hypothesis becomes more pronounced in the direction indicated by an order. Under slightly different assumptions, it is also possible to show that a test based on an arrangement-increasing statistic $T^* = t^*(\mathbf{Z}, \mathbf{y})$ is unbiased against alternatives in which **y** and **u** are positively related; see Rosenbaum (1989a).

In short, the power $\beta(\mathbf{y}, \mathbf{u})$ of a test based on $T^* = t^*(\mathbf{Z}, \mathbf{y})$ increases with the strength of the relationship between $\mathbf{y}$ and $\mathbf{u}$. This yields two qualitative conclusions. First, if concerned about a particular unobserved covariate $\mathbf{u}$, one should search for an unaffected outcome $\mathbf{y}$ that is strongly related to $\mathbf{u}$. Second, if an unaffected outcome $\mathbf{y}$ is available, a test based on $\mathbf{y}$ provides information about unobserved covariates $\mathbf{u}$ with which $\mathbf{y}$ strongly related.

Consider the use of these conclusions in a particular case. Sackett (1979) cataloged and illustrated with examples numerous sources of bias in observational studies. Included were situations in which the search for disease or the inclination to recall or record disease was more intense among subjects exposed to a hazardous agent than among subjects not exposed. The danger, of course, is that more disease will be found among exposed subjects not because the hazardous agent caused the disease, but rather because the agent stimulated intensive search for disease or distorted recall of disease. If $\mathbf{u}$ were an unobserved covariate measuring the inclination to search for or recall disease, then a test for bias due to $\mathbf{u}$ would have highest power if it were based on an unaffected outcome $\mathbf{y}$ strongly related to $\mathbf{u}$. In other words, a good test would be based on records of diseases that could not reasonably be caused by the agent under study but which would likely be detected in diagnostic activities or recalled by subjects. Exposed and control groups should differ with respect to such a $\mathbf{y}$ if there are substantial distortions due differing diagnostic efforts or recall in exposed and control groups.

5.2.4. Did the Test for Bias Reduce Sensitivity to Bias?

In §5.2.2, the difference between treated and control groups in unrelated health conditions appeared reasonably consistent with the absence of hidden bias. The question remains whether these same findings are consistent with hidden biases that would alter the conclusion that methylmercury causes chromosome damage.

In §4.5.4, the effect of mercury on chromosome aberrations was sensitive to hidden bias when $\Gamma = 3$ for a particular value of $\mathbf{u}$, say $\mathbf{u}^\circ$. This $\mathbf{u}^\circ$ contained 1's and 0's and was strongly related to the outcome $\mathbf{r}$ of primary interest, namely, chromosome aberrations. In other words, if treatment assignments were governed by the model (4.2.6) with $(\Gamma, \mathbf{u}) = (3, \mathbf{u}^\circ)$, then there would not be strong evidence that eating contaminated fish causes chromosome damage. For $\Gamma = 3$, this $\mathbf{u}^\circ$ maximized over $\mathbf{u} \in [0, 1]^N$ the significance level of the test for an effect on chromosome aberrations. With a different choice of $\mathbf{u}$, a larger value of Γ would be required to change the conclusion about chromosome aberrations. The distinction between $\Gamma = 1$ and $(\Gamma, \mathbf{u}) = (3, \mathbf{u}^\circ)$ is important because the main conclusions would be different in these two cases.

Are the observed data about unrelated health conditions $\mathbf{y}$ consistent with $(\Gamma, \mathbf{u}) = (3, \mathbf{u}^\circ)$? The unaffected outcome $\mathbf{y}$ can test the hypothesis that $(\Gamma, \mathbf{u}) = (3, \mathbf{u}^\circ)$ in much the same way that it tested the hypothesis that $\Gamma = 1$. If $(\Gamma, \mathbf{u}) = (3, \mathbf{u}^\circ)$ is not rejected, then $\mathbf{y}$ has done nothing to reduce our concern about a bias of the form $(\Gamma, \mathbf{u}) = (3, \mathbf{u}^\circ)$. If $(\Gamma, \mathbf{u}) = (3, \mathbf{u}^\circ)$ is rejected, a point of greatest sensitivity to bias has been rejected as implausible. Rejection is likely if the model with $(\Gamma, \mathbf{u}) = (3, \mathbf{u}^\circ)$ tends to produce large differences in values of y_{si} in treated and control groups, but these large differences are not seen in the data. In short, the question is: Did the test for bias using $\mathbf{y}$ reduce sensitivity to bias?

In §5.2.3, the observed value $T^* = t^*(\mathbf{Z}, \mathbf{y})$ of the test statistic was compared with its distribution when there is no hidden bias, that is, the model $\text{prob}(\mathbf{Z} = \mathbf{z}|\mathbf{m}) = 1/K$ for each $\mathbf{z} \in \Omega$, leading to the exact significance level $|\{\mathbf{z} \in \Omega: t^*(\mathbf{z}, \mathbf{y}) \geq T^*\}|/K$. In fact, any specified value of $(\Gamma, \mathbf{u}^\circ)$ may be tested using $T^* = t^*(\mathbf{Z}, \mathbf{y})$ in much the same way, yielding the one-sided significance level

$$\sum_{\mathbf{z} \in \Omega} [t^*(\mathbf{z}, \mathbf{y}) \geq T^*] \frac{\exp(\gamma \mathbf{z}^T \mathbf{u}^\circ)}{\sum_{\mathbf{b} \in \Omega} \exp(\gamma \mathbf{b}^T \mathbf{u}^\circ)}, \tag{5.2.2}$$

where, as always, $\Gamma = \exp(\gamma)$. Expression (5.2.2) is the probability that the test statistic would exceed its observed value T^* if

$$\text{prob}(\mathbf{Z} = \mathbf{z}|\mathbf{m}) = \frac{\exp(\gamma \mathbf{z}^T \mathbf{u}^\circ)}{\sum_{\mathbf{b} \in \Omega} \exp(\gamma \mathbf{b}^T \mathbf{u}^\circ)}. \tag{5.2.3}$$

For the test statistics in §2.4.2, the exact significance level (5.2.3) may be approximated in large samples using the expectation $\mu^*_{\mathbf{u}^\circ}$ and variance $\sigma^*_{\mathbf{u}^\circ}$ of T^* under the model (5.2.3), and referring the deviate $(T^* - \mu^*_{\mathbf{u}^\circ})/\sigma^*_{\mathbf{u}^\circ}$ to the standard normal distribution. The calculations of the moments of T^* are essentially the same as for T in Chapter 4, except that a different $\mathbf{u}^\circ$ is of interest. An example is given in §5.2.5.

If the test (5.2.2) rejects $(\Gamma, \mathbf{u}^\circ)$, then the test for bias based on the unaffected outcome $\mathbf{y}$ has rejected as implausible the point $(\Gamma, \mathbf{u}^\circ)$ which did most to perturb the inference about the outcome $\mathbf{R}$ of primary interest, so the sensitivity of the primary comparison is reduced. On the other hand, if $(\Gamma, \mathbf{u}^\circ)$ is not rejected, then $(\Gamma, \mathbf{u}^\circ)$ remains plausible in light of the test based on $\mathbf{y}$, so the sensitivity of the primary comparison for $\mathbf{R}$ is unchanged.

As the sample size increases, the test for hidden bias will tend to reject any value $(\Gamma^\circ, \mathbf{u}^\circ)$ for which the expectation $\mu^*_{\mathbf{u}^\circ}$ of T^* differs from the expectation $\mu^*_{\mathbf{u}}$ of T^* at the true $(\Gamma, \mathbf{u})$. Somewhat more precisely, as $N \to \infty$, if $\sigma^*_{\mathbf{u}^*}/\sqrt{N}$ tends to a constant, as is true of the statistics in §2.4.2 under mild conditions that forbid degeneracy, and if $(\mu^*_{\mathbf{u}^\circ} - \mu^*_{\mathbf{u}})/N$ tends to a nonzero constant, then the power of the test for bias based on T^* tends to one. For details, see Rosenbaum (1992).

5.2.5. An Example: Did the Test Using Unrelated Health Conditions Reduce Sensitivity?

In §4.5.4, the association between chromosome damage and eating contaminated fish became sensitive to hidden bias when $\Gamma = 3$ for a $\mathbf{u}^\circ$ containing 21 ones and and 18 zeros, where $\mathbf{u}^\circ$ was ordered in the same way as the percentages of Cu cells in Table 4.5.1. Is this pattern and magnitude of hidden bias, $(\Gamma, \mathbf{u}) = (3, \mathbf{u}^\circ)$, plausible in light of the test for hidden bias in §5.2.3?

The unrelated health conditions in Table 5.2.1 were found to be consistent with the absence of bias, $\Gamma = 1$; specifically, the rank sum was 483.5 with expectation 460, variance 932.4, and standardized deviate $(483.5 - 460)/\sqrt{932.4} = 0.77$. If instead $(\Gamma, \mathbf{u}) = (3, \mathbf{u}^\circ)$, then the rank sum statistic for unrelated health conditions is still 483.5, but its expectation is now 457.5, its variance is 874.3, yielding a deviate of $0.88 = (483.5 - 457.5)/\sqrt{874.3}$. In other words, the observed data on unrelated health conditions are consistent with the absence of hidden bias, $\Gamma = 1$, but they are also quite consistent with a hidden bias $(\Gamma, \mathbf{u}) = (3, \mathbf{u}^\circ)$ that would alter the study's conclusion about the effects of methylmercury on chromosome damage. Though the test found no evidence of hidden bias, it did not reduce the sensitivity to bias of the primary comparison, that is, it did not reject the point $(\Gamma, \mathbf{u}) = (3, \mathbf{u}^\circ)$ of maximum sensitivity.

As noted in §5.2.4, the test for hidden bias based on the unaffected outcome $\mathbf{y}$ has greatest power against an unobserved covariate $\mathbf{u}$ strongly related to $\mathbf{y}$. On the other hand, as seen in Chapter 4, the test of the null hypothesis of no treatment effect on the response $\mathbf{r}$ is most sensitive to bias from an unobserved covariate $\mathbf{u}$ strongly related to $\mathbf{r}$. If $\mathbf{y}$ has only a weak relationship with $\mathbf{r}$, as is true of the relationship between unrelated health conditions and the percent of Cu cells, then the test for bias based on $\mathbf{y}$ will have little power against values of $\mathbf{u}$ that matter most for $\mathbf{r}$. For reducing sensitivity to hidden bias, the ideal $\mathbf{y}$ would be unaffected by the treatment but otherwise would be strongly related to the response $\mathbf{r}$. This ideal $\mathbf{y}$ bears some resemblance to a useful type of covariate, namely, a "pretest score," which is a measure of outcome of interest recorded prior to treatment, so a "pretest" is unaffected by the treatment but is often highly correlated with the outcome.

5.2.6. Should Adjustments Be Made for Unaffected Outcomes?

Should an unaffected outcome be viewed as another covariate? Should adjustments be made for unaffected outcomes in just the same way that adjustments are made for covariates?

The goal is compare subjects who were comparable *prior* to treatment. An outcome is, by definition, measured *after* treatment. Adjustments for unaffected outcomes render people comparable prior to treatment only under special and restrictive circumstances, that is, under assumptions that may be wrong and are often difficult to justify. This is discussed formally and in detail in Rosenbaum (1984b).

A bias visible in an unaffected outcome may only be the faint trace of a much larger imbalance in an unobserved covariate. Indeed, there is a sense in which, if an imbalance in an unaffected outcome is produced by an imbalance in an unobserved covariate, then the covariate imbalance is always at least as large as the outcome imbalance. See Rosenbaum (1989b, §4.3) for formal discussion. Removing the outcome imbalance may do little to remove the imbalance in the unobserved covariate.

5.3. An Outcome for Which the Direction of the Effect Is Known

5.3.1. Using a Nonnegative Effect to Test for Hidden Bias

In many cases, a treatment can have no effect or can have the effect of increasing a particular outcome **Y**, but it cannot plausibly decrease it. In §5.1.2, radiation from nuclear fallout might have no effect on the frequency of childhood cancers other than leukemia, or it might increase them, but it is hard to imagine that fallout is preventing cancers. In Problem 2, occupational exposures to benzene might have no effect on total mortality or it might increase mortality, but it is hard to imagine that benzene exposures prevent death.

In the terminology of §2.5, if the treatment has an effect on **Y**, then **Y** takes the value $\mathbf{y_z}$ if $\mathbf{Z} = \mathbf{z}$ where each $\mathbf{y_z}$ is fixed. There is no interference between units if $y_{zsi} = y_{Tsi}$ whenever $z_{si} = 1$ and $y_{zsi} = y_{Csi}$ whenever $z_{si} = 0$, and there is a nonnegative effect if $y_{Tsi} \geq y_{Csi}$ for all (s, i). Let $T^* = t^*(\mathbf{Z}, \mathbf{Y})$ be an order-preserving statistic, for instance, any of the statistics in §2.4.2.

If the treatment has a nonnegative effect on **Y**, then the statistic T^* will tend to be larger than if the treatment had no effect on **Y**. Compare T^* to its distribution in the absence of hidden bias under the null hypothesis of no effect on **Y**, that is, proceed exactly as in §5.2.1. If T^* is large, falling in the extreme upper tail of this distribution, then two explanations are possible, namely, a positive effect on **Y** or hidden bias. However, if T^* is small, falling in the extreme lower tail of this distribution, then this cannot be due to a nonnegative effect on **Y**, so there is evidence of hidden bias. More precisely, if T^* is small, calculate the tail probability of a value as smaller or smaller than the observed T^* assuming the absence of hidden bias and no effect of the treatment on **Y**. Report this tail probability as the significance level for testing the hypothesis that hidden biases are absent. These highly intuitive ideas are discussed formally in the Appendix, §5.4.

This logic was used in §5.1.2 in connection with childhood cancers other than leukemia in high- and low-exposure counties in Utah. Under the premise that higher levels of radiation from fallout might cause but would not prevent cancers, the finding that high-exposure counties had lower frequencies of other childhood cancers was taken to indicate the presence of hidden bias.

5.3.2. Did the Test Reduce Sensitivity to Bias?

In §5.3.1, an outcome $\mathbf{Y}$ with a nonnegative effect was used to test the hypothesis that there is no hidden bias, that is, $\Gamma = 1$. To use $\mathbf{Y}$ to test the hypothesis H_0: $(\Gamma, \mathbf{u}) = (\Gamma^\circ, \mathbf{u}^\circ)$ with $(\Gamma^\circ, \mathbf{u}^\circ)$ specified, proceed exactly as in §5.2.4, comparing T^* to its distribution determined from (5.3) under the hypothesis of no effect on $\mathbf{Y}$. As in §5.3.1, if T^* is large compared to this distribution, there are two explanations, a positive effect on $\mathbf{Y}$ or an incorrect value of $(\Gamma^\circ, \mathbf{u}^\circ)$. However, if T^* is small compared to this distribution, it cannot be due to a nonnegative effect on $\mathbf{Y}$, so there is evidence against the hypothesized value $(\Gamma^\circ, \mathbf{u}^\circ)$. These considerations are stated formally in the Appendix, §5.4.

As in §5.2.4, the typical use of this procedure is to test whether T^* has rendered implausible the point of greatest sensitivity $(\Gamma^\circ, \mathbf{u}^\circ)$ for the outcome $\mathbf{R}$ of primary interest.

*5.4. Appendix: The Behavior of T^* with Nonnegative Effects

In §5.3, the significance level for the simple hypothesis of no effect on $\mathbf{Y}$ was used to test the composite hypothesis of a nonnegative effect. This section looks at the relevant technical details. Let $T^* = t^*(\mathbf{Z}, \mathbf{Y})$. Under the model (4.2.6), the chance that $T^* \geq c$ is

$$\alpha_c = \sum_{\mathbf{z} \in \Omega} [t^*(\mathbf{z}, \mathbf{y}_\mathbf{z}) \geq c] \frac{\exp(\gamma \mathbf{z}^T \mathbf{u})}{\sum_{\mathbf{b} \in \Omega} \exp(\gamma \mathbf{b}^T \mathbf{u})}. \tag{5.4.1}$$

Let $\mathbf{a}$ be any one fixed treatment assignment, $\mathbf{a} \in \Omega$. Suppose we observed $\mathbf{Z} = \mathbf{a}$ and, consequently, $\mathbf{Y} = \mathbf{y}_\mathbf{a}$, and $T^* = t^*(\mathbf{a}, \mathbf{y}_\mathbf{a})$. The hypothesis that the treatment had no effect on $\mathbf{Y}$ would say that $\mathbf{Y} = \mathbf{y}_\mathbf{a}$ no matter how treatments $\mathbf{Z}$ were assigned. The significance level, say $\alpha_{c,\mathbf{a}}$, for testing this hypothesis would be

$$\alpha_{c,\mathbf{a}} = \sum_{\mathbf{z} \in \Omega} [t^*(\mathbf{z}, \mathbf{y}_\mathbf{a}) \geq c] \frac{\exp(\gamma \mathbf{z}^T \mathbf{u})}{\sum_{\mathbf{b} \in \Omega} \exp(\gamma \mathbf{b}^T \mathbf{u})}. \tag{5.4.2}$$

* This section may be skipped without loss of continuity.

The following proposition says that when the treatment has a nonnegative effect, the statistic T^* is stochastically larger than all of its permutation distributions under the null hypothesis of no effect. In particular, when the treatment has a nonnegative effect, there is at most a 5% chance that T^* will fall in the lower 5% tail of its permutation distribution under the hypothesis of no effect. The proof is the same as the proof of Proposition 2.3 in §2.9.

Proposition 5.2. *If $t^*(\cdot, \cdot)$ is order-preserving and the treatment has a nonnegative effect on* $\mathbf{Y}$, *then* $\alpha_c \geq \alpha_{c,\mathbf{a}}$ for all c and all $\mathbf{a} \in \Omega$.

If $\gamma = 0$, then

$$\alpha_c = \frac{|\{\mathbf{z} \in \Omega: t^*(\mathbf{z}, \mathbf{y}_{\mathbf{z}}) \geq c\}|}{K}, \qquad \alpha_{c,\mathbf{a}} = \frac{|\{\mathbf{z} \in \Omega: t^*(\mathbf{z}, \mathbf{y}_{\mathbf{a}}) \geq c\}|}{K}, \qquad (5.4.3)$$

and $\alpha_{c,\mathbf{a}}$ is the usual significance level for a randomization test.

5.5. Bibliographic Notes

Campbell (1969) and Campbell and Stanley (1963) discuss the strengths and weaknesses of various study designs, including the design in §5.1.2. Disease specificity in §5.1.3 has been widely discussed; see, for example, Yerushalmy, and Palmer (1959), Lilienfeld (1959), Sartwell (1960), Hill (1965), Susser (1973), and Rothman (1986) to sample a range of opinion. The use of known effects to test for hidden bias is discussed in Rosenbaum (1984a, 1989a,b). The link in §5.2.5 between a test for bias using one outcome and a sensitivity analysis for another is discussed in Rosenbaum (1992).

5.6. Problems

5.6.1. Problems: Interpreting Studies with Known Effects

1. Petitti, Perlman, and Sidney (1986) commented on the contradictory results of several studies concerning the relationship between postmenopausal estrogen use and heart disease. After adjusting for several covariates, they found that the risk of death from cardiovascular disease for users of estrogen was half the risk of nonusers; however, they found the same difference for deaths from accidents, homicides, and suicide. They write:

    > There is no biologically plausible reason for a protective effect of postmenopausal estrogen use on mortality from accidents, homicide, and suicide. We believe that our results are best explained by the assumption that postmenopausal estrogen users in this cohort are healthier than those

who had no postmenopausal estrogen use, in ways that have not been quantified and cannot be adjusted for. The selection of healthier women for estrogen use in this population is not necessarily a characteristic shared by other populations.

Do you agree?
Consider also Kreiger, Kelsey, Holford, and O'Connor (1982).

2. Infante, Rinsky, Wagoner, and Young (1977) found a statistically significant excess of leukemia deaths among benzene workers when compared to the general population, but a statistically significant deficit of deaths overall. If one were willing to assume that working with benzene might cause disease or death but would not prevent them, then what would you conclude about the general health of benzene workers compared to the population as a whole? Could this explain the excess of leukemia deaths? Why or why not?

5.6.2. Problems: The Magnitude of the Bias Suggested by a Test for Bias

3. What is the smallest value of Γ such that the expected value of T^* in §5.2.2 would equal the observed value of 483.5 for some $\mathbf{u} \in U$? (*Hint*: Use the methods of Chapter 4. *Answer*: $\Gamma = 1.73$ and $\mathbf{u}$ contains 15 ones and 24 zeros ordered in the same way as $\mathbf{y}$ in Table 5.2.1, that is, $u_i = 0$ when $y_i = 0$ and $u_i = 1$ when $y_i > 0$.)
4. How is the calculation in Problem 3 above related to the range of Hodges–Lehmann estimates of effect that one could calculate for $\mathbf{y}$ using the procedure in §4.5.6? (*Hint*: What is the smallest Γ such that the interval of Hodges–Lehmann estimates includes zero?)
5. Why is there no largest value of Γ such that the expected value of T^* in §5.2.2 equals the observed value of 483.5 for some $\mathbf{u} \in U$? (*Hint*: Review §4.6.5.)

References

Beck, H.L. and Krey, P.W. (1983). Radiation exposures in Utah from Nevada nuclear tests. *Science*, **220**, 18–24.
Campbell, D. (1969). Reforms as experiments. *American Psychologist*, **24**, 409–429.
Campbell, D. and Stanley, J. (1963). *Experimental and Quasi-Experimental Designs for Research.* Chicago: Rand McNally.
D'Abadie, C. and Proschan, F. (1984). Stochastic versions of rearrangement inequalities. In *Inequalities in Statistics and Probability* (ed. Y.L. Tong). Hayward, CA: Institute of Mathematical Statistics, pp. 4–12.
Gastwirth, J. (1988). *Statistical Reasoning in Law and Public Policy.* New York: Academic Press.
Hill, A.B. (1965). The environment and disease: Association or causation? *Proceedings of the Royal Society of Medicine*, **58**, 295–300.
Hollander, M., Proschan, F., and Sethuraman, J. (1977). Functions decreasing in transposition and their applications in ranking problems. *Annals of Statistics*, **5**, 722–733.

Infante, P., Rinsky, R., Wagoner, J., and Young, R. (1977). Leukemia in benzene workers. *Lancet*, July 9, 1977, 76–78.

Kreiger, N., Kelsey, J., Holford, T., and O'Connor, T. (1982). An epidemiologic study of hip fracture in postmenopausal women. *American Journal of Epidemiology*, **116**, 141–148.

Land, C.E. (1979). The hazards of fallout or of epidemiologic research. *New England Journal of Medicine*, **300**, 431–432.

Lilienfeld, A. (1959). On the methodology of investigations of etiologic factors in chronic diseases—Some comments. *Journal of Chronic Diseases*, **10**, 41–46.

Lyon, J.L., Klauber, M.R., Gardner, J.W., and Udall, K.S. (1979). Childhood leukemias associated with fallout from nuclear testing. *New England Journal of Medicine*, **300**, 397–402.

Petitti, D., Perlman, J., and Sidney, S. (1986). Postmenopausal estrogen use and heart disease. *New England Journal of Medicine*, **315**, 131–132.

Rosenbaum, P.R. (1984a). From association to causation in observational studies. *Journal of the American Statistical Association*, **79**, 41–48.

Rosenbaum, P.R. (1984b). The consequences of adjustment for a concomitant variable that has been affected by the treatment. *Journal of the Royal Statistical Society*, Series A, **147**, 656–666.

Rosenbaum, P.R. (1989a). On permutation tests for hidden biases in observational studies: An application of Holley's inequality to the Savage lattice. *Annals of Statistics*, **17**, 643–653.

Rosenbaum, P.R. (1989b). The role of known effects in observational studies. *Biometrics*, **45**, 557–569.

Rosenbaum, P.R. (1992). Detecting bias with confidence in observational studies. *Biometrika*, **79**, 367–374.

Rothman, K. (1986) *Modern Epidemiology*. Boston: Little, Brown.

Sackett, D. (1979). Bias in analytic research. *Journal of Chronic Diseases*, **32**, 51–68.

Sartwell, P. (1960). On the methodology of investigations of etiologic factors in chronic diseases: Further comments. *Journal of Chronic Diseases*, **11**, 61–63.

Skerfving, S., Hansson, K., Mangs, C., Lindsten, J., and Ryman, N. (1974). Methylmercury-induced chromosome damage in man. *Environmental Research*, **7**, 83–98.

Susser, M. (1973). *Causal Thinking in the Health Sciences: Concepts and Strategies in Epidemiology*. New York: Oxford University Press.

Yerushalmy, J., and Palmer, C. (1959). On the methodology of investigations of etiologic factors in chronic diseases. *Journal of Chronic Diseases*, **10**, 27–40.

CHAPTER 6

Multiple Reference Groups in Case-Referent Studies

6.1. Multiple Reference Groups

6.1.1. What Are Multiple Reference Groups?

A case-referent study compares the frequency or intensity of exposure to the treatment among cases and among referents or noncases who are free of the disease; see §3.3. If referents or noncases are selected from several different sources, then the study has several distinct referent groups.

Case-referent studies with several referent groups are common. A few examples follow. Notice, in particular, the types of referent groups used. Gutenshohn, Li, Johnson, and Cole (1975) compared the frequency of tonsillectomy among cases of Hodgkin's disease to two groups of referents, namely, spouses of cases and siblings of cases. The Collaborative Group for the Study of Stroke in Young Women (1973) compared the use of oral contraceptives by cases of stroke among young women aged 15 to 44 with the use of oral contraceptives among two referent groups matched for age and race, namely, women discharged from the same hospital for some other disease and women who were neighbors of the cases. In a study of subacute sclerosing panencephalitis in young children, Halsey, Modlin, Jabbour, Dubey, Eddins, and Ludwig (1980) used playmates of cases as one referent group and children admitted to the same hospital as another. Kreiger, Kelsey, Holford, and O'Connor (1982) compared estrogen use among female cases of hip fracture with estrogen use in two referent groups, namely, trauma and nontrauma referents, that is, other women from the same hospitals who were admitted for traumas such as fractures or sprains other than hip fractures and women who were admitted for disorders that did not involve a physical trauma.

The use of multiple referent groups in case-referent studies is different from the use of multiple control groups. The case group and several referent groups are each exposed to the treatment and each exposed to the control in varying degrees. In contrast, a study with multiple control groups has several groups of subjects who did not receive the treatment. A case-referent study may have a single referent group and several control groups, and one will be discussed in §7.1.2. In principle, a case-referent study could have both multiple referent groups and multiple control groups. Notice that the term "control group" is sometimes used loosely to describe any group used for comparison, but in this book a control group refers specifically to subjects who did not receive the treatment.

6.1.2. An Example: A Study of Reye's Syndrome with Four Referent Groups

Aspirin bottles contain the following warning: "Children and teenagers should not use this medicine for chicken pox or flu symptoms before a doctor is consulted about Reye syndrome, a rare but serious illness reported to be associated with aspirin." Reye's syndrome typically occurs in children under the age of 15 following an upper respiratory tract infection, and it often leads to death from cerebral damage. Hurwitz (1989) surveys the evidence linking Reye's syndrome and aspirin.

Hurwitz, Barrett, Bregman et al. (1985) compared the use of aspirin by cases of Reye's syndrome with its use in four referent groups, namely referents from the case's emergency room, school and community, and inpatients at the case's hospital. This study was not the first to link aspirin with Reye's syndrome, but from a methodological view, it is the most interesting. Earlier studies had convinced the US Surgeon General and the American Academy of Pediatrics to advise against giving children aspirin for chicken pox or flu, but these studies were nonetheless criticized. The US Public Health Service study by Hurwitz, Barrett, Bregman et al. (1985) was an attempt to address the criticisms.

In their study, all subjects, both cases and referents, had to have had a "respiratory, gastrointestinal, or chicken pox antecedent illness within 3 weeks before hospitalization." In other words, the population being described consists of children who have had one of these illness within 3 weeks. The study matched cases to referents on the basis of age, race, date of illness, and the specific type of antecedent illness. In addition, individual referent groups were matched in other ways; specifically, emergency room and inpatient referents came from the same hospital as the case, school referents came from the same school or day care center, and community referents were selected by random digit telephone dialing from the same community as the case. Table 1 in their article shows that cases and referents were similar in terms of age, race, gender, and antecedent illness. (The comparisons that follow are based on the published data and ignore the matching, but

Table 6.1.1. Counts of Aspirin Use by Cases and Referents.

	Aspirin	No aspirin	% Using aspirin
Cases	28	2	93%
Emergency room	7	18	28%
Inpatient	5	17	23%
School	24	17	59%
Community	30	27	53%

for comparisons done both ways, the matched and unmatched analyses agree with each other; compare Tables 6.1.2 and 6.1.3 below and the results reported in the first paragraph of page 852 in Hurwitz, Barrett, Bregman et al. 1985.)

Table 6.1.1 gives the frequencies of aspirin use by case and referent groups in Hurwitz, Barrett, Bregman et al. (1985). Of the 30 cases, 28 had used aspirin. In each referent group, the relative frequency of aspirin use was much lower.

Tables 6.1.2 and 6.1.3 are calculated from Table 6.1.1. Table 6.1.2 gives the odds ratio comparing aspirin use in each pair of groups. For instance, the odds of using aspirin were 9.92 times higher in the case group than in the school control group. Table 6.1.3 gives the usual chi-square statistic with continuity correction for a 2 × 2 table for each pair of groups. Each chi-square has one degree of freedom, and is significant at the 0.05 level or 0.01 level if greater than 3.84 or 6.63, respectively.

Table 6.1.2. Odds Ratios for Aspirin Use.

	School	Community	Emergency room	Inpatient
Cases	9.9	12.6	36.0	47.6
School		1.3	3.6	4.8
Community			2.9	3.8
Emergency room				1.3

Table 6.1.3. Chi-Squares for Aspirin Use.

	School	Community	Emergency room	Inpatient
Cases	9.0	12.9	22.4	24.3
School		0.1	4.6	6.0
Community			3.3	4.6
Emergency room				0.0

Two patterns are apparent in Tables 6.1.2 and 6.1.3. First, cases used aspirin substantially and significantly more than did all four referent groups. If the study were free of biases, Cornfield's (1951) result in §3.3.2 would suggest that aspirin use increases the risk of Reye's syndrome by 9.9 to 47.6 times. The second pattern concerns differences between the referent groups. School and community referents have similar frequencies of aspirin use, the odds ratio being 1.3. Emergency room and inpatient referents have similar frequencies of aspirin use, again with an odds ratio of 1.3. However, school and community referents are three to five times more likely to use aspirin than emergency room or inpatient referents, and three of these four comparisons are significant at the 0.05 level.

The differences in aspirin use among the referent groups are inconsistent with a study that is free of selection and hidden biases. If there were no bias, the frequency of aspirin use in each referent group should differ only by chance from the frequency of aspirin use among all children without Reye's syndrome in the relevant population of children having recently had one of the antecedent illnesses. However, in contrast with the study of nuclear fallout and childhood leukemia in §5.1.2, the differences among the referent groups, while statistically significant, are much smaller than each of the four differences between the cases and each referent group.

6.1.3. Selection Bias and Hidden Bias in Case-Referent Studies

Recall from §3.3 that a synthetic case-referent study begins with the population of M subjects in §3.2.1 and then draws a random sample without replacement of subjects who are cases and a random sample of subjects who are not cases, possibly after stratification on the observed covariates $\mathbf{x}$. In some studies, the sample of cases includes all of the cases in the population; this is a trivial sort of random sampling, but it is random sampling nonetheless. In such a synthetic case-referent study, the reference group is the sample of subjects who are not cases. Synthetic case-referent studies are sometimes conducted and they form a simple model for the situation in which the process of selecting cases and referents neither introduces additional biases nor reduces bias. In a synthetic case-referent study free of hidden bias, §3.3 showed that the odds ratio estimates the population odds ratio and the usual Mantel–Haenszel test appropriately tests the null hypothesis of no effect of the treatment or exposure. Section §4.4 considered the test of no effect and its sensitivity to hidden bias in a synthetic case-referent study in which there is no selection bias.

Most case-referent studies are not synthetic case-referent studies. In particular, a study with several different referent groups is, by definition, not a synthetic case-referent study, since the several reference groups are, by definition, not random samples of subjects who are not cases. The purpose of this section is to discuss case-referent studies in which the reference group is not a random sample or stratified random sample of subjects who are not cases. This section summarizes findings based on the considerations in the Appendix, §6.3, which should be consulted for formal proof.

Hidden bias may be introduced when treatments or exposures are assigned to the M subjects in the population. Selection bias may be introduced when cases and referents are selected into the study from the cases and referents in the population. The relationship between selection and hidden bias can be subtle. A careful choice of referent group may remove a hidden bias that would have been present had the entire population been available for study. A poor choice of referent group may introduce a selection bias into a study that was free of hidden bias, that is, a bias that would not have been present in the entire population.

In the discussion that follows, the model for hidden bias, (4.2.1) or (4.2.2) or (4.2.6), is assumed to hold in the population of M subjects before cases and referents are selected. The intent is to test the null hypothesis of no treatment effect, so the null hypothesis is tentatively assumed to hold for the purpose of testing it. The null hypothesis of no treatment effect says that whether or not a subject is a case is unaffected by the treatment or exposure. More precisely, if $\mathbf{r}$ is N-dimension vector with $r_{si} = 1$ if the ith subject in stratum s is a case and $r_{si} = 0$ if this subject is not a case, then the null hypothesis H_0 says that $\mathbf{r}$ would be unaltered if the pattern of exposures $\mathbf{Z}$ were changed.

The null hypothesis H_0 refers to $\mathbf{r}$, that is, to whether or not a subject is a case. In §6.1.2, $\mathbf{r}$ distinguishes cases of Reye's syndrome from children without Reye's syndrome. In many case-referent studies, other variables besides $\mathbf{r}$ are used in selecting referents, and some of these may be outcomes in the sense that they describe subjects after exposure to the treatment. This turns out to be an important consideration in distinguishing hidden bias and selection bias. As has been true throughout Chapter 4 and Chapter 5, the term hidden bias refers to an unobserved covariate $\mathbf{u}$, that is, a variable describing subjects prior to treatment so that $\mathbf{u}$ is unaffected by the treatment.

Under H_0, if subjects are selected based on quantities that are unaffected by the treatment, such as the response $\mathbf{r}$, the observed covariate $\mathbf{x}$, the unobserved covariate $\mathbf{u}$, or some other unaffected outcome $\mathbf{y}$, then the form of the model (4.2.6) and the value of γ are unchanged. In this situation, if there is no hidden bias in the sense that $\gamma = 0$, then conventional methods such as the Mantel–Haenszel test may be used to test H_0, and if hidden bias is possible in the sense that γ might not equal zero, then the methods in Chapter 4 may be used to appraise sensitivity to bias. In this specific sense, if cases and referents are selected based on quantities that are unaffected by the treatment, there is no selection bias.

Consider instead an outcome that is affected by the treatment, say $\mathbf{Y} = \mathbf{y}_{\mathbf{Z}}$, where as in §5.3 the value of this outcome changes if the treatment assignment **Z** changes. For example, see Problem 1. If subjects are selected using the affected outcome $\mathbf{Y} = \mathbf{y}_{\mathbf{Z}}$, then the selection may introduce a bias where none existed previously, and it may distort the distribution of **Z**, so the model (4.2.6) can no longer be used to study sensitivity to hidden bias. For instance, if the treatment had a positive effect on $\mathbf{Y} = \mathbf{y}_{\mathbf{Z}}$, then selecting referents with high values of Y would produce a referent group in which too many referents had been exposed to the treatment. Under H_0, membership in the referent group should not change with the treatment or exposure. In the hypothetical study in §3.3.3 of smoking and lung cancer, a referent group of cardiac patients would lead to selection bias because smoking causes cardiac disease. Selecting cardiac patients as referents means selecting subjects who have higher frequencies of smoking, more exposure to the treatment, because exposure to the treatment caused some of their cardiac disease.

Although selecting subjects based on quantities unaffected by the treatment does not introduce selection bias in the test of H_0, it may alter the quantity or pattern of hidden bias. That is, even if selection is based on fixed quantities unaffected by the treatment, the selected subjects may differ systematically from the unselected subjects in their values of the unobserved covariate u_{si}. Indeed, an investigator who uses siblings or neighbors as referents is, presumably, trying to reduce hidden bias without introducing selection bias, that is, to produce matched sets that are more homogeneous than the population in terms of certain covariates u that are not explicitly measured, without distorting the frequency of exposure to the treatment. Though often a sensible strategy, an attempt to reduce hidden bias by the choice of referent can at times backfire, increasing rather than reducing the amount of hidden bias, and perhaps this happened with the inpatient referents in §6.1.2. Let u be a binary variable indicating whether or not the adult caring for a child is aware of the advice against aspirin use by the Surgeon General and the American Academy of Pediatrics. It is conceivable that a case tends to differ more from an inpatient referent in term of u than a case would typically differ from a referent selected at random from children who are not cases.

To summarize: Selecting referents based on characteristics unaffected by the treatment does not introduce selection bias in the test of H_0, but it may decrease or increase the quantity of hidden bias. Selecting referents based on characteristics affected by the treatment can introduce selection biases that distort the distribution of treatments or exposures **Z**, thereby invalidating both:

(i) tests that would have been appropriate in the absence of hidden bias; and
(ii) sensitivity analyses that would have been appropriate in the presence of hidden bias.

Again, formal proofs are given in the Appendix, §6.3.

6.2. Matched Studies with Two Referent Groups

6.2.1. An Example: Breast Cancer and Age at First Birth

Table 6.2.1 describes data collected by Lilienfeld, Chang, Thomas and Levin (1976) and reported by Liang and Stewart (1987). The cases in this study were women in Baltimore who were diagnosed with primary malignant breast cancer. Each case was matched to one or two female referents on the basis of age and race. A hospital referent had been admitted to the same hospital as the case. A neighbor referent came from the same neighborhood as the case. There were 409 cases, of whom 195 were matched with both a hospital and a neighbor referent, 164 were matched only to a hospital referent, and 50 were matched only to a neighbor referent. The exposure of interest here is the age of first birth, ≤ 24 or ≥ 25, where women who had never given birth were classified as ≥ 25.

Each count in Table 6.2.1 is a matched set, the total count being 409 matched sets, rather than $(3 \times 195) + (2 \times 164) + (2 \times 50) = 1013$ women. For instance, there are 47 matched sets in which the case and both referents first gave birth at an age ≤ 24. This format is the correct format for presenting matched comparisons because it indicates who is matched to whom, information that is necessary for an appropriate analysis.

For a moment, ignore the distinction between hospital and neighbor referents and compare the age at first birth of cases and referents using the Mantel–Haenszel statistic. There are $S = 409$ strata defined by the matched pairs and matched triples. There are a total of 209 cases who first gave birth at an age of 25 or greater, where 172.3 were expected in the absence of hidden bias, leading to a significance level < 0.0001. If there is no hidden bias, there is strong evidence of a higher risk of breast cancer among women 25 or older at first birth.

Table 6.2.1. Age at First Birth for Cases of Breast Cancer and Matched Referents.

	Hospital referent	Neighbor referent	Case ≤ 24	Case ≥ 25
Referents	≤ 24	≤ 24	47	32
		≥ 25	17	31
	≥ 25	≤ 24	18	21
		≥ 25	8	21
	≤ 24	None	58	53
	≥ 25	None	24	29
	None	≤ 24	17	12
	None	≥ 25	11	10

A first step is to test for hidden bias by comparing the frequency of older first births in the two referent groups. McNemar's test is used to compare the age at first birth in the two referent groups for the 195 matched triples. In the absence of hidden bias, matched hospital and neighbor referents should have similar frequencies of first births at ≤ 24 years of age. There are $87 = 17 + 31 + 18 + 21$ discordant pairs, that is, hospital/neighbor pairs in which one woman first gave birth at an age ≤ 24 and the other at an age ≥ 25. In $48 = 17 + 31$ of these discordant pairs, the hospital referent first gave birth at an age ≤ 24 and in the remaining 39 pairs the neighbor referent first gave birth at an age ≤ 24. McNemar's test compares this 48 versus 39 split of the 87 pairs with a binomial distribution with 87 independent trials each having probability 0.5 of success, finding little or no evidence that the frequency of first births at ages ≤ 24 differs in the two referent groups.

In parallel with the discussion in §5.2.3, it may be shown that the above test comparing the two referent groups has the following properties. First, in testing for hidden bias, the test has the correct level; that is, if $\gamma = 0$, then 5% of the time the test will falsely reject the hypothesis that $\gamma = 0$ with a significance level of 0.05 or less. Second, for $\gamma > 0$, the power of the test increases steadily as the values of u are rearranged to make the two referent groups more dissimilar; that is, the power function of the appropriate one-sided test is arrangement-increasing in the pair of vectors containing the u's and the binary indicators distinguishing hospital and neighbor referents. The test can hope to distinguish $\gamma = 0$ from $\gamma > 0$ if the referent groups are quite different in their typical values of u.

6.2.2. Partial Comparability of Cases and Multiple Referent Groups

All of the case-referent studies discussed in this chapter have the following features in common. A new (or incident) case of disease is made available to the investigator. This case is then matched to one or more referents who are similar in certain ways, for instance, a new patient at the same hospital of the same age and race. Then the case is matched to one or more additional referents who are similar to the case in different ways, for instance, a neighbor of the same age and race. This strategy will be called *partial comparability*. Each referent group is comparable to the case in some ways but not in others. In Gutenshohn, Li, Johnson, and Cole (1975), the cases and their spouses have similar home environments and diets as adults, while cases and siblings have similar childhood environments and diets and have some genetic similarities.

Consider testing the null hypothesis of no treatment or exposure effect, H_0, by comparing cases with two referent groups, the situation with more than two groups being similar. Matched set s contains $n_s \geq 2$ subjects, the

Table 6.2.2. Notation for a Matched Set with Two Referent Groups.

Group size	Case 1	Referent group 1 I_s	Referent group 2 J_s
Subscripts	$(s, 1)$	$(s, 2), (s, 3), \ldots, (s, I_s + 1)$	$(s, I_s + 2), (s, I_s + 3), \ldots, (s, n_s)$

first subject is the case, so $r_{s1} = 1$, the next $I_s \geq 0$ subjects are referents from the first group, and the last $J_s \geq 0$ subjects are referents from the second group, so $n_s = 1 + I_s + J_s$ and $r_{si} = 0$ for $i = 2, \ldots, n_s$. This is summarized in Table 6.2.2 for matched set s. In the example in §6.2.1, $n_s = 3$, $I_s = 1$, $J_s = 1$, for 195 matched sets s; $n_s = 2$, $I_s = 1$, $J_s = 0$, for 164 more matched sets, and $n_s = 2$, $I_s = 0$, $J_s = 1$, for 50 matched sets.

Under H_0 and in the absence of selection bias, a subject does not change from one referent group to another based on exposure to the treatment Z_{si}, nor does exposure to the treatment cause cases of disease, so r_{si} and the order of subjects is fixed under H_0 not varying with $\mathbf{Z}$.

To express partial comparability, write the unobserved covariate u_{si} as a weighted sum of three unobserved covariates

$$u_{si} = \psi_0 v_{0si} + \psi_1 v_{1si} + \psi_2 v_{2si},$$

where

$$\psi_k \geq 0, \qquad 1 = \sum \psi_k, \qquad 0 \leq v_{ksi} \leq 1 \quad \text{for all } k, s, i. \tag{6.2.1}$$

The first thing to notice about (6.2.1) is that it is not a new assumption, but rather a re-expression of the old assumption in §4.2 that $\mathbf{u} \in U$. Any $\mathbf{u} \in U$ may be written as $\mathbf{u} = \psi_0 \mathbf{v}_0 + \psi_1 \mathbf{v}_1 + \psi_2 \mathbf{v}_2$ for some $\mathbf{v}_k \in U$ for $k = 1, 2, 3$, and some $\psi_k \geq 0$, $1 = \sum \psi_k$, and any $\mathbf{u}$ written in this way is an element of U, so (6.2.1) is the same as the assumption that $\mathbf{u} \in U$. So far, (6.2.1) is nothing new.

To express partial comparability, require

$$v_{1si} = v_{1s1} \quad \text{for } i = 2, \ldots, I_s + 1,$$

and

$$v_{2si} = v_{2s1} \quad \text{for } i = I_s + 2, \ldots, n_s, \tag{6.2.2}$$

which says that the referents in group 1 have the same value of v_{1si} as the case and the referents in group 2 have the same value of v_{2si} as the case, though v_{0si} may differ for cases and referents. The first thing to note is that, taken together, (6.2.1) and (6.2.2) are not a new assumption, just a re-expression of the assumption that $\mathbf{u} \in U$. Once again, any $\mathbf{u} \in U$ may be written to satisfy (6.2.1) and (6.2.2) by taking $\psi_0 = 1$, $\psi_1 = 0$, $\psi_2 = 0$, and $\mathbf{v}_0 = \mathbf{u}$, and any $\mathbf{u}$ that satisfies (6.2.1) and (6.2.2) is an element of U. So far, nothing new has been assumed.

The sensitivity analysis considers a range of values of the parameter $(\gamma, \psi_0, \psi_1, \psi_2)$ under the model (4.2.6). Taking $(\gamma, \psi_0, \psi_1, \psi_2) = (\gamma, 1, 0, 0)$ gives exactly the sensitivity analysis in Chapter 4, because in this case (6.2.2) does not restrict $\mathbf{u}$ in any way and $u_{si} = v_{0si}$. In other words, $(\gamma, \psi_0, \psi_1, \psi_2) = (\gamma, 1, 0, 0)$ signifies that partial comparability did nothing to make either type of referent similar to the case in terms of the unobserved covariate u. If instead $(\gamma, \psi_0, \psi_1, \psi_2) = (\gamma, 0, 1, 0)$ then $u_{si} = v_{1si}$ and, using (6.2.2), the case and the first referent group have identical values of u, so there is no hidden bias using the first reference group. In the example of §6.2.1, this would mean that cases and matched hospital referents are comparable in terms of u. In the same way, $(\gamma, \psi_0, \psi_1, \psi_2) = (\gamma, 0, 0, 1)$ signifies that cases and the second referent group have identical values of u, e.g., that cases and neighbors have identical values of u. If $(\gamma, \psi_0, \psi_1, \psi_2) = (\gamma, \frac{1}{3}, \frac{1}{3}, \frac{1}{3})$ then the bias is split equally between the uncontrolled v_0 and the partially controlled v_1 and v_2, so partial comparability has reduced but not eliminated hidden bias for both groups.

In other words, the sensitivity considers several possibilities, including the possibility that partial comparability did nothing to reduce hidden bias, that it eliminated bias for one referent group but did nothing for the other, or that is reduced but did not eliminate bias for one or both groups.

6.2.3. Sensitivity Analysis with Two Matched Referent Groups

The procedure for sensitivity analysis with two reference groups is similar to that in §4.4.4 and §4.4.5 for case-referent studies; however, the sensitivity bounds now reflect the partial comparability of the referent groups. Since the first subject in each matched set is the case, so $r_{s1} = 1$ and $r_{si} = 0$ for $i = 2, \ldots, n_s$, it follows that a sign-score statistic has the form $T = \sum_s d_s \sum_i r_{si} Z_{si} = \sum_s d_s Z_{s1}$. Most often, T would be the Mantel–Haenszel statistic with $d_s = 1$ for each s and $T = \sum_{s1}$ is the number of exposed cases, but the statistic could give different weights d_s to different cases as in §4.4.5.

In the notation of §4.4, $B_s = Z_{s1}$ and $p_s = \text{prob}(Z_{s1} = 1|\mathbf{m})$. To conduct the sensitivity analysis, bounds are needed on p_s under (6.2.1) and (6.2.2), say $P_s^- \leq p_s \leq P_s^+$. For the Mantel–Haenszel statistic, the bounds are used in (4.4.7) for a large sample approximation; alternatively, a formula analogous to (4.3.4) is used for exact calculations. If there are varying weights d_s, then (4.4.8) is used.

The bounds require some preliminary notation. Let $\boldsymbol{\eta}_s^+$ be the vector of dimension $n_s - 1$ with coordinates

$$\boldsymbol{\eta}_s^+ = [\underbrace{\exp(\gamma\psi_1), \ldots, \exp(\gamma\psi_1)}_{I_s} \;\; \underbrace{\exp(\gamma\psi_2), \ldots, \exp(\gamma\psi_2)}_{J_s}]$$

and let $\boldsymbol{\eta}_s^-$ be the vector of the same dimension with coordinates

$$[\underbrace{\exp\{\gamma(\psi_0+\psi_2)\}, \ldots, \exp\{\gamma(\psi_0+\psi_2)\}}_{I_s} \;\; \underbrace{\exp\{\gamma(\psi_0+\psi_1)\}, \ldots, \exp\{\gamma(\psi_0+\psi_1)\}}_{J_s}].$$

The mth elementary symmetric function of an n-dimensional argument $\mathbf{a}$ is the sum of all products of m coordinates of $\mathbf{a}$, that is,

$$\mathrm{SYM}_m(\mathbf{a}) = \sum_{\substack{\mathbf{b}:\, \mathbf{b}^T\mathbf{1}=m \\ b_j \in \{0,1\} \text{ for all } j}} \prod_j a_j^{b_j}.$$

This function is defined for $m = 0, 1, \ldots, n$. For instance, $\mathrm{SYM}_0(\mathbf{a}) = 1$, $\mathrm{SYM}_1(\mathbf{a}) = a_1 + \cdots + a_n$, $\mathrm{SYM}_2(\mathbf{a}) = a_1a_2 + a_1a_3 + \cdots + a_{n-1}a_n$, and $\mathrm{SYM}_n(\mathbf{a}) = a_1a_2\ldots a_n$.

The bounds, $P_s^- \le p_s \le P_s^+$, on $p_s = \mathrm{prob}(Z_{s1} = 1|\mathbf{m})$ are $P_s^- = P_s^+ = 1$ if $m_s = n_s$, $P_s^- = P_s^+ = 0$ if $m_s = 0$, and otherwise if $0 < m_s < n_s$,

$$P_s^+ = \frac{\exp(\gamma)}{\exp(\gamma) + \dfrac{\mathrm{SYM}_{m_s}(\boldsymbol{\eta}_s^+)}{\mathrm{SYM}_{m_s-1}(\boldsymbol{\eta}_s^+)}},$$

and (6.2.3)

$$P_s^- = \frac{1}{1 + \dfrac{\mathrm{SYM}_{m_s}(\boldsymbol{\eta}_s^-)}{\mathrm{SYM}_{m_s-1}(\boldsymbol{\eta}_s^-)}}.$$

These expressions are derived in the Appendix, §6.4.

To illustrate, consider the example in §6.2.1. Consider one of the 195 matched sets with both a hospital and a patient referent, so $n_s = 3$, $I_s = 1$, $J_s = 1$. In this matched set, the number of women who gave birth at age 25 or above could be $m_s = 0, 1, 2$, or 3. We want bounds, P_s^- and P_s^+, on the probability $p_s = \mathrm{prob}(Z_{s1} = 1|\mathbf{m})$ that the case in this matched set was among the women who gave birth at age 25 or above. Given $m_s = 0$, the case certainly did not give birth at age 25 or above, so in this case $P_s^- = P_s^+ = 0$. Similarly, if all of the women in this matched set gave birth at 25 or above, so $m_s = 3$, then $P_s^- = P_s^+ = 1$. The two cases just considered, $m_s = 0$ or $m_s = n_s$, concern concordant matched sets which contribute only a constant to the test statistic and so do not affect the inference. Now $\boldsymbol{\eta}_s^+ = [\exp(\gamma\psi_1), \exp(\gamma\psi_2)]$ and $\boldsymbol{\eta}_s^- = [\exp\{\gamma(\psi_0 + \psi_2)\}, \exp\{\gamma(\psi_0 + \psi_1)\}]$. If $m_s = 1$, then only one of the three women gave birth at 25 or above, so $\mathrm{SYM}_1(\boldsymbol{\eta}_s^+) = \exp(\gamma\psi_1) + \exp(\gamma\psi_2)$, $\mathrm{SYM}_0(\boldsymbol{\eta}_s^+) = 1$, and

$$P_s^+ = \frac{\exp(\gamma)}{\exp(\gamma) + \exp(\gamma\psi_1) + \exp(\gamma\psi_2)},$$

and similarly

$$P_s^- = \frac{1}{1 + \exp\{\gamma(\psi_0 + \psi_2)\} + \exp\{\gamma(\psi_0 + \psi_1)\}}.$$

If $\gamma = 0$, so there is no hidden bias, then $P_s^- = P_s^+ = 1/3$ in the two expressions just given. If partial comparability did nothing to reduce bias, so $\gamma \ne 0$, $\psi_0 = 1$, and $\psi_1 = \psi_2 = 0$, then $P_s^+ = \exp(\gamma)/\{\exp(\gamma) + 2\}$ and $P_s^- = 1/\{1 + 2\exp(\gamma)\}$ in the expressions just given, and these are the bounds from §4.4.4.

In the same way, if two of the three women gave birth at 25 or above, $m_s = 2$, $\text{SYM}_2(\boldsymbol{\eta}_s^+) = \exp\{\gamma(\psi_1 + \psi_2)\}$, so

$$P_s^+ = \frac{\exp(\gamma)}{\exp(\gamma) + \dfrac{\exp\{\gamma(\psi_1 + \psi_2)\}}{\exp(\gamma\psi_1) + \exp(\gamma\psi_2)}},$$

and similarly

$$P_s^- = \frac{1}{1 + \dfrac{\exp\{\gamma(2\psi_0 + \psi_1 + \psi_2)\}}{\exp\{\gamma(\psi_0 + \psi_2)\} + \exp\{\gamma(\psi_0 + \psi_1)\}}}.$$

If $\gamma = 0$, then $P_s^- = P_s^+ = 2/3$ in the two expressions just given. If $\gamma \neq 0$, $\psi_0 = 1$, and $\psi_1 = \psi_2 = 0$, then $P_s^+ = 2 \cdot \exp(\gamma)/\{2 \cdot \exp(\gamma) + 1\}$ and $P_s^- = 2 \cdot \exp(\gamma)/\{2 \cdot \exp(\gamma) + \exp(2\gamma)\}$, as in §4.4.4.

The pairs in §6.2.1 with a single referent have $n_s = 2$. For these pairs, $P_s^- = P_s^+ = 0$ if $m_s = 0$ and $P_s^- = P_s^+ = 1$ if $m_s = 2$. For $n_s = 2$, $m_s = 1$ with just a hospital referent,

$$P_s^- = \frac{\exp(\gamma\psi_1)}{\exp(\gamma\psi_1) + \exp(\gamma)}, \qquad P_s^+ = \frac{\exp(\gamma)}{\exp(\gamma) + \exp(\gamma\psi_1)}; \qquad (6.2.4)$$

see Problem 5. In these expressions, notice that $P_s^- = P_s^+ = 1/2$ if $\psi_1 = 1$. In other words, if there is substantial hidden bias in the sense that γ is large, but if all of that bias is in the variable v_{1si} for which cases and hospital referents are comparable, then there is no bias in pairs containing a case and a hospital referent. Similarly, with just a neighbor referent,

$$P_s^- = \frac{\exp(\gamma\psi_2)}{\exp(\gamma\psi_2) + \exp(\gamma)}, \qquad P_s^+ = \frac{\exp(\gamma)}{\exp(\gamma) + \exp(\gamma\psi_2)}.$$

6.2.4. Example: Sensitivity Analysis for Breast Cancer and Age at First Birth

Table 6.2.3 gives the results of the sensitivity analysis for the study in §6.2.1 which compared cases of breast cancer to hospital and neighbor referents. The sensitivity analysis considers a range of assumptions about the hidden biases affecting each of the referent groups. Among other possibilities, the table considers the situations in which both groups are affected by the same biases, the hospital referents are free of bias but the neighbors are not, or neighbors are free of bias but the hospital referents are not.

The table gives the sensitivity parameters $(\Gamma, \psi_0, \psi_1, \psi_2)$ where $\Gamma = \exp(\gamma)$, together with the upper bound on the significance level for the Mantel–Haenszel test comparing the age at first birth for cases and referents. The lower bound on the significance level is less than 0.0001 in each situation and so is not included in the table. The parameter Γ measures the

Table 6.2.3. Sensitivity Analysis for Breast Cancer and Age at First Birth.

	Interpretation	Γ	ψ_0	ψ_1	ψ_2	Significance level
A	No hidden bias	1	1	0	0	<0.0001
B	No reduction in bias	1.5	1	0	0	0.0197
C	No reduction in bias	2	1	0	0	0.48
D	Hospital referents free of bias	2	0	1	0	0.0007
E	Neighbor referents free of bias	2	0	0	1	0.025
F	Bias partly removed in both groups	2	1/3	1/3	1/3	0.0485
G	Hospital referents free of bias	3	0	1	0	0.0113
H	Neighbor referents free of bias	3	0	0	1	0.36

total quantity of hidden bias, so in this regard, situations C, D, E, and F are similar, and situations G and H are similar. The parameters (ψ_0, ψ_1, ψ_2), where $1 = \psi_0 + \psi_1 + \psi_2$, indicate the degree to which hidden bias is controlled by using referents from the same hospital or neighborhood. With $(\psi_0, \psi_1, \psi_2) = (1, 0, 0)$, the use of hospital and neighbor referents did nothing to make the cases and referents comparable in terms of the unobserved covariate u_{si}—this gives the sensitivity analysis discussed in Chapter 4. With $(\psi_0, \psi_1, \psi_2) = (0, 1, 0)$, the hospital referents are free of hidden bias, because cases and hospital referents have the same value of u_{si}. With $(\psi_0, \psi_1, \psi_2) = (0, 0, 1)$, the neighbor referents are free of hidden bias. With $(\psi_0, \psi_1, \psi_2) = (\frac{1}{3}, \frac{1}{3}, \frac{1}{3})$, the bias is partially reduced in both referent groups.

Of the 409 cases of breast cancer, 209 first gave birth at age 25 or older. In the absence of hidden bias, situation A in Table 6.2.3, there is a single significance level <0.0001 from the usual Mantel–Haenszel test, indicating increased risk of breast cancer. This comparison is insensitive to all patterns of hidden bias of magnitude $\Gamma = 1.5$, situation B, but it becomes sensitive for $\Gamma = 2$, situation C. An unobserved covariate associated with twice the odds of an older first birth could explain the apparently higher risk of breast cancer among such women. However, a bias of magnitude $\Gamma = 2$ or $\Gamma = 3$ could not explain the apparently higher risk if the bias affected only the neighbors and not the hospital referents, situations D and G. A bias of magnitude $\Gamma = 2$ could not explain the increased risk if it affected only hospital referents and not neighbors, situation E, but a larger bias $\Gamma = 3$ with the same pattern could explain the increased risk, situation H.

In short, an unobserved covariate that doubled the odds of a late first birth could explain the observed increase in risk of breast cancer if it affected both referent groups equally. A bias of this magnitude affecting only one referent group could not explain the increased risk.

Table 6.2.4 shows some of the computations leading to situation D in Table 6.2.3, that is, $(\Gamma, \psi_0, \psi_1, \psi_2) = (2, 0, 1, 0)$. This is the situation in which there is hidden bias of magnitude $\Gamma = 2$, but it does not affect hospital

Table 6.2.4. Computations for $(\Gamma, \psi_0, \psi_1, \psi_2) = (2, 0, 1, 0)$.

	n_s	I_s	J_s	m_s	#Matched sets	#Sets with a case ≥25 at first birth	P_s^-	P_s^+
1	3	1	1	0	47	0	0.00	0.00
2	3	1	1	1	67	32	0.25	0.40
3	3	1	1	2	60	52	0.60	0.75
4	3	1	1	3	21	21	1.00	1.00
5	2	1	0	0	58	0	0.00	0.00
6	2	1	0	1	77	53	0.50	0.50
7	2	1	0	2	29	29	1.00	1.00
8	2	0	1	0	17	0	0.00	0.00
9	2	0	1	1	23	12	0.33	0.67
10	2	0	1	2	10	10	1.00	1.00
Total					409	209		

referents. Rows 1–4 in Table 6.2.4 describe the $195 = 47 + 67 + 60 + 21$ matched triples, rows 5–7 describe the $164 = 58 + 77 + 29$ pairs with only a hospital referent, and rows 8–10 describe the $50 = 17 + 23 + 10$ pairs with only a neighbor referent. Rows 1, 4, 5, 7, 8, and 10 describe concordant matched sets in which either all n_s women gave birth at an age ≤ 24 or all gave birth at an age ≥ 25. While these sets may be included in the computations, they contribute only a constant to the test statistic and therefore do not affect the results. Row 2 describes the 67 matched triples in which two women had first births at ages ≤ 24 and one had a first birth at an age ≥ 25. In 32 of these matched sets, the case of breast cancer was the one women whose first birth occurred at an age ≥ 25. In these 67 matched sets, depending on the value of the unobserved covariate $\mathbf{u}$, the probability that the case will have age ≥ 25 at first birth could be as low as $P_s^- = 0.25$ or as high as $P_s^+ = 0.40$. These probabilities are based on the formula in §6.4.3. Summing from $s = 1$ to $s = 409$ in formula (4.4.6) gives the deviate used to obtain the significance level in Table 4.2.3.

Row 6 of Table 4.2.4 concerns the 77 discordant matched pairs in which there is a hospital referent but no neighbor referent. Note that $P_s^- = P_s^+ = \frac{1}{2}$ in these pairs, because the hospital referents are assumed free of bias in this calculation.

*6.3. Appendix: Selection and Hidden Bias

This appendix discusses the impact of certain types of nonrandom selection of subjects for comparison. The section serves two purposes. First, it provides the basis for the assertions in §6.1.3 concerning the relationship be-

* This section may be skipped without loss of continuity.

tween selection bias and hidden bias in case-referent studies. Second, many observational studies entail several comparisons with subsets of the subjects selected in various ways, and the appendix considers the relationship between these different comparisons.

Assume that hidden bias is expressed by the model (4.2.1) or (4.2.2) or (4.2.6). Also, to test the null hypothesis of no treatment effect, H_0, assume that the treatment does not affect the fixed responses **r**.

Divide the units in subclass s into two mutually exclusive and exhaustive groups, G_{Is} and G_{Es}, based on any fixed feature of the units, so $G_{Is} \cup G_{Es} = \{1, \ldots, n_s\}$ and $G_{Is} \cap G_{Es} = \varnothing$. The division may be based on the unobserved covariate **u** or on an unaffected outcome **y** or, under H_0, on the unaffected responses **r** themselves, or on a combination of all three, possibly with the aid of a table of random numbers. Suppose that the units in G_{Is} are included in a comparison but the units in G_{Es} are excluded. Write $\dot{\mathbf{Z}}$ for the coordinates of **Z** in G_{Is}, $s = 1, \ldots, S$, and write $\ddot{\mathbf{Z}}$ for the coordinates in G_{Es}, $s = 1, \ldots, S$, with a similar notation for other quantities. For instance, write

$$\dot{m}_s = \sum_{i \in G_{Is}} Z_{si} \qquad \text{and} \qquad \dot{\mathbf{m}} = \begin{bmatrix} \dot{m}_1 \\ \vdots \\ \dot{m}_S \end{bmatrix},$$

so $\dot{m}_s$ of the $\dot{n}_s$ subjects included in G_{Is} from stratum s received the treatment. Also, $\dot{\Omega}$ contains all vectors with binary coordinates of dimension $\dot{N} = \dot{n}_1 + \cdots + \dot{n}_S$ with $\dot{m}_S$ ones among the $\dot{n}_s$ coordinates for stratum s.

Starting from (4.2.1) or (4.2.2), in parallel with §4.2.3,

$$\mathrm{pr}(\dot{\mathbf{Z}} = \dot{\mathbf{z}} | \dot{\mathbf{m}}) = \frac{\exp(\gamma \dot{\mathbf{z}}^{\mathrm{T}} \dot{\mathbf{u}})}{\sum_{\dot{\mathbf{b}} \in \dot{\Omega}} \exp(\gamma \dot{\mathbf{b}}^{\mathrm{T}} \dot{\mathbf{u}})}, \qquad \dot{\mathbf{u}} \in [0, 1]^{\dot{N}}. \tag{6.3.1}$$

The same distribution (6.3.1) can be obtained beginning with (4.2.6) instead of beginning with (4.2.1) or (4.2.2) by conditioning on **m** and $\ddot{\mathbf{Z}}$. Note that **m** and $\ddot{\mathbf{Z}}$ determine $\dot{\mathbf{m}}$. Using the definition of conditional probability applied to (4.2.6) and simplifying using $\mathbf{z}^{\mathrm{T}}\mathbf{u} = \dot{\mathbf{z}}^{\mathrm{T}}\dot{\mathbf{u}} + \ddot{\mathbf{z}}^{\mathrm{T}}\ddot{\mathbf{u}}$ gives

$$\mathrm{pr}(\dot{\mathbf{Z}} = \dot{\mathbf{z}} | \mathbf{m}, \ddot{\mathbf{Z}}) = \frac{\exp(\gamma \dot{\mathbf{z}}^{\mathrm{T}} \dot{\mathbf{u}})}{\sum_{\dot{\mathbf{b}} \in \dot{\Omega}} \exp(\gamma \dot{\mathbf{b}}^{\mathrm{T}} \dot{\mathbf{u}})}, \qquad \dot{\mathbf{u}} \in [0, 1]^{\dot{N}}. \tag{6.3.2}$$

In other words, the distribution in (6.3.1) and (6.3.2) arises either by starting with (4.2.2) and ignoring the excluded units, or by starting with (4.2.6) and "setting aside" the excluded units by conditioning on their treatment assignments $\ddot{\mathbf{Z}}$ so they no longer enter the permutation distribution.

The models in (6.3.1) and (4.2.6) have the same form and the same value of the parameter γ, but selection of subjects changes **u** to $\dot{\mathbf{u}}$. If there is no hidden bias to start with, that is, if $\gamma = 0$, then there is no hidden bias after selection, and the treatments $\dot{\mathbf{Z}}$ are uniformly distributed on $\dot{\Omega}$, so the methods in Chapter 3 may be used. If there is hidden bias to start with, that is, if $\gamma \neq 0$, then after selection the methods in Chapter 4 may be used to appraise sensitivity to bias. However, as selection changes $\mathbf{u}$ to $\dot{\mathbf{u}}$, the quantity and pattern of hidden bias may be different after selection.

The situation is entirely different if subjects are selected based on a quantity affected by the treatment, say $\mathbf{Y} = \mathbf{y}_\mathbf{Z}$. This may distort the distribution of treatments or exposures $\mathbf{Z}$, so model (6.3.1) does not hold and the methods of Chapter 4 cannot be used. Even if there is no hidden bias in the sense that $\gamma = 0$, after selection based on $\mathbf{y}_\mathbf{Z}$, the distribution of $\dot{\mathbf{Z}}$ will not generally be uniform on $\dot{\Omega}$, so there will be selection bias in testing H_0 though there is no hidden bias.

*6.4. Appendix: Derivation of Bounds for Sensitivity Analysis

This appendix derives the bounds $P_s^- \leq p_s \leq P_s^+$ on $p_s = \text{prob}(Z_{s1} = 1|\mathbf{m})$ in §6.2.3 assuming partial comparability expressed by (6.2.1) and (6.2.2) and $\gamma \geq 0$. The process is slightly different from that in Chapter 4 because the v_{ksi} are subject to the constraint (6.2.2). Write $\zeta_{si} = \exp(\gamma u_{si}) = \exp\{\gamma(\psi_0 v_{0si} + \psi_1 v_{1si} + \psi_2 v_{2si})\}$. Recall the notation in Table 6.2.2. Write

$$\zeta_s = [\underbrace{\zeta_{s2}, \ldots, \zeta_{s,I_s+1}}_{I_s} \quad \underbrace{\zeta_{s,I_s+2}, \ldots, \zeta_{s,n_s}}_{J_s}]$$

for the $n_s - 1$ dimensional vector describing the $n_s - 1$ referents in matched set s. To calculate $\text{prob}(Z_{s1} = 1|\mathbf{m})$, take term s in the product in (4.2.6), and sum over all $\mathbf{z}_s \in \Omega_s$ such that $z_{s1} = 1$, which is a sum over $\binom{n_s - 1}{m_s - 1}$ terms. Then

$$\text{prob}(Z_{s1} = 1|\mathbf{m}) = \frac{\zeta_{s1} \cdot \text{SYM}_{m-1}(\zeta_s)}{\zeta_{s1} \cdot \text{SYM}_{m-1}(\zeta_s) + \text{SYM}_m(\zeta_s)} = \frac{\zeta_{s1}}{\zeta_{s1} + g(\zeta_s)}, \tag{6.4.1}$$

where $g(\mathbf{a}) = \text{SYM}_m\{\mathbf{a}\}/\text{SYM}_{m-1}\{\mathbf{a}\}$ and $g(\mathbf{a})$ is defined for vectors $\mathbf{a}$ with strictly positive coordinates.

In obtaining bounds on $\text{prob}(Z_{s1} = 1|\mathbf{m})$ in (6.4.1), the following fact about ratios of symmetric functions is useful. See Mitrinovic (1970, §2.15.3, Theorem 1, p. 102) for a proof of the first part of Lemma 6.1.

Lemma 6.1. *The ratio $g(\mathbf{a})$ is monotone-increasing in each coordinate of* $\mathbf{a}$. *Hence* $\text{prob}(Z_{s1} = 1|\mathbf{m})$ *is increasing in ζ_{s1} and decreasing in ζ_{si} for $i \geq 2$.*

Because of the constraints (6.2.2), the ζ_{si} are linked and cannot be changed arbitrarily. In particular, changing either v_{1s1} or v_{2s1} affects several ζ_{si}. As a result, the monotonicity in Lemma 1 is not sufficient to determine

* This section may be skipped without loss of continuity.

the bounds $P_s^- \le p_s \le P_s^+$. As will be seen in the proof of Proposition 6.1 below, Lemma 6.2 is needed to determine the extreme values of v_{1s1} or v_{2s1}.

Define $\mathbf{f}(\alpha, \beta)$ as the $n_s - 1$ dimensional vector

$$\mathbf{f}(\alpha, \beta) = [\underbrace{\alpha, \dots, \alpha}_{I_s} \quad \underbrace{\beta, \dots, \beta}_{J_s}].$$

Lemma 6.2. *The function*

$$\frac{\upsilon\alpha\beta}{\upsilon\alpha\beta + g\{\mathbf{f}(\alpha, \beta)\}} \tag{6.4.2}$$

is monotone-increasing in α and in β for $\alpha > 0$, $\beta > 0$, and fixed $\upsilon > 0$.

PROOF. It will be shown that (6.4.2) is monotone-increasing in α, the proof for β being similar. Proving that (6.4.2) is monotone-increasing in α is the same as proving that $g\{\mathbf{f}(\alpha, \beta)\}/\alpha$ is monotone-decreasing in α. Using the definition of SYM_m, it follows that if $\theta > 0$, then $g(\theta\mathbf{a}) = \theta g(\mathbf{a})$. Let $\alpha \le \alpha^*$ so the task is to show $g\{\mathbf{f}(\alpha, \beta)\}/\alpha \ge g\{\mathbf{f}(\alpha^*, \beta)\}/\alpha^*$. Let $\theta = \alpha^*/\alpha$. Then, as required,

$$\frac{g\{\mathbf{f}(\alpha, \beta)\}}{\alpha} = \frac{\theta g\{\mathbf{f}(\alpha, \beta)\}}{\theta\alpha} = \frac{g\{\mathbf{f}(\theta\alpha, \theta\beta)\}}{\theta\alpha}$$
$$\ge \frac{g\{\mathbf{f}(\theta\alpha, \beta)\}}{\theta\alpha} = \frac{g\{\mathbf{f}(\alpha^*, \beta)\}}{\alpha^*},$$

where the inequality follows from the monotonicity of $g(\mathbf{a})$. □

Proposition 6.3. *Under partial comparability expressed by* (6.2.1) *and* (6.2.2),

$$\frac{1}{1 + g[\mathbf{f}\{\exp(\gamma\psi_0 + \gamma\psi_2), \exp(\gamma\psi_0 + \gamma\psi_1)\}]}$$
$$\le \operatorname{prob}(Z_{s1} = 1|\mathbf{m})$$
$$\le \frac{\exp(\gamma)}{\exp(\gamma) + g[\mathbf{f}\{\exp(\gamma\psi_1), \exp(\gamma\psi_2)\}]}. \tag{6.4.3}$$

PROOF. Under the constraint (6.2.2),

$$\exp\{\gamma(\psi_1 v_{1s1} + \psi_2 v_{2s1})\} \le \zeta_{s1} \le \exp\{\gamma(\psi_0 + \psi_1 v_{1s1} + \psi_2 v_{2s1})\}$$
$$\exp(\gamma\psi_1 v_{1s1}) \le \zeta_{si} \le \exp\{\gamma(\psi_0 + \psi_1 v_{1s1} + \psi_2)\}$$
$$\text{for} \quad i = 2, \dots, I_s + 1, \tag{6.4.4}$$
$$\exp(\gamma\psi_2 v_{2s1}) \le \zeta_{si} \le \exp\{\gamma(\psi_0 + \psi_1 + \psi_2 v_{2s1})\}$$
$$\text{for} \quad i = I_s + 2, \dots, n_s.$$

Using Lemma 6.1, $\text{prob}(Z_{s1} = 1|\mathbf{m})$ is increased by raising ζ_{s1} to its upper bound and reducing ζ_{si} for $i \geq 2$ to its lower bound in (6.4.4). If ζ_{s1} is at its upper bound and ζ_{si} for $i \geq 2$ is at its lower bound in (6.4.4), then

$$\text{prob}(Z_{s1} = 1|\mathbf{m}) = \frac{\upsilon\alpha\beta}{\upsilon\alpha\beta + g\{\mathbf{f}(\alpha, \beta)\}} \tag{6.4.5}$$

with $\upsilon = \exp(\gamma\psi_0)$, $\alpha = \exp(\gamma\psi_1 v_{1s1})$, and $\beta = \exp(\gamma\psi_2 v_{2s1})$; hence, using Lemma 6.2, (6.4.5) is maximized when $v_{1s1} = v_{2s1} = 1$, giving the upper bound in (6.4.3) upon using $1 = \psi_0 + \psi_1 + \psi_2$. If ζ_{s1} is at its lower bound and ζ_{si} for $i \geq 2$ is at its upper bound in (6.4.4), then (6.4.5) holds with $\upsilon = \exp\{-\gamma(2\psi_0 + \psi_1 + \psi_2)\}$, $\alpha = \exp\{\gamma(\psi_0 + \psi_1 v_{1s1} + \psi_2)\}$, and $\beta = \exp\{\gamma(\psi_0 + \psi_1 + \psi_2 v_{2s1})\}$, so the minimum value of (6.4.5) occurs with $v_{1s1} = v_{2s1} = 0$. □

6.5. Bibliographic Notes

There are two books about case-referent studies by Breslow and Day (1980) and Schlesselman (1982) and, in addition, most epidemiology texts discuss case-referent studies and selection bias in detail; see Kelsey, Thompson, and Evans (1986), Kleinbaum, Kupper, and Morgenstern (1982), Lilienfeld and Lilienfeld (1980), MacMahon and Pugh (1970), Miettinen (1985a), Rothman (1986), and also the articles by Cornfield (1951), Holland and Rubin (1988), Mantel (1973), Mantel and Haenszel (1959), and Prentice and Breslow (1978). Gastwirth (1988) discusses the studies of Reye's syndrome. The use of more than one reference group is briefly discussed in most texts and is discussed in greater detail by Cole (1979), Fairweather (1987), Kelsey, Thompson, and Evans (1986, pp. 160–163), Liang and Stewart (1987), and Rosenbaum (1987). The technical material in §6.2–§6.4 is largely based on Rosenbaum (1991).

6.6. Problems

6.6.1. A Subject's Knowledge of the Investigator's Hypothesis

1. Weiss (1994) discusses whether or not it is appropriate to exclude from a case-referent study those subjects who were aware of the study's hypothesis. He writes:

> It is likely that for many persons who develop an illness, knowledge of a hypothesis concerning its etiology is obtained *after* the diagnosis.... In some instances, knowledge of the hypothesis could occur more commonly among exposed than nonexposed cases: inquiries by medical personnel into the history of exposure to a possible etiologic factor may be of relatively greater salience to exposed persons....

> [... excluding knowledgeable subjects] will create bias if the acquisition of knowledge of the etiologic hypothesis by cases is related to their exposure status.

Express Weiss's correct argument in the terminology of §6.1.3, that is, in terms of an affected outcome $\mathbf{Y} = \mathbf{y}_Z$ used in selecting subjects.

6.6.2. Bad Tequila in Acapulco

2. In 1985, the *Journal of Chronic Diseases* published an exchange led by Olli Miettinen (1985b) with dissent and discussion by James Schlesselman, Alvan Feinstein, and Olav Axelson. Miettinen writes:

   > As an illustration, consider the hypothetical example of testing the hypothesis that a cause of traveller's diarrhea is the consumption of tequila (a Mexican drink), with cases derived from a hospital in Acapulco, Mexico, over a defined period of time. What might be the proper [referent] group?

   Answer Miettinen's question.

3. Read the exchange mentioned in Problem 2. Compare your answer to the answers given there.

4. In Problem 2, what is the treatment? What is the outcome? What is $\pi_{[j]}$? Consider the binary variable indicating whether a subject was in Acapulco during the defined period of time. Is this binary variable a covariate? How would $\pi_{[j]}$ vary with this binary variable? What would be the consequence of matching on this variable? Would this automatically preclude certain referent groups?

6.6.3. Derivation

5. Obtain (6.2.4) from (6.2.3) using the observation that $\psi_1 = 1 - \psi_0 - \psi_2$ from (6.2.1).

References

Breslow, N. and Day, N. (1980). *The Analysis of Case-Control Studies.* Volume 1 of *Statistical Methods in Cancer Research.* Lyon, France: International Agency for Research on Cancer of the World Health Organization.

Cole, P. (1979). The evolving case-control study. *Journal of Chronic Diseases*, **32**, 15–27.

Collaborative Group for the Study of Stroke in Young Women (1973). Oral contraception and increased risk of cerebral ischemia or thrombosis. *New England Journal of Medicine*, **288**, 871–878.

Cornfield, J. (1951). A method of estimating comparative rates from clinical data: Applications to cancer of the lung, breast and cervix. *Journal of the National Cancer Institute*, **11**, 1269–1275.

Fairweather, W. (1987). Comparing proportion exposed in case-control studies using several control groups. *American Journal of Epidemiology*, **126**, 170–178.

Gastwirth, J. (1988). *Statistical Reasoning in Law and Public Policy*. New York: Academic Press.

Gutensohn, N., Li, F., Johnson, R., and Cole, P. (1975). Hodgkin's disease, tonsillectomy and family size. *New England Journal of Medicine*, **292**, 22–25.

Halsey, N., Modlin, J., Jabbour, J., Dubey, L., Eddins, D., and Ludwig, D. (1980). Risk factors in subacute sclerosing panencephalitis: A case-control study. *American Journal of Epidemiology*, **111**, 415–424.

Herbst, A., Ulfelder, H., and Poskanzer, D. (1971). Adenocarcinoma of the vagina: Association of maternal stilbestrol therapy with tumor appearance in young women. *New England Journal of Medicine*, **284**, 878–881.

Holland, P. and Rubin, D. (1988). Causal inference in retrospective studies. *Evaluation Review*, **12**, 203–231.

Hurwitz, E. (1989). Reye's syndrome. *Epidemiologic Reviews*, **11**, 249–253.

Hurwitz, E.S., Barrett, M.J., Bregman, D., Gunn, W.J., Pinsky, P., Schonberger, L.B., Drage, J.S., Kaslow, R.A., Burlington, D.B., and Quinnan, G.V. (1985). Public health service study on Reye's syndrome and medications. *New England Journal of Medicine*, **313**, 14, 849–857.

Kelsey, J., Thompson, W., and Evans, A. (1986). *Methods in Observational Epidemiology*. New York: Oxford University Press.

Kleinbaum, D., Kupper, L., and Morgenstern, H. (1982). *Epidemiologic Research*. Belmont, CA: Wadsworth.

Kreiger, N., Kelsey, J., Holford, T., and O'Connor, T. (1982). An epidemiologic study of hip fracture in postmenopausal women. *American Journal of Epidemiology*, **116**, 141–148.

Liang, K. and Stewart, W. (1987). Polychotomous logistic regression methods for matched case-control studies with multiple case or control groups. *American Journal of Epidemiology*, **125**, 720–730.

Lilienfeld, A., Chang, L., Thomas, D., and Levin, M. (1976). Rauwolfia derivatives and breast cancer. *Johns Hopkins Medical Journal*, **139**, 41–50.

Lilienfeld, A. and Lilienfeld, D. (1980). *Foundations of Epidemiology* (second edition). New York: Oxford University Press.

MacMahon, B. and Pugh, T. (1970). *Epidemiology: Principles and Methods*. Boston: Little, Brown.

Mantel, N. (1973). Synthetic retrospective studies and related topics. *Biometrics*, **29**, 479–486.

Mantel, N. and Haenszel, W. (1959). Statistical aspects of retrospective studies of disease. *Journal of the National Cancer Institute*, **22**, 719–748.

Miettinen, O. (1985a). *Theoretical Epidemiology*. New York: Wiley.

Miettinen, O. (1985b). The "case-control" study: Valid selection of subjects (with Discussion) *Journal of Chronic Diseases*, **38**, 543–548.

Mitrinovic, D.S. (1970). *Analytic Inequalities*. Berlin: Springer-Verlag.

Prentice, R. and Breslow, N. (1978). Retrospective studies and failure time models. *Biometrika*, **65**, 153–158.

Rosenbaum, P.R. (1987). The role of a second control group in an observational study (with Discussion). *Statistical Science*, **2**, 292–316.

Rosenbaum, P.R. (1991). Sensitivity analysis for matched case-control studies. *Biometrics*, **47**, 87–100.

Rothman, K. (1986). *Modern Epidemiology*. Boston: Little, Brown.

Schlesselman, J. (1982). *Case-Control Studies*. New York: Oxford University Press.

Weiss, N. (1994). Should we consider a subject's knowledge of the etiologic hypothesis in the analysis of case-control studies? *American Journal of Epidemiology*, **139**, 247–249.

CHAPTER 7
Multiple Control Groups

7.1. The Role of a Second Control Group

7.1.1. Observational Studies with More than One Control Group

An observational study has multiple control groups if it has several distinct groups of subjects who did not receive the treatment. In a randomized experiment, every control is denied the treatment for the same reason, namely, the toss of a coin. In an observational study, there may be several distinct ways that the treatment is denied to a subject. If these several control groups have outcomes that differ substantially and significantly, then this cannot reflect an effect of the treatment, since no control subject received the treatment. It must reflect, instead, some form of bias.

Multiple control groups arise in several ways. In some contexts, a subject may become a control either because the treatment was not offered or because, though offered, it was declined. Some years ago, the Educational Testing Service sought to evaluate the impact of the College Board's Advanced Placement (AP) Program which offers high-school students the opportunity to earn college credit for advanced courses taken in high school. In such a study, a student may become a control in either of two ways: the student may decline to participate in the AP Program, or the student may attend a school that does not offer the AP Program. What hidden biases might be present with these two sources of controls? Schools offering the AP Program may have greater resources or have more college bound students than other schools, and these characteristics are often found in school districts with higher family incomes. Possibly, students in schools offering the AP Program would have better outcomes even if the AP Program itself had no effect. On the other hand, if the AP Program is available in a school, students who decline to participate may be less motivated for academic work than those who do participate. Here, too, if the AP Program had no effect, greater motivation among AP participants might possibly lead to better outcomes. The use of both control groups provides little protection in this study because the most plausible hidden biases lead to patterns of responses that resemble treatment effects. In the end, a single control group was used, namely, matched students in the same school who declined to participate.

A related but different situation arises with compensatory programs, that is, programs such as Head Start that are intended to assist disadvantaged students. Campbell and Boruch (1975) discuss biases that may affect studies of compensatory programs. As with the AP Program, if a compensatory program is offered, students who decline to participate might be less motivated. In contrast with the AP Program, admission to a compensatory program often depends on need, the program being offered to students with the greatest need. After matching or adjustment for observed covariates, such as test scores and grades prior to admission into the program, if the compensatory program had no effect, one might expect that program participates, being better motivated, would outperform students who decline to participate, but would not perform as well as students who were not offered the program because of less need.

Generally, the goal is to select control groups so that, in the observed data, an actual treatment effect would have an appearance that differed from the most plausible hidden bias. Arguably, this is true in the case of the compensatory program but not in the case of the AP Program.

Multiple control groups also arise when controls are selected from several existing groups of subjects who did not receive the treatment. Seltser and Sartwell (1965) studied the possibility that X-ray exposure was causing deaths among radiologists. They compared members of the Radiological Society of North America, a professional association of radiologists, to four other specialty societies. In two of these four societies, the American Academy of Ophthalmology and Otolaryngology and the American Association of Pathologists and Bacteriologists, most society members would normally use X-rays vastly less often than radiologists do. For the period 1945–1954, Seltser and Sartwell (1965, Table 7) calculated age-adjusted mortality rates per 1000 person years of 16.4 for the radiologists, 12.5 for the pathologists and bacteriologists, and 11.9 for the ophthalmologists and otolaryngologists. In other words, in these two control groups the mortality rates are similar, as they should be in the absence of hidden bias. Both groups had substantially lower mortality than among the radiologists, consistent with a possible effect of radiation exposure.

If subjects who receive negligible doses of a treatment are distinguished from those who actively avoid or abstain from the treatment, then two control groups are produced. In §7.1.2, an example will be discussed that distinguishes abstention from alcohol and extremely low consumption of alcohol.

Multiple control groups are sometimes formed from several groups that received different treatments that were believed to be the same in ways that matter for the outcome under study. When formed in this way, differences among the control groups may reflect either hidden bias or unanticipated differences in the effects of the control treatments. The problems in §7.4.1 discuss an example in which the treatment is contraception using the intrauterine device, and there are five comparison groups defined by other forms of contraception.

As noted in Chapter 6, a case-referent study may have one or more referent groups and one or more control groups. A referent is not a case. A control did not receive the treatment. A case-referent study of alcohol consumption might have two referent groups, say neighbors and siblings of cases, and two control groups, say abstainers and those who are not abstainers but who report virtually no alcohol consumption. Such a design would offer several opportunities to look for selection and hidden biases.

7.1.2. An Example: Are Small Doses of Alcohol Beneficial?

Petersson, Trell, and Kristenson (1982) reviewed six studies suggesting that moderate consumption of alcohol may have beneficial effects in preventing some cardiac disease, and then they conducted a similar study with the additional feature that there were two control or "no alcohol" groups. Alcohol use was graded based on the response to ten questions. Nine of the questions described situations and asked about drinking in relation to those situations. One question asked about alcohol abstention. One control group consisted of men who described themselves as abstaining from all consumption of alcohol. The other control group consisted of men who did not say they abstained from alcohol, but who answered "no" to all nine questions about drinking behavior.

Between 1974 and 1979, the study obtained questionnaire responses from 7725 male residents of Malmö, Sweden. By 1980, there had been 127 deaths among these 7725 men. Each death was matched with two referents who were alive in 1980, the referents being matched to the death based on age and date of entry into the study. In principle, the analysis should make use of the matching, but this is not possible from the published data, so for this example, the matching will be ignored. This study is, almost, a synthetic case-referent study in the sense of §3.3.2; see also §6.1.3. Petersson, Trell, and Kristenson (1982)'s data are given in Table 7.1.1. (In this table, their 0 is called "None," their 1 is called "Lowest," their 2 is called "Middle" and their ≥ 3 is called "Highest.")

Table 7.1.1. Alcohol and Mortality.

	Group	Dead	Alive
Control groups	Abstain	7	6
	None	27	71
Treated groups	Lowest	14	38
	Middle	13	20
	Highest	21	20

Table 7.1.2 contains odds ratios comparing the groups in Table 7.1.1. In this table, the odds for the row are divided by the odds for the column, so for None/Abstain the odds ratio is (27/71)/(7/6) = 0.33. Table 7.1.3 gives the chi-square statistics for a 2 × 2 table, with the continuity correction; they have one degree of freedom and are significant at 0.05 or 0.025 if greater than 3.84 or 5.02, respectively. For instance, 2.60 is the chi-square for the 2 × 2 table Abstain/None × Dead/Alive.

When High alcohol consumption is compared to None or Low, the odds ratios are above two and significant at 0.05, consistent with greater mortality among those who consume larger quantities of alcohol. However, the High group does not differ significantly from the Abstain group, and the odds ratio is actually slightly less than one. In fact, the odds ratios suggest the greatest mortality is found among abstainers, though the differences are not quite statistically significant. In particular, comparing the two control groups, the odds of death are slightly more than three times greater in the

Table 7.1.2. Odds Ratios for Alcohol and Mortality.

	Abstain	None	Low	Middle
None	0.33			
Low	0.32	0.97		
Middle	0.56	1.71	1.76	
High	0.90	2.76	2.85	1.62

Table 7.1.3. Chi-Squares for Alcohol and Mortality.

	Abstain	None	Low	Middle
None	2.6			
Low	2.3	0.0		
Middle	0.31	1.1	0.9	
High	0.02	6.1	4.8	0.6

Abstain group than in the None group, though again this difference is not significant at 0.05. Notice that there are only 13 men in the Abstain group.

Of the abstainers, Petersson, Trell, and Kristenson write: "Most of these men, however, had chronic disease as the reason for their abstention, or even a past history of alcoholism. Increased mortality in nondrinkers may create a false impression of a preventive effect of any versus no daily drinking in relation to general and cardiovascular health." In other words, though earlier studies had claimed a beneficial effect of moderate alcohol consumption, Petersson, Trell, and Kristenson are raising doubts about whether that is an actual effect of alcohol, because moderate consumption appears better than abstention but no better and perhaps worse than negligible doses.

7.1.3. Selecting Control Groups: Systematic Variation and Bracketing

The goal in selecting control groups is to distinguish treatment effects from the most plausible systematic biases. If this is to be done successfully, the pattern of outcomes anticipated if the treatment has an effect must differ from the pattern anticipated from a hidden bias. The principles of "systematic variation" and "bracketing" are intended to ensure that this is so.

In a thoughtful discussion of control groups in behavioral research, Campbell (1969) quotes Bitterman's (1965) discussion of control by systematic variation. Bitterman's work is in a challenging field, comparative psychology, which seeks to "study the role of the brain in learning [by comparing] the learning of animals with different brains" [Bitterman (1965, p. 396)]. Bitterman (1965, pp. 399–400) writes:

> Another possibility to be considered is that the difference between the fish and rat which is reflected in these curves is not a difference in learning at all, but a difference in some confounded variable—sensory, motor, or motivational. Who can say, for example, whether the sensory and motor demands made upon the two animals in these experiments were exactly the same? Who can say whether the fish were just as hungry as the rats?...
>
> I do not, of course, know how to arrange a set of conditions for the fish which will make sensory and motor demands exactly equal to those which are made upon the rat in some given experimental situation. Nor do I know how to equate drive level or reward value in the two animals. Fortunately, however, meaningful comparisons are still possible, because for *control by equation* we may substitute what I call control by *systematic variation*. Consider, for example, the hypothesis that the difference between the curves ... is due to a difference, not in learning, but in degree of hunger. The hypothesis implies that there is a level of hunger at which the fish *will* show progressive improvement, and, put in this way, the hypothesis becomes easy to test. If, despite the widest possible variation in hunger, progressive improvement fails to appear in the fish, we may reject the hunger hypothesis.

In the terminology of this book, Bitterman's "control by equation" is analogous to controlling confounding due to an observed covariate **x** by matching subjects with the same or similar values of **x** as in §3—then differences in outcomes cannot be due to differences in **x**. "Control by systematic variation" concerns a variable u which is not recorded. If, however, two control groups can be formed so that u is much higher in one group than in the other, and if the groups do not differ materially in their outcomes, then this is consistent with the claim that differences in u are not responsible for differences in outcomes observed between treated and control groups. The principle of systematic variation leads to the following advice. When designing an observational study, identify the most plausible hidden biases and find control groups that sharply vary their levels.

A step beyond systematic variation of u is "bracketing," as discussed by Campbell (1969). Bracketing seeks two control groups such that, in the first group, u tends to be higher than in the treated group and, in the second, u tends to be lower than in the treated group. The goal is two control groups that are further apart from each other in terms of u than they are from the treated group. Possibly this is true of the example in §6.1.1 of a compensatory educational program. In the absence of a program effect, matched controls who decline to participate might be expected to underperform the program group, while matched controls who are not eligible because of less need might be expected to outperform the program group. On the other hand, with a large positive treatment effect and no hidden bias, the program group would tend to outperform both matched control groups. Bracketing yields a study design in which treatment effects and plausible biases are likely to have different appearances in observable data.

7.2. Comparing Outcomes in Two Control Groups

7.2.1. A Model for Assignment to One of Several Groups

In earlier chapters, the notation and model described two groups, a treated and a control group. This section discusses a model for assignment to one of two or more groups. As will be seen, it is closely connected both to the model discussed in earlier chapters and to randomized experiments with two or more groups. Specifically, the model generalizes (4.2.2) to permit more than two groups. Also, when there is no hidden bias, the model includes the distribution of treatment assignments in completely randomized experiments and randomized block experiments.

As always, there are n_s subjects in stratum s defined by the observed covariate $\mathbf{x}_s$, with $N = n_1 + \cdots + n_S$, and the ith of these n_s subjects has an unobserved covariate u_{si}, with $0 \le u_{si} \le 1$. If each subject falls in one of K

groups, $K \geq 2$, rather than a single treated group or a single control group, write $G_{sik} = 1$ if the ith subject in stratum s is in group k, and write $G_{sik} = 0$ otherwise, with $1 = \sum_{k=1}^{K} G_{sik}$ for each (s, i). With fixed covariates $\mathbf{x}_s$ and u_{si}, consider the following multinomial logit model (Cox, 1970, §7.5) for $(G_{si1}, \ldots, G_{siK})$:

$$\operatorname{prob}(G_{sik} = 1) = \frac{\exp\{\xi_k(\mathbf{x}_s) + \delta_k u_{si}\}}{\sum_{j=1}^{K} \exp\{\xi_j(\mathbf{x}_s) + \delta_j u_{si}\}}, \tag{7.2.1}$$

where the N vectors $(G_{si1}, \ldots, G_{siK})$ are mutually independent.

Suppose that we wish to compare two specific groups, j and k, $j \neq k$, and so select all subjects with $G_{sij} + G_{sik} = 1$, that is, all subjects who belong to one of these two groups. Then

$$\begin{aligned}\operatorname{prob}(G_{sij} = 1 \mid G_{sij} + G_{sik} = 1) &= \frac{\exp\{\xi_j(\mathbf{x}_s) + \delta_j u_{si}\}}{\exp\{\xi_j(\mathbf{x}_s) + \delta_j u_{si}\} + \exp\{\xi_k(\mathbf{x}_s) + \delta_k u_{si}\}} \\ &= \frac{\exp\{\kappa(\mathbf{x}_s) + \gamma u_{si}\}}{1 + \exp\{\kappa(\mathbf{x}_s) + \gamma u_{si}\}},\end{aligned}$$

where

$$\kappa(\mathbf{x}_s) = \xi_j(\mathbf{x}_s) - \xi_k(\mathbf{x}_s), \qquad \gamma = \delta_j - \delta_k, \qquad \text{and} \quad 0 \leq u_{si} \leq 1, \tag{7.2.2}$$

which is identical in form to (4.2.2). In other words, if the model (7.2.1) describes assignment to all K groups, then when comparing any two fixed groups, say j and k, the model and methods of earlier chapters may be used directly, recognizing only that the $\kappa(\mathbf{x}_s)$ and the γ appropriate for comparing two groups j and k will be different from the $\kappa(\mathbf{x}_s)$ and the γ for comparing two other groups j' and k'. In particular, if there are only $K = 2$ groups, then models (7.2.1) and (4.2.2) are essentially the same.

Write $m_{sk} = \sum_i^{n_s} G_{sik}$ for the number of subjects assigned to group k in stratum s. Also write $\boldsymbol{\delta} = (\delta_1, \ldots, \delta_K)^{\mathrm{T}}$, so there is no hidden bias if $\boldsymbol{\delta} = \mathbf{0}$.

7.2.2. The Model and Randomized Block Experiments with Several Groups

This section, which may be skipped, describes the relationship between the model (7.2.1) and a randomized block experiment. The conclusion is that, if there is no hidden bias in the sense that $\boldsymbol{\delta} = \mathbf{0}$, then the distribution of assignments G_{sik} under (7.2.1) is the same as in a randomized block experiment if one conditions on a sufficient statistic for the $\xi_k(\mathbf{x}_s)$. This is analogous to the conclusion in §3.2 for two groups. The notation introduced and results in this section will not be used in later sections. The section is intended solely as further motivation for the model (7.2.1).

Write $\mathbf{G}$ for the $N \times K$ matrix containing the G_{sik}, and write $\mathbf{M}$ for the $S \times K$ matrix containing the $m_{sk} = \sum_i^{n_s} G_{sik}$. Conditioning on $\mathbf{M}$ fixes the number of subjects in group k in stratum s. Then

$$\text{prob}(\mathbf{G} = \mathbf{g} | \mathbf{M}) = \frac{\exp(\mathbf{u}^T \mathbf{g} \boldsymbol{\delta})}{\sum^* \exp(\mathbf{u}^T \mathbf{g}^* \boldsymbol{\delta})}, \tag{7.2.3}$$

where $\sum^*$, is a sum is over all $\prod \binom{n_s}{m_{s1} \ldots m_{sK}}$ possible $\mathbf{g}^*$ such that:

(i) $g_{sik}^* = 0$ or $g_{sik}^* = 1$;
(ii) $1 = \sum_k g_{sik}^*$; and
(iii) $m_{sk} = \sum_i^{n_s} g_{sik}^*$ for each s, k.

If there is no hidden bias in the sense that $\boldsymbol{\delta} = \mathbf{0}$, then (7.2.3) is constant or uniform, assigning the same probability to each possible $\mathbf{g}$. If there is just one stratum, $S = 1$, then this uniform distribution of treatment assignments is the same as the distribution used to derive the Kruskal–Wallis (1952) test. If there are several strata, $S \geq 2$, but each $m_{sk} = 1$, so one subject gets each treatment in each stratum, then this uniform distribution is the same as the distribution used to derive Friedman's (1937) test. These two standard tests are discussed in many texts, for instance, Hollander and Wolfe (1973) or Lehmann (1975).

In short, the model (7.2.1) is familiar in two senses. First, as shown in §7.2.1, in comparing groups two at a time, the model leads back to the methods used to compare treated and control groups in earlier chapters. Second, when all groups are considered together, if there is no hidden bias, the model leads back to familiar randomization tests comparing several groups.

7.2.3. Power of the Test Comparing Outcomes in Two Control Groups

Consider a study in which group $k = 1$ is the treated group and groups $k = 2, \ldots, K$ are the $K - 1$ control groups, with $K > 2$. What does it mean to say that group 1 is a treatment group and groups $k = 2, \ldots, K$ are control groups? The answer *defines* multiple control groups, distinguishing a study with multiple control groups from a study which simply has several different treatments. Consider the response R_{si} of the ith subject in stratum s under the K possible group assignments for this subject. To say that groups $k = 2, \ldots, K$ are control groups is to say that the response of this subject, R_{si}, would be the same no matter which control group received this subject, though R_{si} might be higher or lower if the subject received the treatment by being assigned to group 1. Notice carefully that this is a statement about the responses of individual subjects under different treatments, not a statement about the observed responses in the several groups. This is, again, the definition of multiple control groups. If the response of an individual would change depending upon the control group to which that subject was assigned, then the differences between the control groups affect the response, so they are not really control groups but rather active treatments with varied effects.

Pick two control groups, $j \geq 2$ and $k \geq 2$, $j \neq k$. Suppose that $G_{sik} + G_{sij} = 1$ so the ith subject in stratum s is in one of these two control groups. Then the response R_{si} of this subject is the same if $G_{sik} = 1$ or if $G_{sij} = 1$, that is, conditionally given that $G_{sik} + G_{sij} = 1$, the response R_{si} is fixed. Also, $\text{prob}(G_{sik} = 1 | G_{sij} + G_{sik} = 1)$ is given by (7.2.2), which has the form used in Chapter 4 to compare two stratified or matched groups. Focus attention on the subjects in these two groups, setting aside subjects in other groups. Write $\boldsymbol{u}$ and $\boldsymbol{r}$ for the vectors of dimension $\sum_s (m_{sj} + m_{sk})$ containing the u_{si} and R_{si} for subjects in groups j and k, and write $\boldsymbol{Z}$ for the vector of the same dimension where $Z_{si} = 1$ if $G_{sij} = 1$ and $Z_{si} = 0$ if $G_{sik} = 1$. Note that $\boldsymbol{r}$ is fixed as $\boldsymbol{Z}$ varies since groups j and k are control groups. Also, let Ω be the set of possible values of $\boldsymbol{Z}$, so each $\boldsymbol{z} \in \Omega$ has m_{sj} ones and m_{sk} zeros among its $m_{sj} + m_{sk}$ coordinates for stratum s.

Contrast the responses of the subjects in groups j and k using an arrangement-increasing statistic $T = t(\boldsymbol{Z}, \boldsymbol{r})$, for instance any of the statistics in §2.4.2. For example, if $S = 1$ then T might be the rank sum statistic comparing responses in groups j and k, or if there are matched sets with one subject from each group in each matched set, then T might be the signed rank or McNemar statistic comparing the pair of subjects from groups j and k. Let

$$\alpha = \frac{1}{|\Omega|} \sum_{\boldsymbol{z} \in \Omega} [t(\boldsymbol{Z}, \boldsymbol{r}) \geq a]$$

and

$$\beta(\boldsymbol{r}, \boldsymbol{u}) = \sum_{\boldsymbol{z} \in \Omega} [t(\boldsymbol{Z}, \boldsymbol{r}) \geq a] \frac{\exp(\gamma \boldsymbol{z}^{\mathrm{T}} \boldsymbol{u})}{\sum_{\boldsymbol{b} \in \Omega} \exp(\gamma \boldsymbol{b}^{\mathrm{T}} \boldsymbol{u})},$$

where $\gamma = \delta_j - \delta_k$ as in (7.2.2). Using arguments exactly parallel to those in §5.2.3, the following points are established. All statements about probabilities refer to conditional probabilities given the $G_{sik} + G_{sij}$.

- *The comparison of responses in two control groups tests the null hypothesis that there is no hidden bias.* A test which rejects the hypothesis of no hidden bias—that is, the hypothesis that $\boldsymbol{\delta} = \boldsymbol{0}$—when $T \geq a$ has level α. For instance, the rank sum test or the signed rank test would have their usual null distribution discussed in Chapter 2. This is true because $\boldsymbol{\delta} = \boldsymbol{0}$ implies $\gamma = 0$.
- *The test can have power only if there is systematic variation of the unobserved covariate.* The power of the test is $\beta(\boldsymbol{r}, \boldsymbol{u})$, which can differ from α only if $\gamma = \delta_j - \delta_k \neq 0$. If $\delta_j > \delta_k$ then group j will tend to have higher values of u than group k, and similarly if $\delta_j < \delta_k$ then group j will tend to have

lower values of u, but if $\delta_j = \delta_k$ they will tend to have the same distribution of u. In the terminology of §7.1.3, a comparison of control groups j and k can hope to detect a hidden bias only if these two groups exhibit systematic variation of u.

- *The power of this test for hidden bias is greatest if the unobserved covariate u and the responses of controls are strongly related.* This is highly desirable, since an unobserved covariate strongly related to the response is one that does the most to distort inferences about treatment effects. More precisely, the power function $\beta(\boldsymbol{r}, \boldsymbol{u})$ is arrangement-increasing. This is analogous to Proposition 5.1 in §5.2.3.
- *Bracketing.* There is bracketing in the sense of §7.1.3 if $\delta_j > \delta_1 > \delta_k$ for in this case the distribution of u in the treated group tends to fall below the distribution in group j but above that in group k.
- *Unbiased tests.* If the unobserved covariate and the response under the controls are positively related, then the test is an unbiased test of H_0: $\delta_j = \delta_k$ against H_A: $\delta_j > \delta_k$; for specifics, see Rosenbaum (1989a, §6).

In short, the behavior of the power function $\beta(\boldsymbol{r}, \boldsymbol{u})$ provides some formal justification for the principles of systematic variation and bracketing. See also §7.4.3 for the relationship between the power $\beta(\boldsymbol{r}, \boldsymbol{u})$ and the sensitivity of tests for treatment effects.

7.3. Bibliographic Notes

The use of multiple control groups in observational studies is discussed in detail by Campbell (1969) and Rosenbaum (1984, 1987). Another use of multiple control groups is to examine the impact of measurement procedures separated from effects of the treatment; see Solomon (1949). The power and unbaisedness of tests for hidden bias in §7.2.3 are discussed in Rosenbaum (1989a). The Advanced Placement Program example in §7.1 is discussed in Rosenbaum (1987). See also the Bibliographic Note, §6.5, concerning multiple referent groups in case-referent studies.

7.4. Problems

7.4.1. Problems: IUD and Ectopic Pregnancy

1. Rossing, Daling, Voigt, Stergachis, and Weiss (1993) conducted a case-referent study of the possible increase in the risk of a tubal or ectopic pregnancy caused by use of an intrauterine device as a contraceptive. Can you spot an important issue that will arise in such a study? (*Hint*: An increased risk compared to *what*?)

2. Rossing et al. (1993, p. 252) write:

> There is considerable evidence to indicate that, among women who conceive a pregnancy, users of an intrauterine device (IUD) are more likely than nonusers to have implantation occur outside the uterus ... [In studies of] nonpregnant, sexually active women ... both increased and decreased risks have been reported. In the two studies in which the comparison group was restricted to noncontracepting, nonpregnant women, however, current IUD users were observed to be at a reduced risk of ectopic pregnancy.

   Why do the studies disagree with each other?

3. Rossing et al. (1993, p. 252) collected data on female members of the Group Health Cooperative of Puget Sound who developed an ectopic pregnancy between 1981 and 1986; these were the cases. Referents were females from the same cooperative, selected using age and county of residence, with certain exclusions. Table 7.4.1 gives their data comparing cases and referents for IUD users and five other contraceptive groups. What conclusion do you draw?

Table 7.4.1. Tubal Pregnancy and IUD.

Contraceptive method	Cases	Referents
IUD	18	60
Oral	13	202
Barrier	19	263
Sterilization	31	153
Rhythm, withdrawal, other	11	33
None	157	120

4. Table 7.4.1 compares IUD users to five comparison groups. Are these control groups in the sense defined in §7.2.3? Why or why not?

7.4.2. Problems: Dose and Response

5. The example in §7.1.2 had two control groups and it also had three treated groups at three doses or levels of alcohol consumption. Lilienfeld and Lilienfeld (1980, p. 309) write:

> If a factor is of causal importance in a disease, then the risk of developing the disease should be related to the degree of exposure to the factor; that is, a *dose-response relationship* should exist.

   Ignoring the Abstain group, does there appear to be increasing risk of death with increasing alcohol consumption in Table 7.1.2?

6. Excluding the Abstain group, some of the odds ratios in Table 7.1.2 differ significantly and others do not; see Table 7.1.3. Given this, what would you say about the presence or absence of a dose response relationship. Compare your thoughts with the sensible recommendations of Maclure and Greenland (1992, p. 103).

7. Concerning dose and response, Weiss (1981, p. 488) writes:

 > ... one or more confounding factors can be related closely enough to both exposure and disease to give rise to [a dose response relationship] in the absence of cause and effect.

 Suggest a simple linear regression model with a continuous dose Z, a continuous response R, and an unobserved covariate u such that a dose response relationship is produced in the absence of a treatment effect. That is, build a simple model illustrating Weiss's point.

8. Find an exponential family model relating discrete or continuous doses Z to an unobserved covariate u which reduces to the logit model (4.2.2) when there are just two doses. Rosenbaum (1989b)

7.4.3. Power of the Test for Bias and Sensitivity to Bias

9. Suppose the treatment has no effect, and that there are two control groups and a binary response $\mathbf{r}$. Suppose further that the one-sided Mantel–Haenszel statistic is used twice. First, in a test for hidden bias, the two control groups are compared to each other. Second, in a test for a treatment effect, the treated group is compared to the two control groups, ignoring the distinction between the two control groups. Define the notation so each test rejects when its test statistic is large in the positive direction. What value $\mathbf{u} \in U$ of the unobserved covariate leads to greatest sensitivity in the test for a treatment effect in §4.4.6? What value $\boldsymbol{u}$ of the unobserved covariate leads to greatest conditional power $\beta(\boldsymbol{r}, \boldsymbol{u})$ in the test for hidden bias in §7.2.3? What is the relationship between $\mathbf{u}$ and $\boldsymbol{u}$? Why is this highly desirable?

10. Continuing Question 9: How would you investigate whether the test for bias has reduced sensitivity to bias? That is: How would you conduct an analysis similar to that in §5.2.4? What value $\boldsymbol{u}^{\circ}$ would be tested? How is $\boldsymbol{u}^{\circ}$ related to the value that maximizes the power $\beta(\boldsymbol{r}, \boldsymbol{u})$? Why is this, too, highly desirable?

References

Bitterman, M. (1965). Phyletic differences in learning. *American Psychologist*, **20**, 396–410.

Campbell, D. (1969). Prospective: Artifact and Control. In *Artifact in Behavioral Research* (eds., R. Rosenthal and R. Rosnow). New York: Academic Press, pp. 351–382.

Campbell, D. and Boruch, R. (1975). Making the case for randomized assignment to treatments by considering the alternatives: Six ways in which quasi-experimental evaluations in compensatory education tend to underestimate effects. In: *Evaluation and Experiment* (eds., C. Bennett and A. Lumsdaine). New York: Academic Press, pp. 195–296.

Campbell, D. and Stanley, J. (1963). *Experimental and Quasi-Experimental Designs for Research.* Chicago: Rand McNally.

Cox, D.R. (1970). *The Analysis of Binary Data.* London: Methuen.

Friedman, M. (1937). The use of ranks to avoid the assumption of normality implicit in the analysis of variance. *Journal of the American Statistical Association*, **32**, 675–701.

Hollander, M. and Wolfe, D. (1973). *Nonparametric Statistical Methods*. New York: Wiley.

Kruskal, W. and Wallis, W. (1952). Use of ranks in one-criterion variance analysis. *Journal of the American Statistical Association*, **47**, 583–621.

Lehmann, E.L. (1975). *Nonparametrics: Statistical Methods Based on Ranks*. San Francisco: Holden-Day.

Lilienfeld, A., Chang, L., Thomas, D., and Levin, M. (1976). Rauwolfia derivatives and breast cancer. *Johns Hopkins Medical Journal*, **139**, 41–50.

Lilienfeld, A. and Lilienfeld, D. (1980). *Foundations of Epidemiology* (second edition). New York: Oxford University Press.

Maclure, M. and Greenland, S. (1992). Tests for trend and dose-response: Misinterpretations and alternatives. *American Journal of Epidemiology*, **135**, 96–104.

Petersson, B., Trell, E., Kristenson, H. (1982). Alcohol abstention and premature mortality in middle aged men. *British Medical Journal*, **285**, 1457–1459.

Rosenbaum, P.R. (1984). From association to causation in observational studies. *Journal of the American Statistical Association*, **79**, 41–48.

Rosenbaum, P.R. (1987). The role of a second control group in an observational study (with Discussion). *Statistical Science*, **2**, 292–316.

Rosenbaum, P.R. (1989a). On permutation tests for hidden biases in observational studies: An application of Holley's inequality to the Savage lattice. *Annals of Statistics*, **17**, 643–653.

Rosenbaum, P.R. (1989b). Sensitivity analysis for matched observational studies with many ordered treatments. *Scandinavian Journal of Statistics*, **16**, 227–236.

Rossing, M., Daling, J., Voigt, L., Stergachis, A., and Weiss, N. (1993). Current use of an intrauterine device and risk of tubal pregnancy. *Epidemiology*, **4**, 252–258.

Seltser, R. and Sartwell, P. (1965). The influence of occupational exposure to radiation on the mortality of American radiologists and other medical specialists. *American Journal of Epidemiology*, **81**, 2–22.

Solomon, R. (1949). An extension of control group design. *Psychological Bulletin*, 137–150.

Weiss, N. (1981). Inferring causal relationships: Elaboration of the criterion of "dose-response." *American Journal of Epidemiology*, **113**, 487–490.

CHAPTER 8

Coherence and Focused Hypotheses

8.1. Coherent Associations

8.1.1. What Is Coherence?

The 1964 US Surgeon General's report, *Smoking and Health* (Bayne-Jones et al. 1964, p. 20), lists five criteria for judgment about causality, the fifth being "the coherence of the association." A single sentence defines coherence (Bayne-Jones et al. 1964, p. 185): "A final criterion for the appraisal of causal significance of an association is its coherence with known facts in the natural history and biology of the disease." There follows a long discussion of the many ways in which the association between smoking and lung cancer is coherent. Per capita consumption of cigarettes had, at that time, been increasing, and the incidence of lung cancer was also increasing. Men, at that time, smoked much more than women and had a much higher incidence of lung cancer. And so on. To this, Sir Austin Bradford Hill (1965, p. 10) adds: "... I regard as greatly contributing to coherence the histopathological evidence from the bronchial epithelium of smokers and the isolation from cigarette smoke of factors carcinogenic for the skin of laboratory animals." The pattern of associations in §1.2 between smoking and cardiovascular disease would also be described as coherent. Coherence is discussed by Susser (1973, pp. 154–162) and more critically by Rothman (1986, p. 19). MacMahon and Pugh 1970, p. 21) use the phrase "consonance with existing knowledge" in place of coherence. Coherence is related to Fisher's "elaborate theory," as discussed in §1.2.

Typically, coherence is defined briefly, if at all, and then illustrated with examples, often compelling examples. This chapter offers a definition of coherence. The definition is reasonably close to traditional usage of the term, and it has consequences for design and inference. The purpose is to distinguish coherence from efforts to detect hidden biases, as described in Chapters 5, 6, and 7. Though both are useful, coherence and detection differ in their logic and their evidence.

In detecting hidden biases, the goal is to collect data so that an actual treatment effect would be visibly different from the most plausible hidden biases. Control groups, referent groups, or outcomes with known effects are selected with this goal in mind; see, in particular, §5.2.3, §6.2.2, §7.1, and §7.2.3. The many tables in Campbell and Stanley (1963) concern the ability of various research designs to distinguish treatment effects from biases of various kinds. Detection concerns distinguishing biases and effects.

Coherence is different. There is no reference to particular biases, no assurance that certain specific biases would be visibly different from treatment effects. A coherent pattern of associations is one that is, at each of many points, in harmony with existing knowledge of how the treatment should behave if it has an effect. In a coherent association, an elaborate theory describing the treatment and its effects closely fits equally elaborate data. What does such a close fit say about hidden biases?

8.1.2. Focused Hypotheses About Treatment Effects

Sir Karl Popper writes:

> It is easy to obtain confirmations, or verifications, for nearly every theory —if we look for confirmations .... Confirmations should count only if they are the result of *risky predictions*; that is to say, if, unenlightened by the theory in question, we should have expected an event that was incompatible with the theory—an event that would have refuted the theory [Popper, 1965, p. 36] ... as scientists we do not seek highly probable theories but ... powerful and improbable theories [Popper, 1965, p. 58].
>
> A theory will be said to be better corroborated the more severe the [evaluations] it has passed.... [An evaluation is] more severe the greater the probability of failing it (the absolute or prior probability as well as the probability in the light of what I call our 'background knowledge', that is to say, knowledge which, by common agreement, is not questioned while [evaluating] the theory under investigation) [Popper, 1983, p. 244].

In §1.1, Fisher's elaborate theories are risky predictions because the form of the anticipated treatment effect is tightly and narrowly specified. If such an elaborate theory is confirmed at each of many opportunities, then one senses there is more dramatic evidence against the null hypothesis of no treatment effect and in favor of the elaborate theory of an effect. Indeed, the statistical theory supports this view in the sense that hypothesis tests against ordered alternatives often provide more dramatic evidence against a null hypothesis; see, for instance, Jonckheere (1954), Page (1963), Barlow, Bartholomew, Bremner, and Brunk (1972), and Robertson, Wright, and Dykstra (1988). In Chapter 4, dramatic evidence was typically less sensitive to hidden bias. Perhaps a coherent association, an association that matches an elaborate theory, is less sensitive to hidden bias.

8.1.3. Focused Hypotheses Expressed in Terms of Partial Orders: An Example

Many focused hypotheses can be expressed in terms of an anticipated partial ordering of the responses, as discussed in §2.8. Typically, there are several responses, that is, R_{si} is no longer a single number for each subject, but rather R_{si} records several responses for each subject. The focused hypothesis often expresses the thought that one pattern of responses is closer to what is anticipated from a treated subject and another pattern is closer to what is anticipated from a control.

Recall from §4.5.4 the study of possible chromosome damage due to eating fish contaminated with methylmercury, that is, the study by Skerfving, Hansson, Mangs, Lindsten, and Ryman (1974). There were two groups, a treated or exposed group of subjects who ate large quantities of contaminated fish, and a control group that is believed not to have eaten contaminated fish. Table 8.1.1 presents three outcomes from this study, namely the level of mercury found in the blood in ng/g, the percent of cells exhibiting chromosome abnormalities, and the percent of C_u cells exhibiting a particular type of abnormality. This study has a single stratum, $S = 1$, so the s subscript is omitted. The second column in this table is the indicator of the treatment group, $Z_i = 1$ if exposed to methylmercury and $Z_i = 0$ if not exposed.

If eating contaminated fish is causing chromosome damage, then one would expect:

(i) higher levels of mercury in the blood of exposed subjects;
(ii) more chromosome abnormalities in total;
(iii) more C_u cells; and
(iv) the greatest chromosome damage in subjects with the highest levels of mercury in their blood.

If some of these predictions were confirmed and others were not, the results might be described as incoherent, and we might doubt that the treatment is actually the cause of differences seen. In the study of nuclear fallout in §5.1.2, doubts about causality arose because increases in childhood leukemia were accompanied by decreases in other childhood cancers, an incoherent result.

Table 8.1.1. Mercury in Fish and Chromosome Damage: Poset Rank Scores.

i	Group	Mercury	% Abnormal	% Cu	Rank score
1	0	5.3	8.6	2.7	−8
2	0	15	5	0.5	−13
3	0	11	8.4	0	−13
4	0	5.8	1	0	−25
5	0	17	13	5	14
6	0	7	5	0	−20
7	0	8.5	1	0	−25
8	0	9.4	2.3	1.3	−16
9	0	7.8	2	0	−24
10	0	12	6.4	1.8	−7
11	0	8.7	7	0	−15
12	0	4	1.7	0	−26
13	0	3	4	1	−20
14	0	12.2	1.8	1.8	−14
15	0	6.1	2.8	0	−23
16	0	10.2	4.7	3.1	−2
17	1	100	6.4	0.7	−1
18	1	70	9.2	4.6	16
19	1	196	3.6	0	−7
20	1	69	3.7	1.7	−6
21	1	370	14.2	5.2	35
22	1	270	7	0	−5
23	1	150	13.5	5	24
24	1	60	21.5	9.5	19
25	1	330	9	2	18
26	1	1100	11	3	26
27	1	40	8	1	−4
28	1	100	9.2	3.5	16
29	1	70	8	2	7
30	1	150	14	5	24
31	1	200	11.9	5.5	26
32	1	304	10	2	17
33	1	236	6.6	3	17
34	1	178	13	4	23
35	1	41	0	0	−17
36	1	120	6	2	5
37	1	330	13.1	2.2	23
38	1	62	0	0	−16
39	1	12.8	5.3	2	−3

The response R_i of subject i is a triple recording mercury level, total aberrations, C_u aberrations. Compare subjects $i = 1$ and $i = 5$ in Table 8.1.1, both of whom are controls. Subject $i = 5$ has higher values all three responses than subject $i = 1$. Comparing subjects $i = 1$ and $i = 5$, it is subject $i = 5$ whose responses more closely resemble those anticipated from a treated subject. Now compare subject 5 to subject 21. Subject 21 has higher responses on all three variables than subject 5, so subject 21 has responses that more closely resemble those anticipated from a treated subject. Finally, compare subject 5 to subject 17. Here, subject 17 has more mercury in the blood but fewer chromosome aberrations in total and of type C_u. It is not possible to say which of these response triples more closely resembles that anticipated from the treatment. The response triples are partially but not totally ordered. The possible values of the response form a partially ordered set or poset. For a single response, say level of mercury, for any two subjects, either one has a higher level of mercury than the other or else they have equal levels of mercury—real numbers are totally ordered. In contrast, for response triples there is another possibility, as in the case of subjects 5 and 17, the responses may not be comparable—response triples are partially ordered. Partial orders are discussed more formally in §8.2.

The ranks in the last column of Table 8.1.1 assign high values to individuals whose responses resemble those anticipated from a treated subject. The calculation of the ranks is discussed in §8.2, but for now consider how the ranks behave. Notice that subject 21 has the highest rank and, in fact, no other subject is higher on all three responses than subject 21, though some subjects are higher on one or two responses. No other subject has responses that more closely resemble those anticipated from the treatment. If subject 21 were found in the control group it would work strongly against the claim of a treatment effect, while if found in the treated group it would work strongly in favor of a treatment effect. The lowest rank score is for subject $i = 12$, being low on all three outcomes. Finding subject 12 in the control groups supports the claim of a treatment effect. Subject $i = 17$ has a score of -1, nearly neutral, because this subject has a high level of mercury but only typical levels of chromosome aberrations. Subject 17 pulls in the direction of the null hypothesis of no treatment effect whether found in the treated group or in the control group.

Most of the large ranks are found in the treated group and most of the negative ranks are found in the control group, a finding that tends to support the claim that the treatment has its anticipated effects. There are individuals who work against this claim, for instance, control subject $i = 5$ with a positive score of 14 and treated subject $i = 35$ with a negative score -17. Subject 5 did not eat contaminated fish but had many chromosome aberrations, while subject 15 ate large quantities of contaminated fish but had no chromosome aberrations. This example will be discussed further in §8.2.

8.2. Sensitivity of a Test for a Coherent Association

8.2.1. Calculating Poset Rank Scores and a Poset Statistic

The rank scores in Table 8.1.1 were calculated using the method in §2.8.4, which will now be described in the context of the example of mercury and chromosome damage. Then the null hypothesis of no effect will be tested against a coherent alternative. Write **R** for the matrix whose N rows are the vector responses for the N subjects under study.

There is a partial order $\lesssim$ on the set of possible values of the response. In the example, the response R_i is a 3-tuple giving (level of mercury, % abnormal cells, % C_u cells) and one response R_j is smaller, or at least no larger, than another response R_i, written $R_j \lesssim R_i$, if each coordinate of R_j is less than or equal to each coordinate of R_i. This partial order satisfies the three conditions that define a partially ordered set, namely, for all values a, b, and c in the set:

(i) $a \lesssim a$;
(ii) $a \lesssim b$ and $b \lesssim a$ imply $a = b$; and
(iii) $a \lesssim b$ and $b \lesssim c$ imply $a \lesssim c$.

A totally ordered set has the additional property that

(iv) $a \lesssim b$ or $b \lesssim a$ for every a and b;

however, this is true for real numbers but not for triples of real numbers, as seen with subjects 5 and 17 in Table 8.1.1.

Let L_{ij} indicate whether subject i has a higher response than subject j, or a lower response, or an equal or noncomparable response; that is, let

$$L_{ij} = \begin{cases} 1 & \text{if } R_j \lesssim R_i \text{ with } R_i \neq R_j, \\ -1 & \text{if } R_i \lesssim R_j \text{ with } R_i \neq R_j, \\ 0 & \text{otherwise.} \end{cases}$$

For instance, $L_{1,5} = -1$, $L_{5,1} = 1$, $L_{5,21} = -1$, $L_{1,1} = 0$, $L_{5,17} = 0$. Notice that $L_{ii} = 0$ for all i.

The rank score for subject i in Table 8.1.1 is

$$q_i = \sum_{j=1}^{n} L_{ij},$$

so q_i is the number of subjects with lower responses minus the number with higher responses, ignoring subjects with the same or noncomparable responses. For instance, for subject $i = 21$ in Table 8.1.1, there are 35 subjects with responses lower than that of subject i, no subjects with higher responses, three subjects with noncomparable responses, namely, $j = 24, j =$

26, and $j = 31$, and one subject with the same response, namely, the subject $i = 21$ itself, so $q_i = 35$. A large positive rank q_i indicates most subjects had lower responses than subject i, and a large negative rank q_i indicates most subjects had higher responses. A rank q_i near zero has two possible interpretations, either a subject in the middle, or an incongruous subject who is not comparable to most others perhaps because of high responses on some coordinates and low responses on others.

The statistic

$$T = t(\mathbf{Z}, \mathbf{R}) = \sum_{i=1}^{n} \sum_{j=1}^{n} Z_i(1 - Z_j)L_{ij} \tag{8.2.1}$$

is the number of times a treated subject had a higher response than a control minus the number of times a control had a higher response than a treated subject, where higher refers to the partial order $\lesssim$. In Table 8.1.1, there are $m = 23$ exposed subjects and $n - m = 16$ controls, so there are $23 \times 16 = 368$ ways to compare a treated subject to a control. It follows that T could be as large as 368 if all treated subjects had higher responses than all controls, or as small as -368 if all controls had higher responses than all treated subjects. In fact, $T = 237$, so treated subjects had higher responses than controls far more often than controls had higher responses than treated subjects, in fact, 237 more times.

As shown in §2.8.4, using the device of Mantel (1967), the statistic T may be written as the sum of the rank scores in the treated group,

$$T = t(\mathbf{Z}, \mathbf{R}) = \sum_{i=1}^{n} Z_i q_i.$$

In other words, T is a sum statistic, as defined in §2.4.3. Also, using (2.8.3), it follows that $0 = \sum_{i=1}^{m} q_i$. Therefore, using Proposition 2.2, in the absence of hidden bias, if the treatment has no effect, $E(T) = 0$ and

$$\operatorname{var}(T) = \frac{m(n-m)}{n(n-1)} \sum_{i=1}^{n} q_i^2. \tag{8.2.2}$$

For the example, (8.2.2) is 3019, leading to a standardized deviate of $(237 - 0)/\sqrt{3019} = 4.3$, which gives an approximate one-sided significance level less than 0.0001 using the Normal distribution. If the data in Table 8.1.1 were from a randomized experiment, there would be strong evidence against the null hypothesis of no treatment effect, evidence in the direction indicated by the partial order $\lesssim$.

The poset statistic (8.2.1) generalizes the tests of Wilcoxon (1945) and Mann and Whitney (1947); see Problem 1. A second generalization is discussed and contrasted with (8.2.1) in the problems in §8.5.1. For properties of this test against coherent alternatives, see the problems in §8.5.2.

Table 8.2.1. Sensitivity of the Test for a Coherent Association.

e^{γ}	Deviate	Bound
1	4.26	<0.0001
2	3.40	0.0003
3	2.95	0.0016
10	1.79	0.04
12	1.63	0.05
14	1.50	0.07

8.2.2. Sensitivity Analysis

Table 8.2.1 is a sensitivity analysis for the poset statistic T for the mercury contamination data. Because T is a sum statistic, the sensitivity analysis is conducted using the general method in §4.5.2 with the scores q_i in Table 8.1.1. Table 8.2.1 gives the upper bound on the significance level for T, the lower bound being less than 0.0001 in all cases.

The sensitivity analysis indicates that the null hypothesis of no effect becomes plausible for $\Gamma = e^{\gamma} \geq 12$. Very large hidden biases would be required to explain away the entire pattern of associations consistent with the predicted ordering of responses.

8.2.3. Strengths and Limitations of Coherence

As an argument, coherence has two strengths. First, a test against a coherent alternative penalizes a data set that exhibits nonsensical associations, making it more difficult to detect an effect in this case. Second, evidence of an effect may be dramatic in either of two ways: the effect may appear dramatic in size, or it may make numerous predictions each of which is confirmed in data. A test against a coherent alternative provides a basis for appraising the latter possibility.

Coherence also has two significant limitations. First, a coherent pattern of associations may result from a coherent pattern of hidden biases. For instance, in §8.2.2, if fish contaminated with methylmercury were also contaminated with other substances, then mercury in the blood and chromosome damage might be much more common among subjects who consume the fish, even if mercury is not the cause of chromosome damage. As discussed in §8.1.1, unlike efforts to detect hidden bias, coherence is concerned with the pattern of associations that is consistent with a treatment effect, but it does not take active steps to ensure that this pattern differs from what is expected from the most plausible hidden biases. Contrast coherence with "control by systematic variation" in §7.1.3 and §7.2.3. If control groups are selected to systematically vary a relevant unobserved covariate then, as seen in §7.2.3, there are good prospects that any hidden bias that may be due to this covariate will be revealed when the control groups are compared. In contrast, the test against a coherent alternative comes with no promise that it can assist in distinguishing a treatment effect from any particular hidden bias.

The second limitation of coherence is apparent when Table 8.2.1 is compared with the separate sensitivity analysis for mercury levels and C_u cells in §4.5.4. Taken alone, the comparison for mercury levels was highly insensitive to hidden bias, being significant even for $\Gamma = 35$. Taken alone, the comparison for C_u cells became sensitive to bias for $\Gamma = 3$. Table 8.2.1 indicates that the entire pattern of associations consistent with a treatment effect is highly insensitive to hidden bias. Although this is true, it may say that there is strong evidence, insensitive to hidden bias, asserting that eating fish contaminated with mercury raises the level of mercury in the blood, together with some evidence, more sensitive to bias, that mercury causes chromosome damage. In general, a test against a coherent alternative concerns evidence against the null hypothesis of no effect and in the direction of a coherent pattern of associations. The evidence against the null hypothesis of no effect of any kind may be strong and insensitive to hidden bias, but the evidence about individual outcomes may be quite varied, as is true in the current example. Because outcomes differ in their consequences for policy, it is rarely sufficient to know that there is at least some effect in one of the outcomes. A sensitivity analysis against a coherent alternative should generally be accompanied by a sensitivity analysis for individual outcomes.

In short, a sensitivity analysis for a test against a coherent alternative is useful in providing an appraisal of the evidence provided by the confirmation of numerous predictions. However, such an analysis is not a substitute for sensitivity analyses for individual outcomes nor for active steps to detect hidden bias.

*8.3. Appendix: Arrangement-Increasing Functions of Matrices

In this section, the definition of arrangement-increasing functions is extended to certain partial orders and, in particular, the statistic (8.2.1) as used in §8.2.1 is seen to be arrangement-increasing with respect to a partial order. Section 2.4.3 discussed arrangement-increasing functions $f(\mathbf{a}, \mathbf{b})$ of two stratified N-dimensional vectors $\mathbf{a}, \mathbf{b}$, that is, two vectors indexed by $s = 1, \ldots, S$, $i = 1, \ldots, n_s$, $N = n_1 + \cdots + n_S$. Such a function was defined by two properties. First, it was permutation-invariant in the sense that interchanging two coordinates from the same stratum in both vectors does not alter the value of the function. Second, interchanging two coordinates from the same

* This section may be skipped without loss of continuity.

stratum in one vector so their order disagrees with the order in the other vector decreases, or at least does not increase, the value of the function.

These two properties of arrangement-increasing functions may be stated formally using "interchange vectors" $\boldsymbol{\iota}_{sij}$. There is an interchange vector for each s, $s = 1, \ldots, S$, and each i, j with $1 \leq i \leq j \leq n_s$, and $\boldsymbol{\iota}_{sij}$ is the N-dimensional column vector with a one in the ith coordinate for stratum s, a negative one in the jth coordinate for stratum s, and a zero in all other coordinates. Let $\mathbf{I}$ denote the $N \times N$ identity matrix. When an N-dimensional vector $\mathbf{b}$ is multiplied by the matrix $\mathbf{I} - \boldsymbol{\iota}_{sij}\boldsymbol{\iota}_{sij}^{\mathrm{T}}$ the result $(\mathbf{I} - \boldsymbol{\iota}_{sij}\boldsymbol{\iota}_{sij}^{\mathrm{T}})\mathbf{b}$ is $\mathbf{b}$ with its ith and jth coordinates of stratum s interchanged. Then $f(\cdot, \cdot)$ is permutation invariant if, for each $\mathbf{a}$, and $\mathbf{b}$ in the domain of $f(\cdot, \cdot)$, for every interchange vector $\boldsymbol{\iota}_{sij}$, $f(\mathbf{a}, \mathbf{b}) = f[(\mathbf{I} - \boldsymbol{\iota}_{sij}\boldsymbol{\iota}_{sij}^{\mathrm{T}})\mathbf{a}, (\mathbf{I} - \boldsymbol{\iota}_{sij}\boldsymbol{\iota}_{sij}^{\mathrm{T}})\mathbf{b}]$. A permutation invariant $f(\cdot, \cdot)$ is arrangement-increasing if $\mathbf{a}^{\mathrm{T}}\boldsymbol{\iota}_{sij}\boldsymbol{\iota}_{sij}^{\mathrm{T}}\mathbf{b} \geq 0$ implies $f(\mathbf{a}, \mathbf{b}) \geq f[\mathbf{a}, (\mathbf{I} - \boldsymbol{\iota}_{sij}\boldsymbol{\iota}_{sij}^{\mathrm{T}})\mathbf{b}]$.

In the same way, let $\mathbf{A}$ and $\mathbf{B}$ be two matrices each with N stratified rows numbered $s = 1, \ldots, S$, $i = 1, \ldots, n_s$. Let $f(\mathbf{A}, \mathbf{B})$ be a real-valued function of these matrix arguments. Then $f(\cdot, \cdot)$ is permutation invariant if simultaneously interchanging two rows of $\mathbf{A}$ and the same two rows of $\mathbf{B}$ in the same subclass does not change $f(\mathbf{A}, \mathbf{B})$, that is, if for every $\mathbf{A}$ and $\mathbf{B}$ in the domain of $f(\cdot, \cdot)$, and for every interchange vector, $f(\mathbf{A}, \mathbf{B}) = f[(\mathbf{I} - \boldsymbol{\iota}_{sij}\boldsymbol{\iota}_{sij}^{\mathrm{T}})\mathbf{A}, (\mathbf{I} - \boldsymbol{\iota}_{sij}\boldsymbol{\iota}_{sij}^{\mathrm{T}})\mathbf{B}]$. Write $\mathbf{C} \geq \mathbf{0}$ if $c_{km} \geq 0$ for each k, m. A permutation-invariant function $f(\cdot, \cdot)$ is arrangement-increasing if $\mathbf{A}^{\mathrm{T}}\boldsymbol{\iota}_{sij}\boldsymbol{\iota}_{sij}^{\mathrm{T}}\mathbf{B} \geq \mathbf{0}$ implies $f(\mathbf{A}, \mathbf{B}) \geq f[\mathbf{A}, (\mathbf{I} - \boldsymbol{\iota}_{sij}\boldsymbol{\iota}_{sij}^{\mathrm{T}})\mathbf{B})$. In other words, if rows (s, i) and (s, j) of $\mathbf{A}$ and $\mathbf{B}$ are ordered in the same way, then interchanging these two rows of $\mathbf{B}$ reduces, or at least does not increase, $f(\mathbf{A}, \mathbf{B})$.

Let $\mathbf{R}$ be an $N \times p$ matrix of p-dimensional responses for N subjects, with the usual partial order $\lesssim$ saying that subject (s, i) has a lower response than subject (s, j) if each of the p responses for subject (s, i) is less than or equal to each of the p responses for subject (s, j). This is the partial order used in §8.1.3 and the first partial order in §2.8.3. With this partial order, the statistic $t(\mathbf{Z}, \mathbf{R})$ in (2.8.2) and (8.2.1) is arrangement-increasing.

Hollander, Proschan, and Sethuraman (1977) discuss properties of arrangement-increasing functions of two vectors, and many of these extend to arrangement-increasing functions of two matrices, sometimes with small modifications; see Rosenbaum (1991, §2). For instance, using an extension of their composition theorem, Proposition 5.1 in §5.2.3 may be proved for poset statistics. Specifically, if the unaffected outcome is a matrix $\mathbf{y}$ of q responses and $t^*(\mathbf{z}, \mathbf{y})$ is arrangement-increasing as a function of the vector $\mathbf{z}$ and matrix $\mathbf{y}$, such as the statistic (2.8.2), then the power $\beta(\mathbf{y}, \mathbf{u})$ is arrangement-increasing as a function of the matrix $\mathbf{y}$ and the vector $\mathbf{u}$. A different application arises when there is underlying and unobserved dose of treatment, such as the amount of mercury consumed in the example in §8.1.3; see Rosenbaum (1991, §4) for a practical example.

8.4. Bibliographic Notes

Coherence is discussed with varying terminology and varying degrees of enthusiasm by Hill (1965), MacMahon and Pugh (1970, p. 21), Susser (1973, pp. 154–162), Rothman (1986, p. 19), among others. There is a large literature on statistical methods for order restricted inference, though the most widely used methods are for totally ordered outcomes rather than partially ordered outcomes, that is, they concern dose-response relationships rather than coherence of multiple outcomes. See Jonckheere (1954) and Page (1963) for nonparametric methods for totally ordered outcomes; these tests are also discussed in the textbook by Hollander and Wolfe (1973). For comprehensive surveys of order restricted inference with extensive bibliographies, see Barlow, Bartholomew, Bremner, and Brunk (1972), and Robertson, Wright, and Dykstra (1988). Maclure and Greenland (1992) are critical of potential misinterpretation of tests for trend and suggest careful terminology for discussing such tests. The statistic (8.2.1) is discussed in Rosenbaum (1994) with reference to coherence; it generalizes the Mann and Whitney (1947) statistic and is related to the ideas in Mantel (1967). Poset statistics and arrangement-increasing functions of matrices are discussed in Rosenbaum (1991).

8.5. Problems

8.5.1. Contrasting Two Generalizations of the Rank Sum Test for Posets

1. Suppose the outcomes R_i are distinct real numbers and the order $\lesssim$ is the ordinary inequality $\leq$ of real numbers. What are the rank scores q_i in §8.2.1 in this case? How do they relate to conventional ranks? What is the relationship between the statistic (8.2.1) and Wilcoxon's rank sum test?

2. Problem 1 showed that the statistic (8.2.1) is a generalization of Wilcoxon's rank sum test. A second generalization for multivariate outcomes such as those in Table 8.1.1 is to separately rank each outcome and form a combined rank for each subject by summing the ranks for the several outcomes for each subject. In other words, separately rank each of the three columns in Table 8.1.1 and then sum the three ranks in row i to form a score for subject i. The statistic is the sum of the scores for treated subjects. Show that both of these statistics are arrangement-increasing in the sense of §8.3.

3. Compare the two generalizations of the rank sum statistic in (8.2.1) and Problem 2. What would happen to each statistic if, in Table 8.1.1, the mercury levels for the 23 treated subjects were permuted among themselves, without permuting the

other two outcomes? For instance, what would happen to the test statistics if the mercury levels of 60 and 1100 for subjects $i = 24$ and $i = 26$ were interchanging leaving the rest of the data unchanged? How is the answer related to prediction (iv) in §8.1.3?

4. Continuing Problem 3, which of the two generalizations would seem to be preferable if there are many outcomes? Why?

5. How would one conduct a sensitivity analysis for the second generalization in Problem 2?

8.5.2. A Poset Statistic Under Alternative Hypotheses

6. Consider the quantity

$$\tilde{L}_{sij} = \begin{cases} 1 & \text{if } r_{Csj} \lesssim r_{Tsi} \text{ with } r_{Tsi} \neq r_{Csj}, \\ -1 & \text{if } r_{Tsi} \lesssim r_{Csj} \text{ with } r_{Tsi} \neq r_{Csj}, \\ 0 & \text{otherwise.} \end{cases}$$

Write the statistic (2.8.2) in terms of $\tilde{L}_{sij}$. Rosenbaum (1994)

7. Use the representation in Problem 6 to show that the statistic (2.8.2) is order preserving. What does Proposition 2.4 say about tests against coherent alternatives?

References

Barlow, R., Bartholomew, D., Bremner, J., and Brunk, H. (1972). *Statistical Inference Under Order Restrictions.* New York: Wiley.

Bayne-Jones, S., Burdette, W., Cochran, W., Farber, E., Fieser, L., Furth, J., Hickman, J., LeMaistre, C., Schuman, L., Seevers, M. (1964). *Smoking and Health: Report of the Advisory Committee to the Surgeon General of the Public Health Service.* Washington, DC: US Department of Health, Education, and Welfare.

Campbell, D. and Stanley, J. (1963). *Experimental and Quasi-Experimental Designs for Research.* Chicago: Rand McNally.

Hill, A.B. (1965). The environment and disease: Association or causation? *Proceedings of the Royal Society of Medicine*, **58**, 295–300.

Hollander, M., Proschan, F., and Sethuraman, J. (1977). Functions decreasing in transposition and their applications in ranking problems. *Annals of Statistics*, **5**, 722–733.

Hollander, M. and Wolfe, D. (1973). *Nonparametric Statistical Methods.* New York: Wiley.

Jonckheere, A. (1954). A distribution-free k-sample test against ordered alternatives. *Biometrika*, **41**, 133–145.

Maclure, M. and Greenland, S. (1992). Tests for trend and dose-response: Misinterpretations and alternatives. *American Journal of Epidemiology*, **135**, 96–104.

MacMahon, B. and Pugh, T. (1970). *Epidemiology: Principles and Methods.* Boston: Little, Brown.

Mann, H. and Whitney, D. (1947). On a test of whether one of two random variables is stochastically larger than the other. *Annals of Mathematical Statistics*, **18**, 50–60.

Mantel, N. (1967). Ranking procedures for arbitrarily restricted observations. *Biometrics*, **23**, 65–78.

Page, E. (1963). Ordered hypotheses for multiple treatments: A significance test for linear ranks. *Journal of the American Statistical Association*, **58**, 216–230.

Popper, K. (1965). *Conjectures and Refutations*. New York: Harper & Row.

Popper, K. (1983). *Realism and the Aim of Science*. Totowa, NJ: Rowman and Littlefield.

Robertson, T., Wright, F.T., and Dykstra, R.L. (1988). *Order Restricted Statistical Inference*. New York: Wiley.

Rosenbaum, P.R. (1991). Some poset statistics. *Annals of Statistics*, **19**, 1091–1097.

Rosenbaum, P.R. (1994). Coherence in observational studies. *Biometrics*, **50**, 368–374.

Rothman, K. (1986). *Modern Epidemiology*. Boston: Little, Brown.

Skerfving, S., Hansson, K., Mangs, C., Lindsten, J., and Ryman, N. (1974). Methylmercury-induced chromosome damage in man. *Environmental Research*, **7**, 83–98.

Susser, M. (1973). *Causal Thinking in the Health Sciences*. New York: Oxford University Press.

Wilcoxon, F. (1945). Individual comparisons by ranking methods. *Biometrics*, **1**, 80–83.

CHAPTER 9

Constructing Matched Sets and Strata

9.1. Introduction: Propensity Scores, the Form of Optimal Strata, the Construction of Optimal Matched Samples

This chapter discusses the construction of matched sets or strata when there are several, perhaps many, observed covariates **x**. There are three topics: the propensity score, the form of an optimal stratification, and the construction of optimal matched sets. This introduction summarizes the main issues and findings.

As the number p of covariates increases, it becomes difficult to find matched pairs with the same or similar values of **x**. Even if each covariate is a binary variable, there will be 2^p possible values of **x**, so with $p = 20$ covariates there are more than a million possible values of **x**. If there are hundreds or thousands of subjects and $p = 20$ covariates, it is likely that many subjects will have unique values of **x**. For this reason, it is sometimes said that matching is not useful when there are many covariates. Actually, the problem is not with matching as a technique but rather with specific objective of obtaining matched pairs or sets which are homogeneous in **x**. If the objective is defined in other ways, then matching and stratification are not difficult with many covariates.

There are two objectives of matching and stratification besides matched sets that are homogeneous in **x**. First, if there is no hidden bias so that it suffices to adjust for **x**, then strata or matched sets are desired that permit use of the conventional methods in Chapter 2. Second, whether or not there is hidden bias, one would like to compare treated and control groups with similar distributions of **x**, even if matched individuals have differing values of **x**. The second objective is called covariate balance.

The propensity score is a device for constructing matched sets or strata when $\mathbf{x}$ contains many covariates. If the true propensity score were known, both objectives in previous paragraph would be attained by matching or stratifying on the propensity score, a single covariate. That is, if strata or matched sets are formed that are homogeneous in the propensity score, even if they are heterogeneous in $\mathbf{x}$, then the methods of Chapter 2 are appropriate in the absence of hidden bias, and the observed covariates $\mathbf{x}$ will tend to balance whether or not there is hidden bias. In the case of the first objective, this was demonstrated in §3.2.5; see also §9.2.1 below. Covariate balance is demonstrated in §9.2.2. In practice, the propensity score is not known and must be estimated. One approach was discussed in §3.5. For matching and stratification, use of estimated propensity scores is illustrated in §9.2.3 and simulation results are reviewed in §9.4.7.

Section 9.3 discusses optimal stratification. As it turns out, under quite general conditions, the form of an optimal stratification is always the same. This implies that in searching for a good stratification, the search may be confined to stratifications of this form because one is sure to be optimal. This optimal form is called a full matching. It is a matched sample in which each matched set contains either a treated subject and one or more controls or else a control subject and one or more treated subjects. Pair matching is not optimal, and neither is matching with a fixed number of controls. In the simulation results described in §9.4.7, full matching is found to be substantially better than matching with a fixed number of controls. It is easy to see why this happens. When the treated and control groups have different distributions of the observed covariates $\mathbf{x}$, there are regions of $\mathbf{x}$ values with many treated subjects and few controls, and other regions with many controls and few treated subjects. Forcing every treated subject to have the same number of controls creates some poor matched sets.

The construction of optimal matched samples is discussed in §9.4. Various types of matching are illustrated and compared, including matching with a fixed number of controls, matching with a variable number of controls, full matching and balanced matching. Network flow techniques are used to obtain optimal matched samples, and optimal matching is contrasted with a commonly used alternative, greedy matching.

9.2. The Propensity Score

9.2.1. Definition of the Propensity Score

The propensity score is the conditional probability of receiving the treatment given the observed covariates $\mathbf{x}$. Recall from §3.2 the model for treatment assignment with $\pi_{si} = \text{prob}(Z_{si} = 1)$ and $0 < \pi_{si} < 1$. For all n_s subjects in stratum s, define the *propensity score* to be

$$\lambda(\mathbf{x}_s) = \frac{\sum_{i=1}^{n_s} \pi_{si}}{n_s}.$$

Since each π_{si} satisfies $0 < \pi_{si} < 1$, it follows that $0 < \lambda(\mathbf{x}_s) < 1$ for each s.

The propensity score $\lambda(\mathbf{x}_s)$ has the following operational interpretation. Pick a subject at random from stratum s, each subject having probability $1/n_s$ of being selected; then this random subject receives the treatment with probability $\lambda(\mathbf{x}_s)$. That is, pick subject (s, i) with probability $1/n_s$, where (s, i) receives the treatment with probability π_{si}, so the marginal probability that a random subject will receive the treatment is $\sum_{i=1}^{n_s} \pi_{si}/n_s = \lambda(\mathbf{x}_s)$.

The propensity score has two useful properties. The first, the more important of the two properties, was discussed in §3.2.5 and it applies when there is no hidden bias, that is, when $\pi_{si} = \lambda(\mathbf{x}_s)$ for every s and i. If there is no hidden bias, then one need not form strata or matched sets that are homogeneous in $\mathbf{x}_s$; it suffices to obtain strata or matched sets that are homogeneous in $\lambda(\mathbf{x}_s)$. If there is no hidden bias, if the strata are homogeneous in $\lambda(\mathbf{x}_s)$, then the conditional distribution of treatment assignments is uniform, and the statistical methods in Chapter 2 for a randomized experiment may be used; see §3.2.5 for details. Since $\mathbf{x}_s$ may be of high dimension, but $\lambda(\mathbf{x}_s)$ is a number, it is often much easier to find subjects with similar values of $\lambda(\mathbf{x}_s)$ than with similar values of $\mathbf{x}_s$. When there is no hidden bias, when there is only overt bias due to $\mathbf{x}_s$, it suffices to adjust for the propensity score $\lambda(\mathbf{x}_s)$.

The second property applies whether or not there is hidden bias, that is, it applies even if $\pi_{si} \neq \lambda(\mathbf{x}_s)$. Strata or matched sets that are homogeneous in $\lambda(\mathbf{x}_s)$ tend to balance $\mathbf{x}_s$ in the sense that treated and control subjects in the same stratum or matched set tend to have the same distribution of $\mathbf{x}_s$. In an experiment, randomization tends to balance all covariates, observed and unobserved, in the sense that treated and control groups tend to have the same distribution of covariate values. In an observational study, strata or matched sets that are homogeneous in the propensity score $\lambda(\mathbf{x}_s)$ tend to balance the observed covariates $\mathbf{x}_s$, though there may be imbalances in unobserved covariates. The balancing property is demonstrated in the next section.

9.2.2. Balancing Properties of the Propensity Score

Pick a value of the propensity score, say Λ, such that there is at least one s with $\lambda(\mathbf{x}_s) = \Lambda$. There may be several strata s with this same value of the propensity score, $\lambda(\mathbf{x}_s) = \Lambda$, so write n_Λ for the total number of subjects in these strata, that is,

$$n_\Lambda = \sum_{s:\lambda(\mathbf{x}_s)=\Lambda} n_s = \sum{}^* n_s.$$

where $\sum^*$ denotes a sum over all s such that $\lambda(\mathbf{x}_s) = \Lambda$. Pick a subject at random from these strata, that is, pick a subject (s, i) from $\{(s, i): \lambda(\mathbf{x}_s) = \Lambda\}$ where each such subject has the same probability of being selected, namely, $1/n_\Lambda$. Write Z and $\mathbf{X}$ for the treatment assignment and observed covariate for this random subject, that is, if the subject selected is (s, i) then $Z = Z_{si}$ and $\mathbf{X} = \mathbf{x}_s$.

What is the probability that $Z = 1$? To emphasize that the probability refers to a subject randomly selected from the n_Λ subjects with $\lambda(\mathbf{x}_s) = \Lambda$, write the probability as $\text{prob}\{Z = 1 | \lambda(\mathbf{X}) = \Lambda\}$. If $\lambda(\mathbf{x}_s) = \Lambda$, then the random subject comes from stratum s with probability n_s/n_Λ, so

$$\text{prob}\{Z = 1 | \lambda(\mathbf{X}) = \Lambda\} = \sum{}^* \frac{n_s \lambda(\mathbf{x}_s)}{n_\Lambda} = \sum{}^* \frac{n_s \Lambda}{n_\Lambda} = \Lambda,$$

where again $\sum^*$ denotes a sum over all s such that $\lambda(\mathbf{x}_s) = \Lambda$.

This same conclusion, $\text{prob}\{Z = 1 | \lambda(\mathbf{X}) = \Lambda\} = \Lambda$, may be expressed in a slightly different way. If $\lambda(\mathbf{x}_s) = \Lambda$, then subject (s, i) is selected with probability $1/n_\Lambda$, and receives the treatment with probability π_{si}; so

$$\text{prob}\{Z = 1 | \lambda(\mathbf{X}) = \Lambda\} = \frac{1}{n_\Lambda} \sum{}^* \sum_{i=1}^{n_s} \pi_{si} = \sum{}^* \frac{n_s \lambda(\mathbf{x}_s)}{n_\Lambda} = \Lambda.$$

Proposition 9.1 below describes the balancing property of the propensity score. It says that a treated and a control subject with the same value of the propensity score have the same distribution of the observed covariate $\mathbf{X}$. This means that in a stratum or matched set that is homogeneous in the propensity score, treated and control subjects may have differing values $\mathbf{X}$, but the differences will be chance differences rather than systematic differences. Again, this balancing concerns observed covariates, but unlike a randomized experiment, the propensity score does not typically balance unobserved covariates. More precisely, Proposition 9.1 says the following. Pick a value of the propensity score Λ and pick one subject at random from among the n_Λ subjects with this value of the propensity score; then for this subject, treatment assignment Z is independent of the value of the covariate $\mathbf{X}$ given the value of the propensity score $\lambda(\mathbf{X}) = \Lambda$. The proposition is due to Rosenbaum and Rubin (1983).

Proposition 9.1. *If* $\lambda(\mathbf{x}_s) = \Lambda$, *then*

$$\text{prob}\{\mathbf{X} = \mathbf{x}_s | \lambda(\mathbf{X}) = \Lambda, Z = 1\} = \text{prob}\{\mathbf{X} = \mathbf{x}_s | \lambda(\mathbf{X}) = \Lambda, Z = 0\}.$$

Proof. If $\lambda(\mathbf{x}_s) = \Lambda$, then by Bayes' theorem,

$$\text{prob}\{\mathbf{X} = \mathbf{x}_s | \lambda(\mathbf{X}) = \Lambda, Z = 1\}$$
$$= \frac{\text{prob}\{Z = 1 | \lambda(\mathbf{X}) = \Lambda, \mathbf{X} = \mathbf{x}_s\}\ \text{prob}\{\mathbf{X} = \mathbf{x}_s | \lambda(\mathbf{X}) = \Lambda\}}{\text{prob}\{Z = 1 | \lambda(\mathbf{X}) = \Lambda\}}.$$

Now, $\text{prob}\{Z = 1 | \lambda(\mathbf{X}) = \Lambda, \mathbf{X} = \mathbf{x}_s\} = \text{prob}\{Z = 1 | \mathbf{X} = \mathbf{x}_s\} = \lambda(\mathbf{x}_s) = \Lambda$, and $\text{prob}\{Z = 1 | \lambda(\mathbf{X}) = \Lambda\} = \Lambda$, so

$$\text{prob}\{\mathbf{X} = \mathbf{x}_s | \lambda(\mathbf{X}) = \Lambda, Z = 1\} = \text{prob}\{\mathbf{X} = \mathbf{x}_s | \lambda(\mathbf{X}) = \Lambda\},$$

proving the result. □

9.2.3. Matching or Stratifying on an Estimated Propensity Score: An Example

In practice, the propensity score $\lambda(\mathbf{x})$ is unknown. In this section, the propensity score $\lambda(\mathbf{x})$ is estimated using a logit model and the estimate is used in place of the true propensity score.

The example concerns a comparison of coronary bypass surgery and medical or drug therapy in the treatment of coronary artery disease; it is from Rosenbaum and Rubin (1984). There were $N = 1515$ subjects, of whom 590 were surgical patients and 925 were medical patients. The vector $\mathbf{x}$ contained 74 observed covariates describing hemodynamic, angiographic, laboratory and exercise test results, together with information about patient histories. Each of these 74 covariates showed a statistically significant imbalance in treated and control groups; contrast this with randomized experiment in §2.1.

These 74 covariates were controlled using five strata formed from an estimated propensity score. The propensity score was estimated using a linear logit model that predicted treatment assignment Z from the observed covariates $\mathbf{x}$; see Cox (1970) for detailed discussion of logit models. The model included some interactions and quadratic terms selected by a sequential process described in Rosenbaum and Rubin (1984). The 1515 patients were divided into five strata, each containing 303 patients, based on the estimated propensity score. The stratum with the highest estimated probabilities of surgery contained 69 medical patients and 234 surgical patients, while the stratum with the lowest estimated probabilities of surgery contained 277 medical patients and 26 surgical patients. The theory in §9.2.2 suggests that, had the true propensity scores been used in place of estimates, medical and surgical patients in the same stratum should have similar distributions of the 74 observed covariates.

Table 9.2.1 shows the imbalance in five of the 74 variables before and after stratification on the propensity score. The column labeled "Before" contains the square of the usual two-sample t-statistic, that is the F-ratio, comparing the means of each variable among medical and surgical patients. The values in this column are large indicating the medical and surgical patients exhibit statistically significant differences on each of these variables, and in fact on each of the 74 covariates. The last two columns give F-ratios from a two-way 2×5 analysis of variance performed for each variable, the two factors being medical versus surgical and the five propensity score

Table 9.2.1. Covariate Imbalance Before and After Stratification on an Estimate of the Propensity Score: F-Statistics for Selected Covariates.

Variable	Before	After: Main effect	After: Interaction
Abnormal left ventricle contraction	51.8	0.4	0.9
Progressing chest pain	43.6	0.1	1.4
Left main stenosis	22.1	0.3	0.2
Cardiomegaly	25.0	0.2	0.0
Currently being treated with long acting nitrates	31.4	0.1	2.2

strata. The column labeled "After: Main Effect" is the F-ratio for the main effect of medical versus surgical. The column labeled "After: Interaction" is the F-ratio for the two-way interaction. None of the F-ratios in Table 9.2.1 is significant at the 0.05 level, and most of the F-ratios are less than one, so there is no indication of systematic imbalances in these covariates. For all 74 covariates, only one of the $2 \times 74 = 148$ F-ratios was significant at the 0.05 level. In a randomized experiment, $0.05 \times 148 = 7.4$ F-ratios significant at 0.05 would have been expected by chance alone. Evidently, these 74 observed covariates are more closely balanced within strata than would have been expected in a randomized experiment with five strata. However, randomization would balance both observed and unobserved covariates, while stratification on the propensity score cannot be expected to balance unobserved covariates.

In this case, a fairly coarse stratification on an estimate of the propensity score did tend to balance 74 observed covariates. Using these five strata, the medical and surgical patients were compared with respect to outcomes such as survival and pain relief. Adjusting for the five strata, there was little or no difference in survival. See Rosenbaum and Rubin (1984) for detailed results.

9.3. Optimal Strata

9.3.1. Evaluating Stratifications Based on the Average Remaining Distance

This section characterizes the form of an optimal stratification. Optimality is defined in terms of a distance between each treated subject and each control, and the goal is a stratification that minimizes a weighted average of the distances within each stratum. Under mild conditions, the optimal stratification has a simple form developed in §9.3.4.

Initially, there are two sets of subjects, the treated subjects are in a set A and the controls are in a set B, with $A \cap B = \emptyset$. The initial number of treated subjects is $|A|$ and the number of controls is $|B|$, where $|\cdot|$ denotes the number of elements of a set. For each $a \in A$ and each $b \in B$, there is a distance, δ_{ab} with $0 \le \delta_{ab} \le \infty$. The distance measures the difference between a and b in terms of their observed covariates, say $\mathbf{x}_a$ and $\mathbf{x}_b$; however, it need not be a distance in the sense used to define a metric space, and it is not required to have any properties besides being nonnegative. An infinite distance, $\delta_{ab} = \infty$, indicates that $\mathbf{x}_a$ and $\mathbf{x}_b$ are so different that it is forbidden to place a and b in the same stratum.

The nature of the distance δ_{ab} is not important in characterizing the form of an optimal stratification. Many distances have been proposed. Cochran and Rubin (1973) consider the following distances:

(i) the categorical distance, if $\mathbf{x}_a = \mathbf{x}_b$, then $\delta_{ab} = 0$, otherwise $\delta_{ab} = \infty$;
(ii) caliper distance, if $|x_{ai} - x_{bi}| \le c_i$ for each coordinate i of $\mathbf{x}$, then $\delta_{ab} = 0$, otherwise $\delta_{ab} = \infty$, where the c_i are given constants;
(iii) quadratic distances, $\delta_{ab} = (\mathbf{x}_a - \mathbf{x}_b)^{\mathrm{T}}\mathbf{D}(\mathbf{x}_a - \mathbf{x}_b)$ for some matrix $\mathbf{D}$, including the Mahalanobis distance in which $\mathbf{D}$ is the inverse of a sample variance–covariance matrix of the $\mathbf{x}$'s; and
(iv) the squared difference along a linear discriminant.

See also Carpenter (1977) and Rubin (1980) for discussion of the Mahalanobis distance. Smith, Kark, Cassel, and Spears (1977) standardized the coordinates of $\mathbf{x}$ and take δ_{ab} equal to the squared length of the standardized difference $\mathbf{x}_a - \mathbf{x}_b$; this is a quadratic distance with $\mathbf{D}$ equal to a diagonal matrix of reciprocals of sample variances. Similar measures have been defined with the coordinates of $\mathbf{x}$ replaced by their ranks. Rosenbaum and Rubin (1985) define three distances that use the propensity score and other coordinates of $\mathbf{x}$. Again, the nature of the distance is not important in §9.3 and §9.4.

A *stratification* $(A_1, \ldots, A_S; B_1, \ldots, B_S)$ with S strata consists of S nonempty, disjoint subsets of A and S nonempty, disjoint subsets of B, so $|A_s| \ge 1$, $|B_s| \ge 1$, $A_s \cap A_{s'} = \emptyset$ for $s \ne s'$, $B_s \cap B_{s'} = \emptyset$ for $s \ne s'$, $A_1 \cup \cdots \cup A_S \subseteq A$, $B_1 \cup \cdots \cup B_S \subseteq B$. For $s = 1, \ldots, S$, stratum s consists of the treated units from A_s and the controls from B_s. Notice that a stratification may discard some units, that is, it may happen that $|A_1 \cup \cdots \cup A_S| < |A|$ or $|B_1 \cup \cdots \cup B_S| < |B|$. Write $\alpha = |A_1 \cup \cdots \cup A_S|$ and $\beta = |B_1 \cup \cdots \cup B_S|$, so the stratification includes α treated units and β controls, and call (α, β) the *size* of the stratification.

A *pair matching* is a stratification $(A_1, \ldots, A_S; B_1, \ldots, B_S)$ in which $|A_S| = |B_S| = 1$ for each s. A *matching with multiple controls* is a stratification $(A_1, \ldots, A_S; B_1, \ldots, B_S)$ in which $|A_s| = 1$ for each s. A *full matching* is a stratification $(A_1, \ldots, A_s; B_1, \ldots, B_S)$ in which $\min(|A_s|, |B_s|) = 1$ for each s, so a stratum consists of a single treated subject and one or more controls or else a single control and one or more treated subjects.

9.3.2. A Small Example: Nuclear Power Plants at New and Existing Sites

As an illustration, Table 9.3.1 describes 26 light water nuclear power plants built in the United States. The data are due to W.E. Mooz and are reported by Cox and Snell (1981). Seven of these plants were built on the site of an existing nuclear plant; they are plants $A = \{3, 5, 9, 18, 20, 22, 24\}$ and they form the columns of the table. The other nineteen plants were built at a new site; they are $B = \{1, 2, 4, 6, 7, 8, 10, 11, 12, 13, 14, 15, 16, 17, 19, 21, 23, 25, 26\}$ and they form the rows of the table. The numerical labels in A and B correspond to those used by Cox and Snell (1981). Excluded are six "partial turnkey" whose costs may contain hidden subsidies. The comparison of interest is the cost of plants built at new or existing sites, adjusting for covariates related to the cost. The example is intended solely as a small illustration of matching and stratification.

The covariate **x** is two dimensional and gives the year the construction permit was issued and the capacity of the power plant. The values in the

Table 9.3.1. Distances Between Covariates Describing Nuclear Power Plants.

		Plants built at a previous site A						
		3	***5**	**9**	**18**	**20**	***22**	**24**
Plants built at a new site B	***1**	28	24	10	7	17	20	14
	2	0	3	18	28	20	31	32
	***4**	3	0	14	24	16	28	29
	***6**	22	22	18	8	32	35	30
	7	14	10	4	14	18	20	18
	8	30	27	12	2	26	29	24
	***10**	17	14	5	10	20	22	17
	11	28	26	11	6	18	20	16
	***12**	26	24	9	12	12	14	9
	13	28	24	10	0	24	26	22
	14	20	16	14	24	0	12	12
	15	22	19	12	22	2	9	10
	16	23	20	5	4	20	22	17
	***17**	26	23	14	24	6	5	6
	19	21	18	22	32	7	15	16
	21	18	16	10	20	4	12	14
	23	34	31	16	18	14	9	4
	25	40	37	22	16	20	11	8
	26	28	25	28	38	14	12	17

* Plant built in the northeastern United States.

table are distances δ_{ab} between power plants with $a \in A$ and $b \in B$. The distance is formed as follows. The two covariates in $\mathbf{x}$ were ranked separately from 1 to 26 with average ranks used for ties. The δ_{ab} is the sum of the absolute values of the two differences in ranks for $\mathbf{x}_a$ and $\mathbf{x}_b$. For instance, $\delta_{3,2} = 0$ because plants 3 and 2 were started in the same year and had the same capacity, so their year covariates and their capacity covariates were each assigned the same tied rank, and the sum of the absolute differences in their tied ranks was zero.

The boxed distances describe a matching that is optimal among all matchings in which each plant in A is matched to two plants in B. For instance, plant 3 is matched to plant 2 and plant 21. Notice that plant 4 is better than plant 21 as a match for plant 3, but if plant 4 were matched to plant 3, then plant 5 would not receive its best match. The match described by the boxes is optimal in that it minimizes the total distance within pairs among all matchings with two controls. The construction of an optimal matching is discussed in §9.4.

To illustrate the notation in §9.3.1, the optimal match in Table 9.3.1 is the stratification of size $(\alpha, \beta) = (7, 14)$ with $S = 7$ strata $(A_1, \ldots, A_7; B_1, \ldots, B_7) = (\{3\}, \{5\}, \ldots, \{24\}; \{2, 21\}, \{4, 7\}, \ldots, \{23, 25\})$, so the first stratum consists of $A_1 = \{3\}$ together with $B_1 = \{2, 21\}$. The goal in §9.3 is to characterize optimal stratifications, and in §9.4 to construct them.

9.3.3. Evaluating Stratifications Based on the Average Distance Within Strata

A good stratification would place similar subjects in the same stratum. The distances between treated and control subjects in the same stratum would be small. Let $\delta(A_s, B_s)$ be the average of the $|A_s| \times |B_s|$ distances δ_{ab} with $a \in A_s$ and $b \in B_s$. For instance, in Table 9.3.1, $\delta(A_1, B_1) = \delta(\{3\}, \{2, 21\}) = (\delta_{3,2} + \delta_{3,21})/2 = (0 + 18)/2 = 9$. If $|A_s| = 2$ and $|B_s| = 5$ then $\delta(A_s, B_s)$ would be an average of ten distances.

To find an optimal stratification, a numerical criterion is needed that evaluates a stratification $(A_1, \ldots, A_S; B_1, \ldots, B_S)$, combining the S distances $\delta(A_s, B_s)$, $s = 1, \ldots, S$ into a single number. For this purpose, introduce a weight function, $w(\cdot, \cdot)$. The distances within the S strata will be combined with weights $w(|A_s|, |B_s|)$, so the weights are a function of the sizes of the treated and control groups within each stratum. The weight function $w(\cdot, \cdot)$ is assumed to be strictly positive and finite, and it is defined for strictly positive integer arguments. Define the *distance* Δ for a stratification $(A_1, \ldots, A_S; B_1, \ldots, B_S)$ to be

$$\Delta = \sum_{s=1}^{S} w(|A_s|, |B_s|)\delta(A_s, B_s).$$

There are several natural choices of weight function. Let the stratification $(A_1, \ldots, A_S; B_1, \ldots, B_S)$ be of size (α, β). One weight function is $w(|A_s|, |B_s|) = |A_s|/\alpha$, that is, the weight is the proportion of the α treated subjects who fall in stratum s. In this case, Δ has a simple interpretation. From the stratification, pick one of the α treated subjects at random, and pick a control at random from the stratum containing the selected treated subject; then Δ is the expected distance between these two subjects. Another weight function is $w(|A_s|, |B_s|) = |B_s|/\beta$ with a similar interpretation. Still another weight function is $w(|A_s|, |B_s|) = (|A_s| + |B_s|)/(\alpha + \beta)$. With this weight function, pick one of the $\alpha + \beta$ subjects at random and pick a comparison subject at random in the same stratum but from the other treatment group; then Δ is the expected distance between these subjects. For these three weight functions, $1 = \sum w(|A_s|, |B_s|)$, so Δ is truly a weighted average of the $\delta(A_s, B_s)$. A different weight function is $w(|A_s|, |B_s|) = |A_S| \times |B_s|$; then Δ is the total of all distances within strata.

Distinguish three types of weight function. They are defined in terms of the impact of removing a pair of subjects from one stratum to form a new stratum comprised of just that pair. Does the total weight increase, decrease or stay the same when a pair is separated? A weight function *favors large strata* if $w(p, q) < w(p-1, q-1) + w(1, 1)$ for all integers $p \geq 2$, $q \geq 2$, it *favors small strata* if $w(p, q) > w(p-1, q-1) + w(1, 1)$, and it is *neutral* if $w(p, q) = w(p-1, q-1) + w(1, 1)$. A weight function that favors large strata increases the total weight when a pair is separated, so there is a penalty for increasing the number of strata, increasing Δ even if the distances are not changed. Similarly, a weight function that favors small strata creates a reward for increasing the number of strata. A neutral weight function neither rewards nor penalizes the creation of additional strata. The three weight functions $|A_s|/\alpha$, $|B_s|/\beta$, and $(|A_s| + |B_s|)/(\alpha + \beta)$ are all neutral, while the weight function $|A_s| \times |B_s|$ favors small strata.

Notice that $\Delta = \infty$ if there is a stratum s and a pair of subjects $a \in A_s$ and $b \in B_s$ with $\delta_{ab} = \infty$. In words, if a comparison of a and b is forbidden, then placing a and b in the same stratum yields a stratification with an infinite distance Δ. If $\Delta = \infty$ call the stratification *unacceptable*, but if $\Delta < \infty$ call it *acceptable*.

9.3.4. The Structure of Optimal Strata

Consider a fixed data set, a fixed size (α, β) for the stratification, and a fixed weight function $w(\cdot, \cdot)$. A stratification of size (α, β) and distance Δ is *optimal* if there is no other stratification of size (α, β) with a strictly smaller distance. It may happen that $\Delta = \infty$ for an optimal stratification, but in this case there is no acceptable stratification.

A *refinement* of a stratification $(A_1, \ldots, A_S; B_1, \ldots, B_S)$ is another stratification $(\tilde{A}_1, \ldots, \tilde{A}_{\tilde{S}}; \tilde{B}_1, \ldots, \tilde{B}_{\tilde{S}})$ of the same size with $\tilde{S} \geq S$ such that for each $\tilde{s}$, $\tilde{s} = 1, \ldots, \tilde{S}$, there exists an s such that $\tilde{A}_{\tilde{s}} \subseteq A_s$ and $\tilde{B}_{\tilde{s}} \subseteq B_s$. In other words, a refinement subdivides strata.

The following proposition says that if there is no penalty for creating additional strata, then any stratification can be refined into a full matching which is as good or better. Recall that all of the weight functions discussed in §9.3.3 were either neutral or favored small strata. Proposition 9.2 is proved in the Problems in §9.6.2.

Proposition 9.2. *Consider a weight function $w(\cdot, \cdot)$ that is either neutral or favors small strata. If the stratification $(A_1, \ldots, A_S; B_1, \ldots, B_S)$ has distance Δ_0 and is not a full matching, then it has a refinement which is a full matching and has distance no greater than Δ_0.*

For weight functions that are neutral or favor small strata, Proposition 9.2 has several consequences. First, if there is an acceptable stratification of size (α, β), then there is an acceptable full matching of size (α, β). Second, for each fixed size (α, β), there is a full matching that is an optimal stratification. In searching for an optimal stratification, it suffices to confine the search to full matchings because one of them is sure to be optimal. As will be seen in §9.3.5, neither statement is true for pair matching or for matching with multiple controls.

Under the conditions of Proposition 9.2, it can happen that the refinement is only as good as the original stratification, but no better. For this to happen, the distances δ_{ab} must satisfy a large number of linear equations. This will often happen when there are many distances δ_{ab} that are equal to zero. However, it may be shown that if the covariates $\mathbf{x}$ came from a multivariate normal distribution and the distance is the Mahalanobis distance, then with probability one the equations will not be satisfied, and the full matching will be strictly better than the original stratification. In other words, in the multivariate normal case, with probability one, a stratification which is not a full matching is not optimal. For proof, see Rosenbaum (1991, §4).

9.3.5. Pair Matching Is Not Optimal

Unlike an optimal full matching, an optimal pair matching is not generally an optimal stratification. The distance matrix in Table 9.3.2 demonstrates this. There are $|A| = 3$ treated subjects, $A = \{a, b, c\}$, $|B| = 3$ controls, $B = \{1, 2, 3\}$, and the distances take two values, $\omega > \varepsilon$.

For the neutral weight function $w(|A_s|, |B_s|) = (|A_s| + |B_s|)/(\alpha + \beta)$, the optimal stratification and optimal full matching is $(A_1, A_2; B_1, B_2) = (\{a, b\}, \{c\}; \{1\}, \{2, 3\})$ with distance $\Delta = (3\varepsilon + 3\varepsilon)/6 = \varepsilon$. The optimal pair

Table 9.3.2. Hypothetical Distance Matrix.

		A		
		a	b	c
B	1	ε	ε	ω
	2	ω	ω	ε
	3	ω	ω	ε

matching is $(A_1, A_2, A_3; B_1, B_2, B_3) = (\{a\}, \{b\}, \{c\}; \{1\}, \{2\}, \{3\})$ with distance $(2\varepsilon + 2\omega + 2\varepsilon)/6 > \varepsilon$. By letting ω increase, the difference between optimal full matching and optimal pair matching may be made arbitrarily large.

In the same way, the optimal matching with multiple controls is not generally an optimal stratification.

9.4. Optimal Matching

9.4.1. Greedy Matching Versus Optimal Matching

In §9.4, the task is to build a matched sample from a matrix of distances δ_{ab} between treated subjects $a \in A$ and controls $b \in B$. Various types of matching will be considered, including pair matching, matching with multiple controls, balanced matching and full matching. Here, optimal refers to minimizing the total distance within matched sets.

The first impulse is to use what is known as a greedy algorithm. A greedy algorithm divides a large decision problem into a series of simpler decisions each of which is handled optimally, and makes those decisions one at a time without reconsidering early decisions as later ones are made. One greedy algorithm for matching finds the smallest distance δ_{ab} for $a \in A$ and $b \in B$, calls this (a, b) the first matched pair, removes a from A and b from B, and repeats the process to find the next pair. Greedy algorithms do solve a small class of problems optimally, but the matching problem is not a member of that class.

In principle, greedy can perform very poorly as a matching algorithm. To see this, consider the distance matrix in Table 9.4.1 with two treated subjects and two controls, where ε is finite. Greedy pairs a with 1 at a cost of 0 and then is forced to pair b with 2 at a cost of ∞. The optimal match pairs a with 2 and b with 1 for a cost of 2ε. In principle, greedy can be arbitrarily poor compared to optimal matching.

Table 9.4.1. A Small Distance Matrix.

		A	
		a	*b*
B	1	0	ε
	2	ε	∞

The situation in Table 9.4.1 is not far fetched when some pairings are forbidden through the use of infinite distances, as is true with category matching and caliper matching. Suppose one were matching on age, requiring matched pairs to have ages that differ by less than 5 years. As with the caliper distance in §9.3.1, there is an infinite distance δ for any pair whose ages differ by more than 5 years; however, if the age difference is less than or equal to 5 years, the distance is the absolute difference in ages. If 1 and *a* are both 50 years old, 2 is 47 years old, and *b* is 53 years old, then Table 9.4.1 results with $\varepsilon = 3$. Greedy would pair the 50 year olds, and then it would have no acceptable match for the 53 year old. There are also theoretical results suggesting greedy can be very poor in large problems. See, for instance, Walkup (1979). In §9.4.7, greedy and optimal matching are compared by simulation when covariates are multivariate normal.

If greedy were used to match each treated plant to two controls in Table 9.3.1, it would begin by pairing plant 3 to plant 2 at a cost of $\delta_{3,2} = 0$ units of distance, then plant 5 to 4 at a cost of $\delta_{5,4} = 0$, and so on. Table 9.4.2 compares greedy and optimal matching step by step.

In Table 9.4.2, greedy does well at the beginning when there is little competition for controls. In selecting the first eleven pairs, greedy and optimal matching have selected the same pairs and produced the same total distance. At step 12, greedy misses a small opportunity because it never reconsiders previous decisions. Greedy adds the pair (22, 26) at a cost of $\delta_{22,26} = 12$ units of distance. Instead, at step 12, optimal matching deleted pair (20, 15) and added pairs (20, 21) and (22, 15) at a cost of $-\delta_{20,15} + \delta_{20,21} + \delta_{22,15} = -2 + 4 + 9 = 11$. As the final matched pairs are selected and the competition for controls intensifies, greedy misses additional opportunities. In the end, the greedy match gives a total distance $(79 - 71)/71 = 11\%$ higher than necessary.

9.4.2. Optimal Matching and Minimum Cost Flow in a Network

Optimal matching is known to be equivalent to finding a minimum cost flow in a certain network, a problem that has been extensively studied and

Table 9.4.2. Step-by-Step Comparison of Greedy and Optimal Matching.

	Greedy		Optimal		
Step	Add	Total distance	Delete	Add	Total distance
1	(3, 2)	0		(3, 2)	0
2	(5, 4)	0		(5, 4)	0
3	(18, 13)	0		(18, 13)	0
4	(20, 14)	0		(20, 14)	0
5	(18, 8)	2		(18, 8)	2
6	(20, 15)	4		(20, 15)	4
7	(9, 7)	8		(9, 7)	8
8	(24, 23)	12		(24, 23)	12
9	(22, 17)	17		(22, 17)	17
10	(9, 10)	22		(9, 10)	22
11	(24, 25)	30		(24, 25)	30
12	(22, 26)	42	(20, 15)	(20, 21) (22, 15)	41
13	(5, 21)	58	(9, 7)	(9, 16) (5, 7)	52
14	(3, 19)	79	(22, 15) (20, 21)	(22, 26) (20, 15) (3, 21)	71

for which good algorithms exist. The current section sketches a few of the general ideas about network flow. These ideas are applied to matching problems in subsequent sections. A attractive, detailed, modern discussion of this subject is given by Bertsekas (1991); he also provides FORTRAN code in an Appendix and offers to provide the program on diskette for a nominal fee. See also Papadimitriou and Steiglitz (1982), Tarjan (1983), and Rockafellar (1984).

A network is a directed graph, that is, a set V of vertices and a set E of directed edges consisting of ordered pairs of elements of V. Later the vertices in the set V will include subjects available for matching, that is, V will contain $A \cup B$, and the set E of edges will include an edge (a, b) with $a \in A$ and $b \in B$ if a may be matched to b, that is, if $\delta_{ab} < \infty$. However, at times it will be convenient to include in V and E certain other vertices and edges, so for the moment V is any set of vertices and E is any set of ordered pairs of elements of V. A network (V, E) is depicted by drawing a dot for each vertex $v \in V$, and for each edge $e = (v_1, v_2) \in E$ an arrow from vertex v_1 to vertex v_2.

As an illustration, Figure 9.4.1 is a network for the matching problem in Table 9.4.1. The four vertices are $V = \{a, b, 1, 2\}$ and the three edges are $E = \{(a, 1), (a, 2), (b, 1)\}$. There is no edge $(b, 2)$ because $\delta_{b2} = \infty$, so b cannot be matched to 2.

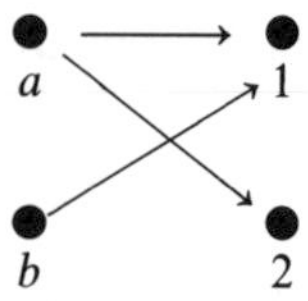

Vertices $V = \{a, b, 1, 2\}$
Edges $E = \{(a, 1), (a, 2), (b, 1)\}$

Figure 9.4.1. Matching network for Table 9.4.1.

Network flow theory was originally concerned with the movement of material from one vertex to another along the edges. One might think of a railroad network or a telephone network. A *flow* in a network assigns a number $\text{FLOW}(i, j)$ to each edge $(i, j) \in E$ signifying that $\text{FLOW}(i, j)$ units of material are to flow from vertex i to vertex j directly across edge (i, j). Network flow optimization finds the best flow for all $(i, j) \in E$ subject to various requirements. Each edge (i, j) must carry at least $\text{MIN}(i, j)$ units of flow and at most $\text{MAX}(i, j)$ units of flow, that is, $\text{MIN}(i, j) \le \text{FLOW}(i, j) \le \text{MAX}(i, j)$ for all $(i, j) \in E$, where $\text{MIN}(i, j)$ and $\text{MAX}(i, j)$ are given numbers for each $(i, j) \in E$. For each vertex $i \in V$, the divergence is the total flow out from i minus the total flow into i, that is,

$$\text{DIV}(i) = \sum_{j:(i,j) \in E} \text{FLOW}(i, j) - \sum_{k:(k,i) \in E} \text{FLOW}(k, i).$$

If there were a warehouse at vertex i, then $\text{DIV}(i) > 0$ would signify that the stock in the warehouse is being depleted, $\text{DIV}(i) < 0$ would signify that stock is building up, and $\text{DIV}(i) = 0$ would signify that stock is arriving and leaving at the same rate. There are specified limits on the divergence at each vertex, $\text{DMIN}(i) \le \text{DIV}(i) \le \text{DMAX}(i)$.

Shipping a unit of flow from i to j along (i, j) costs $\text{COST}(i, j)$. The total cost of a flow is $\sum_{(i,j) \in E} \text{FLOW}(i, j)\,\text{COST}(i, j)$. The minimum cost flow problem is to find a flow, $\text{FLOW}(i, j)$ for $(i, j) \in E$, which minimizes the total cost subject to the constraints, that is,

$$\text{minimize} \quad \sum_{(i,j) \in E} \text{FLOW}(i, j)\,\text{COST}(i, j), \tag{9.4.1}$$

subject to

$$\text{MIN}(i, j) \le \text{FLOW}(i, j) \le \text{MAX}(i, j) \qquad \text{for all} \quad (i, j) \in E,$$

and

$$\text{DMIN}(i) \le \text{DIV}(i) \le \text{DMAX}(i) \qquad \text{for} \quad i \in V.$$

See Bertsekas (1991, Exercise 1.6, p. 19) for discussion of the relationship between this definition of the minimum cost flow problem and other equivalent definitions.

An *integer flow* is a flow in which FLOW(i, j) is an integer for each $(i, j) \in E$. Integer flows arise naturally when it is impossible to divide a unit in half for shipping. Also, in matching problems, only integer flows make sense because whole units must be matched. If all of the capacity constraints, MIN(i, j), MAX(i, j), DMIN(i), and DMAX(j) are integers, then whenever there is a minimum cost flow there is also an integer minimum cost flow (e.g., Tarjan, 1983, p. 110). Notice that the costs COST(i, j) need not be integers. Throughout this chapter, MIN(i, j), MAX(i, j), DMIN(i), DMAX(j) will be integers, and the solution to the minimum cost flow problem is assumed to be one of the integer solutions. This happens automatically with commonly used algorithms.

Good algorithms exist for solving the minimum cost flow problem. In particular, there is an algorithm for optimal matching which requires computational effect that grows no faster than the cube of the number of subjects to be matched; see Papadimitriou and Steiglitz (1982, Theorem 11.1, p. 250) or Tarjan (1983, Theorem 8.13, p. 110). For comparison, the conventional way of multiplying two square matrices with one row per subject requires effort that grows as the cube of the number of subjects. In other words, the rate of growth in the difficulty of these two problems is similar. Bertsekas (1991, §5) compares the performance of several algorithms on a Macintosh Plus computer. In particular, with his favored algorithm for pair matching, he reports solving in less than 6 seconds optimal pair matching problems with 5000 treated subjects, 5000 controls, and 25000 permitted matched pairs (i.e., $|A| = 5000$, $|B| = 5000$, $|E| = |\{(a, b)\colon a \in A, b \in B, \delta_{ab} < \infty\}| = 25000$).

The remainder of §9.4 indicates how to set up various matching problems as minimum cost flow problems. Once cast as minimum cost flow problems, available algorithms provide optimal solutions.

9.4.3. Matching with a Fixed Number of Controls

In matching with a fixed number of controls, each of the treated subjects in A is to be matched with a fixed number, say $k \geq 1$, of controls from B to minimize the total of the $k|A|$ distances between treated subjects and their matched controls. This becomes a minimum cost flow problem with the following identifications.

The vertices are the subjects, $V = A \cup B$. The edges link treated subjects to controls for which the distance is finite, $E = \{(a, b)\colon a \in A, b \in B, \delta_{ab} < \infty\}$. See Figure 9.4.1. The cost is simply the distance, COST$(a, b) = \delta_{ab}$ for all $(a, b) \in E$. The bounds on the flow are MIN$(a, b) = 0$ and MAX$(a, b) = 1$ for all $(a, b) \in E$. The bounds on the divergences are DMIN$(a) =$ DMAX$(a) = k$ for all $a \in A$, and DMIN$(b) = -1$, DMAX$(b) = 0$ for all $b \in B$.

Suppose that there is an integer flow FLOW(a, b) that satisfies the constraints in (9.4.1). Since MIN$(a, b) = 0$ and MAX$(a, b) = 1$ and the FLOW(a, b) is an integer, it follows that FLOW$(a, b) = 0$ or FLOW$(a, b) = 1$ for all $(a, b) \in E$. Since DMIN(a) = DMAX$(a) = k$, it follows that for each $a \in A$, FLOW$(a, b) = 1$ for exactly k subjects $b \in B$. Also, since DMIN$(b) = -1$ and DMAX$(b) = 0$, for each $b \in B$, there is at most one $a \in A$ such that FLOW$(a, b) = 1$. In other words, a flow satisfies the constraints if and only if $\{(a, b)$: FLOW$(a, b) = 1\}$ is a matching of each treated subject to exactly k controls. A minimum cost flow is a matching that minimizes the total distance among all matchings that assign k controls to every treated unit. If no integer flow satisfies the constraints, then there is no acceptable matching with k controls, where an acceptable matching was defined in §9.3.3.

The boxed power plants in Table 9.3.1 are an optimal match with $k = 2$ controls. Notice that there is competition among treated plants for the same control. For instance, control 7 was matched to treated plant 5 at a distance of $\delta_{5,7} = 10$, though 7 is closer to plant 9, etc. See also Table 9.4.2.

9.4.4. Matching with a Variable Number of Controls

An alternative to matching each treated subject $a \in A$ to k controls $b \in B$ is to require each treated subject to have at least k_{min} controls and at most k_{max} controls, with a total of h controls in the matched sample. In this case, the optimization chooses the best number of controls for each treated subject. To produce a matching, the parameters k_{min}, k_{max} and h should satisfy $k_{\text{min}} \geq 1$ and $k_{\text{max}} \leq |B| - |A| + 1$, and $|A| k_{\text{min}} \leq h \leq |A| k_{\text{max}}$.

Optimal matching of h controls with at least k_{min} controls and at most k_{max} controls becomes a minimum cost flow problem in the following way. Add to the vertex set a new vertex called the SINK, so $V = A \cup B \cup \{\text{SINK}\}$. Add an edge from each $b \in B$ to the sink, so $E = \{(a, b)\colon a \in A, b \in B, \delta_{ab} < \infty\} \cup \{(b, \text{SINK})\colon b \in B\}$. See Figure 9.4.2 which adds a SINK to Figure 9.4.1.

The SINK is needed to ensure that optimal matching does not reduce the cost by matching fewer than h controls. Set DMIN$(a) = k_{\text{min}}$, DMAX$(a) = k_{\text{max}}$ for all $a \in A$, DMIN(b) = DMAX$(b) = 0$ for all $b \in B$, and DMIN(SINK) = DMAX(SINK) = $-h$. Set MIN$(b$, SINK$) = 0$, MAX$(b$, SINK$) = 1$, and COST$(b$, SINK$) = 0$ for all $b \in B$. Define all other quantities as in §9.4.3. In an integer FLOW, each of h controls must send

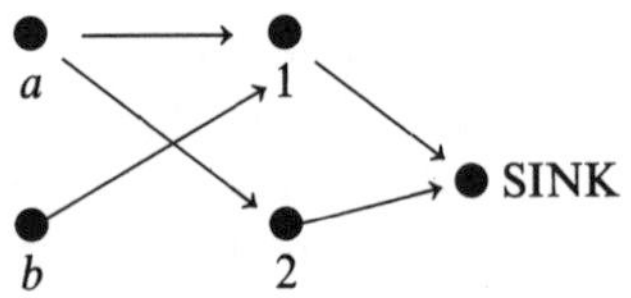

Figure 9.4.2. A network with a sink.

Table 9.4.3. Three Matched Samples with a Variable Number of Controls.

Treated unit	$k_{min} = 1, k_{max} = 13$	$k_{min} = 1, k_{max} = 4$	$k_{min} = 2, k_{max} = 3$
3	2	2	2, 6
5	4	4	4, 7
9	7, 10, 12	1, 7, 10, 16	1, 10, 16
18	1, 6, 8, 11, 13, 16	6, 8, 11, 13	8, 11, 13
20	14, 15, 19, 21	14, 15, 19, 21	14, 19, 21
22	17, 26	17, 26	15, 17, 26
24	23, 25	12, 23, 25	12, 23, 25
Total distance	87	91	118

one unit of flow to the SINK, and each of these controls must receive its unit of flow from different treated unit. An integer FLOW satisfies the constraints in (9.4.1) if and only if $\{(a, b): a \in A, b \in B, \text{FLOW}(a, b) = 1\}$ is a matching in which each treated unit has at least k_{min} and at most k_{max} controls. A minimum cost flow is an optimal matching.

Table 9.4.3 gives optimal matched samples with a variable number of controls for three choices of k_{min} and k_{max}. The table identifies the controls matched to each treated plant. For instance, in the first match with $k_{min} = 1$ and $k_{max} = 13$, treated unit 9 is matched to controls 7, 10, and 12. Every control is matched to some treated unit, so $h = 19$, unlike the match in §9.4.3 where some controls were discarded. The first match in Table 9.4.3 has $k_{min} = 1$ and $k_{max} = 13$, so it requires only that every treated plant have at least one control. This first match assigns one control to treated unit 3 and six controls to treated unit 18. The last match has $k_{min} = 2$ and $k_{max} = 3$. Remember that there are $|B| = 19$ controls and $|A| = 7$ treated units. Since $19/7 = 2.71$, to match with $k_{min} = 2$ and $k_{max} = 3$ is to require the number of controls to be as nearly constant as possible.

Table 9.4.3 also gives the total distance within matched sets for each of the three matched samples. The match with $k_{min} = 1$ and $k_{max} = 4$ has a total distance only slightly greater than the match with $k_{min} = 1$ and $k_{max} = 13$, and both are much better than the match with $k_{min} = 2$ and $k_{max} = 3$. A substantial price was paid in giving more than one control to treated units 3 and 5.

9.4.5. Optimal Full Matching

Recall that in a full matching, each matched set contains a treated unit and one or more controls or a control unit and one or more treated units. This section finds a full matching that minimizes the total distance. Assume at first that $\delta_{ab} > 0$ for each $a \in A$ and $b \in B$; this assumption can be removed, as indicated at the end of this section.

Optimal full matching uses the network in §9.4.3 without a SINK but with the following changes. Set DMIN(a) = 1 and DMAX(a) = $|B|$ for each $a \in A$ and DMIN(b) = $-|A|$ and DMAX(b) = -1 for each $b \in B$. Define other quantities as in §9.4.3. Then an integer FLOW which satisfies the constraints in (9.4.1) must have at least one unit of flow leaving each treated unit, since DMIN(a) = 1, and it must have at least one unit of flow entering each control, since DMAX(b) = -1, so every treated unit is connected to one or more controls and every control is connected to one or more treated units. However, the set $\{(a, b)\colon a \in A, b \in B, \text{FLOW}(a, b) = 1\}$ may not define a full matching because a treated unit a might be connected to several controls including b but b might be connected to several treated units. This is not a problem, however, if FLOW has minimum cost, as will now be demonstrated.

Find a minimum cost integer FLOW. It will now be shown that because FLOW has minimum cost, the set $\{(a, b)\colon a \in A, b \in B, \text{FLOW}(a, b) = 1\}$ does define a full matching. Suppose not, so there are $a, a^* \in A$ and $b, b^* \in B$ such that FLOW(a, b) = 1, FLOW(a^*, b) = 1, and FLOW(a^*, b^*) = 1. Define a new flow called BETTER such that BETTER(a^*, b) = 0 and BETTER(i, j) = FLOW(i, j) for all $(i, j) \neq (a^*, b)$. Then BETTER satisfies the constraints and it has a cost that is lower by $\delta_{a^*,b} > 0$, so FLOW is not optimal, and there is a contradiction. Hence, a minimum cost integer FLOW is an optimal full matching.

It was assumed at the beginning of this section that $\delta_{ab} > 0$ for all $a \in A$ and $b \in B$. This assumption will now be removed. Suppose that $\delta_{ab} = 0$ for d pairs (a, b) with $a \in A$ and $b \in B$. Construct a new minimum cost flow problem identical to the original problem except that $\delta_{ab} = \varepsilon > 0$ for these d pairs. The minimum cost flow for the new problem is a full matching and the total cost is at most $d\varepsilon$ higher than the minimum cost of the original problem. Since ε was arbitrary, it follows that for sufficiently small ε, the minimum cost flow for the new problem is also a minimum cost flow for the original problem.

As it turns out, the first match in Table 9.4.3 is an optimal full matching.

9.4.6. Optimal Balanced Matching

In the nuclear power plant example in Table 9.3.1, two of the treated plants were constructed in the northeastern United States, specifically plants 5 and 22, and six of the control plants were constructed in the northeast. Construction costs, the outcome, are generally high in the northeast, so it would be desirable to match on this covariate in addition to the date of construction and the capacity of the plant. In the minimum distance match with two controls, 2/7 = 29% of the treated plants were constructed in the northeast, but only 3/14 = 21% of the controls were constructed there. The matched sample would be balanced if 4/14 = 29% of the controls came from the northeast. This section finds the matched sample which has the minimum total distance among all balanced samples with two controls per treated unit.

In general, suppose that units are divided into C categories, $c = 1, \ldots, C$, that m_c treated subjects fall in category c, and that at least km_c are available from category c. The task is to match each treated unit with k controls so that a total of km_c controls fall in category c, and the total distance is as small as possible.

The following network is used. Introduce C new vertices, $\text{SINK}_1, \ldots, \text{SINK}_C$. The vertices are the subjects together with the sinks, $V = A \cup B \cup \{\text{SINK}_1, \ldots, \text{SINK}_C\}$. The edges E are of two types. There is an edge linking each treated subject $a \in A$ to each control $b \in B$ for which the distance is finite, $\delta_{ab} < \infty$. These edges have $\text{COST}(a, b) = \delta_{ab}$. There is also an edge linking each control in category c to SINK_c, so each control is linked to exactly one sink. These edges connecting $b \in B$ to a sink have COST equal to zero. The set E of edges is the union of these two sets of edges. All edges have $(i, j) \in E$ have $\text{MIN}(i, j) = 0$ and $\text{MAX}(i, j) = 1$. The bounds on the divergences are $\text{DMIN}(a) = \text{DMAX}(a) = k$ for all $a \in A$, and $\text{DMIN}(b) = \text{DMAX}(b) = 0$ for all $b \in B$, and $\text{DMIN}(\text{SINK}_c) = \text{DMAX}(\text{SINK}_c) = -km_c$ for $c = 1, \ldots, C$.

Suppose that there is an integer flow $\text{FLOW}(a, b)$ that satisfies the constraints in (9.4.1). Then the set $\{(a, b)\colon a \in A, b \in B, \text{FLOW}(a, b) = 1\}$ defines a balanced matched sample with k controls for each treated subject. To see this, note the following. A total of k units of flow leave each treated unit $a \in A$ because $\text{DMIN}(a) = \text{DMAX}(a) = k$, so each treated unit is paired with k controls. No control $b \in B$ can be matched to more than one treated unit because $\text{DMIN}(b) = \text{DMAX}(b) = 0$ and $\text{MAX}(b, \text{SINK}_c) = 1$ for exactly one c, so unit b must pass on all the flow it receives and can pass on at most one unit of flow. The sample is balanced because $\text{DMIN}(\text{SINK}_c) = \text{DMAX}(\text{SINK}_c) = -km_c$, so there must be km_c controls from category c. A minimum cost integer flow therefore defines a balanced matched sample of minimum total distance.

Table 9.4.4 compares optimal balanced matching with greedy matching and optimal matching for the example in Table 9.3.1. All three matched samples have two controls for each treated plant. The greedy and optimal match are from §9.4.3. The optimal balanced match forces 29% of the control plants to come from the northeast because 29% of the treated plants are from the northeast.

In Table 9.4.4, optimal balanced matching corrects the imbalance in plants from the northeast with a distance 5% = (74.5 − 71)/71 higher than the unbalanced optimal match. The optimal balanced match is better than the greedy match both in being balanced and having a smaller total distance.

Table 9.4.4. Comparison of Greedy, Optimal, and Optimal Balanced Matching.

Treated unit	Greedy	Optimal	Optimal balanced
3	2, 19	2, 21	2, 10*
5*	4*, 21	4*, 7	4*, 7
9	7, 10*	10*, 16	12*, 16
18	8, 13	8, 13	8, 13
20	14, 15	14, 15	14, 21
22*	17*, 26	17*, 26	15, 17*
24	23, 25	23, 25	23, 25
Total distance	79	71	74.5
% Treated in northeast	29	29	29
% Control in northeast	21	21	29

* Plant constructed in the northeastern United States.

9.4.7. Comparison of Matching Methods Using Simulation

In a simulation study, Gu and Rosenbaum (1993) compared the methods described in this chapter. This section outlines some of the findings, though the paper should be consulted for detailed discussion, numerical results, and specifics.

The simulation considered multivariate normal covariates $\mathbf{x}$ with common covariance matrix in the treated and control groups but different mean vectors. The number of covariates or dimension of $\mathbf{x}$ was 2, 5 or 20. In each matched sample, there were $|A| = 50$ treated subjects. The number of controls available for matching was $|B| = 50$, 100, 150, or 300. For each matching situation considered, 50 different matched samples were constructed.

Three distances δ_{ab} were used. The first was the absolute difference in propensity scores estimated using a logit model. The second was the Mahalanobis distance. The third method, $M + P$, used a caliper on the propensity score, so $\delta_{ab} = \infty$ if a and b had propensity scores differing by more than 0.6 times the standard deviation of the propensity scores, and otherwise δ_{ab} was the Mahalanobis distance. This third distance was proposed by Rosenbaum and Rubin (1985).

Two types of measure were used to judge the quality of a matched sample. The first was the average distance within matched sets. The average distance is the total distance that was minimized by optimal matching divided by the number of distances summed to form the total. The second was the balance of the matched groups, that is, the difference in covariate mean vectors in treated and control groups after matching. When the matched sets have unequal sizes, as is true for some matching methods, balance is defined in terms of the average over matched sets of the average difference between treated and control subjects within matched sets. Different distances δ_{ab} may be compared in terms of balance but not in terms of total distance, since the difference in propensity scores is not commensurate with the Mahalanobis distance.

The main findings follow:

- When there were 20 covariates, matching on the propensity score produced greater covariate balance than matching using the Mahalanobis distance. When there were two covariates, the methods were similar with no consistent winner. The combined distance, $M + P$, often fell between the other methods but rarely performed poorly.
- When a fixed number $k|A|$ of controls are matched, full matching was found to be much better than matching k controls to each treated subject. That is, the method of §9.4.5 was much better than the method of §9.4.3 when the same total number of controls were matched. This was true for both the average distance and covariate balance.
- Optimal matching was sometimes much better than greedy matching and sometimes only marginally better. Optimal matching was much better than greedy at minimizing the average Mahalanobis distance when there were few controls to choose from, $|B| = 50$ or $|B| = 100$, and the number of covariates was 2 or 5. Optimal matching was only slightly better than greedy at minimizing the propensity distance. Optimal matching was no better than greedy at producing covariate balance.

Some caveats are needed. The comparisons of different distances refer only to covariate balance, because the distances themselves are not commensurate. The comparison of full matching and matching with a fixed number of controls refers only to average distance and balance, measures of bias. An optimal full matching may contain matched sets of extremely unequal sizes, and these may be unattractive for reasons of precision or aesthetics. In practice, with a given data set, a full matching, a matching with a fixed number of controls, and several compromises may be compared, as in §9.4.4.

9.5. Bibliographic Notes

The discussion in §9.2 of the balancing property of the propensity score is based on Rosenbaum and Rubin (1983, 1984). The discussion of optimal stratification in §9.4 is from Rosenbaum (1991). There is a vast literature on network optimization and its relationship to matching. Some early references are Kuhn (1955) and Ford and Fulkerson (1962), and modern discussions are given by Papadimitriou and Steiglitz (1982), Tarjan (1983), Rockafellar (1984), and Bertsekas (1991). The material and examples in §9.4 are from Rosenbaum (1989), except for §9.4.5 which is from Rosenbaum (1991). Section 9.4.7 summarizes the simulation study in Gu and Rosenbaum (1993). Other simulations of greedy matching with a single control are given in Rubin (1973, 1979, 1980).

9.6. Problems

9.6.1. Aspects of Full Matching

1. Find a distance matrix δ_{ab} and a stratification $(A_1, \ldots, A_S; B_1, \ldots, B_S)$ such that the refinement in Proposition 9.2 is as good as but no better than the stratification.

2. Let $x_1, \ldots, x_n$ be independent random variables with distribution $F(\cdot)$ and let $\tilde{x}_1, \ldots, \tilde{x}_n$ be independent random variables with distribution $\tilde{F}(\cdot)$ where the x's are independent of the $\tilde{x}$'s. Suppose that we match x's to $\tilde{x}$'s to minimize the distance $\delta_{ab} = |x_a - \tilde{x}_b|$. Consider both a pair matching of each x to a single $\tilde{x}$ and a full matching. Within each matched set, calculate the average x and the average $\tilde{x}$ and difference these two averages. Each matched set produces one such difference; average the differences across matched sets and call the result B. Then B is a measure of covariate balance. Note that B is is computed from certain $x_a - \tilde{x}_b$, not from $|x_a - \tilde{x}_b|$. Suppose that $F(\cdot)$ and $\tilde{F}(\cdot)$ are two beta distributions on the interval $[0, 1]$ with different means. How does B behave as $n \to \infty$ for pair matching and full matching? Suppose that $F(\cdot)$ is uniform on $[0, 1]$ and $\tilde{F}(\cdot)$ is uniform on $[1, 2]$. How does B behave as $n \to \infty$ for pair matching and full matching? (*Hint*: Does pair matching affect B?) What does this suggest about what full matching can and cannot do?

9.6.2. Proof of the Proposition Characterizing the Form of an Optimal Stratification

3. Problems 3, 4, 5, 6, and 7 prove Proposition 9.2, and the notation of that proposition is used. Why can we assume, without loss of generality, that $\Delta_0 < \infty$? Why can we assume that there is some s such that $|A_s| \geq 2$ and $|B_s| \geq 2$? Fix that s.

4. Pick any two units $a \in A_s$ and $b \in B_s$. Suppose that (a, b) are separated from stratum s to form a new stratum $S + 1$ with $A_{S+1} = \{a\}$ and $B_{S+1} = \{b\}$. Let $\theta(a, b)$ be the difference between the distance Δ_0 for the original stratification and the distance for the new stratification in which (a, b) are in their own strata. Express $\theta(a, b)$ in terms of $w(|A_s|, |B_s|)$, $w(|A_s| - 1, |B_s| - 1)$, $w(1, 1)$, $\delta(A_s, B_s)$, $\delta(A_s - \{a\}, B_s - \{b\})$, and δ_{ab}. Here, $A_s - \{a\}$ means the set A_s with a removed.

5. What is the average of δ_{ab} over all $a \in A_s$ and $b \in B_s$? What is the average of $\delta(A_s - \{a\}, B_s - \{b\})$ over $a \in A_s$ and $b \in B_s$?

6. Show that the average of $\theta(a, b)$ over $a \in A_s$ and $b \in B_s$ is nonnegative. (*Hint*: Use Problem 5 and the assumption that the weight function is either neutral or favors small strata.)

7. Conclude that there is some $a \in A_s$ and $b \in B_s$ such that $\theta(a, b) \geq 0$. Separate this (a, b) to form a new stratum. How does repeating the process prove Proposition 9.2? (A proof of Proposition 9.2 along these lines is given in Rosenbaum (1991, §4)).

References

Bertsekas, D. (1991). *Linear Network Optimization: Algorithms and Codes.* Cambridge MA: MIT Press.

Carpenter, R. (1977). Matching when covariables are normally distributed. *Biometrika*, **64**, 299–307.

Cochran, W.G. (1968). The effectiveness of adjustment by subclassification in removing bias in observational studies. *Biometrics*, **24**, 205–213.

Cochran, W.G. and Rubin, D.B. (1973). Controlling bias in observational studies: A review. *Sankya*, Series A, **35**, 417–446.

Cox, D.R. (1970). *The Analysis of Binary Data.* London: Methuen.

Cox, D.R. and Snell, E.J. (1981). *Applied Statistics: Principles and Examples.* London: Chapman & Hall.

Ford, L. and Fulkerson, D. (1962). *Flows in Networks.* Princeton, NJ: Princeton University Press.

Gu, X.S. and Rosenbaum, P.R. (1993). Comparison of multivariate matching methods: Structures, distances and algorithms. *Journal of Computational and Graphical Statistics*, **2**, 405–420.

Kuhn, H.W. (1955). The Hungarian method for the assignment problem. *Naval Research Logistics Quarterly*, *2*, 83–97.

Papadimitriou, C. and Steiglitz, K. (1982). *Combinatorial Optimization.* Englewood Cliffs, NJ: Prentice-Hall.

Rockafellar, R.T. (1984). *Network Flows and Monotropic Optimization.* New York: Wiley.

Rosenbaum, P.R. (1984). Conditional permutation tests and the propensity score in observational studies. *Journal of the American Statistical Association*, **79**, 565–574.

Rosenbaum, P. (1989). Optimal matching for observational studies. *Journal of the American Statistical Association*, **84**, 1024–1032.

Rosenbaum, P.R. (1991). A characterization of optimal designs for observational studies. *Journal of the Royal Statistical Society*, Series B, **53**, 597–610.

Rosenbaum, P. and Rubin, D. (1983). The central role of the propensity score in observational studies for causal effects. *Biometrika*, **70**, 41–55.

Rosenbaum, P. and Rubin, D. (1984). Reducing bias in observational studies using subclassification on the propensity score. *Journal of the American Statistical Association*, **79**, 516–524.

Rosenbaum, P. and Rubin, D. (1985). Constructing a control group using multivariate matched sampling methods that incorporate the propensity score. *American Statistician*, **39**, 33–38.

Rubin, D.B. (1973). The use of matched sampling and regression adjustment to remove bias in observational studies. *Biometrics*, **29**, 185–203.

Rubin, D.B. (1979). Using multivariate matched sampling and regression adjustment to control bias in observational studies. *Journal of the American Statistical Association*, **74**, 318–328.

Rubin, D.B. (1980). Bias reduction using Mahalanobis metric matching. *Biometrics*, **36**, 293–298.

Smith, A., Kark, J., Cassel, J., and Spears, G. (1977). Analysis of prospective epidemiologic studies by minimum distance case-control matching. *American Journal of Epidemiology*, **105**, 567–574.

Tarjan, R. (1983). *Data Structures and Network Algorithms*. Philadelphia: Society for Industrial and Applied Mathematics.

Walkup, D. (1979). On the expected value of a random assignment problem. *SIAM Journal on Computing*, **8**, 440–442.

CHAPTER 10

Some Strategic Issues

10.1. What Are Strategic Issues?

By and large, the discipline of statistics is concerned with the development of correct and effective research designs and analytical methods, together with supporting theory. Here, correct and effective refer to formal properties of the designs and methods. The first nine chapters discussed issues of this sort. In contrast, a strategic issue concerns the impact that an empirical investigation has on its intended audience. Often, the audience is not focused on statistical technique and theory, may have limited training in statistics, and may be comprised of laymen, that is, laymen with respect to their knowledge of statistics. When this is so, strategic issues may arise in which formal properties are weighed against impact on the intended audience.

Should strategic issues be considered at all? Why, after all, should laymen manage the conduct of science? Why should the decisions of laymen govern the interpretation of empirical research? Why should laymen sit as judge and jury on statistical analyses?

The answer, of course, is that public and corporate managers, governors, judges, and juries are, typically, not professional statisticians. The person responsible for a decision hopes and expects to find the evidence somewhat convincing, not merely to hear that an expert finds the evidence convincing. Often, these hopes and expectations are reasonable, and often they determine the impact of an empirical investigation.

10.2. Strategic Issues in Observational Studies: Some Specific Suggestions

This brief, final chapter discusses a few strategic issues in the design and conduct of observational studies. Motivation and elaboration of these issues is found in the chapters or sections indicated within parentheses.

• *Strategic issues are more than communication and presentation.* Strategic issues affect the choice of research design and analytical methods.

• *Design observational studies.* A recent report of the National Academy of Sciences (Meyer and Fienberg 1992, p. 106) concludes: "Care in design and implementation will be rewarded with useful and clear study conclusions.... Elaborate analytical methods will not salvage poor design or implementation of a study."

Exert as much experimental control as is possible. Use the same measurement techniques in treated and control groups (§1.2). Carefully consider the process that selects individuals into the study—selection can introduce or eliminate biases (§6.1.3, §6.3). Anticipate hidden biases that pose the greatest threat to the study. Actively collect data that can reveal those biases if they are present (Chapters 5, 6, and 7).

• *Focus on simple comparisons.* Tukey (1986) writes: "increase impact of results on consumers .. by focusing on meaningful results (e.g., simple comparisons)". Cox (1958, p. 11) lists "simplicity" as one of the five "requirements for a good experiment" and writes: "This is a very important matter which must be constantly borne in mind...". Peto, Pike, Armitage, Breslow, Cox, Howard, Mantel, McPherson, Peto, and Smith (1976, §4, p. 590) make a similar point.

Simplicity is of greater importance in observational studies. Comparisons are often subject to genuine ambiguity or credible challenge. Issues of this kind are far more likely to be resolved if they are not compounded by unnecessary complications. A complex analysis can often be divided into several simple analyses, each of which can be challenged and debated separately. Writing within economics, Blaug (1992, p. 245) offers good advice for all fields in saying that empirical work should be judged "on the basis of the likely validity of the results reported and not on the technical sophistication of the techniques employed."

• *Compare subjects who looked comparable prior to treatment.* The most direct, most compelling way to address overt biases is to compare treated and control groups that looked comparable prior to treatment in terms of *observed* covariates. Susser (1973, §7) calls this "simplifying the conditions of observation." The matching and stratification methods in Chapter 9 can often produce matched pairs or sets or strata that balance many *observed* covariates.

• *Use sensitivity analyses to delimit discussions of hidden biases due to unobserved covariates.* Even with the greatest care, undetected hidden bias is a legitimate concern in an observational study. However, claims about hidden biases do not become credible merely because the covariates involved were not observed. The issue is explored through sensitivity analyses (Chapter 4). Sensitivity analyses for unobserved covariates are likely to have greatest impact if they build upon standard statistical techniques applied to simple comparisons of treated and control groups that appear comparable in terms of observed covariates.

References

Blaug, M. (1992). *The Methodology of Economics* (second edition). New York: Cambridge University Press.

Cox, D.R. (1958). *The Planning of Experiments*. New York: Wiley.

Meyer, M. and Fienberg, S., eds. (1992). *Assessing Evaluation Studies: The Case of Bilingual Education Strategies*. Washington, DC: National Academy Press.

Peto, R., Pike, M., Armitage, P., Breslow, N., Cox, D., Howard, S., Mantel, N., McPherson, K., Peto, J., and Smith, P. (1976). Design and analysis of randomized clinical trials requiring prolonged observation of each patient, I: Introduction and design. *British Journal of Cancer*, **34,** 585–612.

Susser, M. (1973). *Causal Thinking in the Health Sciences*. New York: Oxford University Press.

Tukey, J. (1986). Sunset salvo. *The American Statistician*, **40**, 72–76.

Index

Springer Series in Statistics

(continued from p. ii)

Reinsel: Elements of Multivariate Time Series Analysis.
Reiss: A Course on Point Processes.
Reiss: Approximate Distributions of Order Statistics: With Applications to Nonparametric Statistics.
Rieder: Robust Asymptotic Statistics.
Rosenbaum: Observational Studies.
Ross: Nonlinear Estimation.
Sachs: Applied Statistics: A Handbook of Techniques, 2nd edition.
Särndal/Swensson/Wretman: Model Assisted Survey Sampling.
Seneta: Non-Negative Matrices and Markov Chains, 2nd edition.
Shao/Tu: The Jackknife and Bootstrap.
Siegmund: Sequential Analysis: Tests and Confidence Intervals.
Tanner: Tools for Statistical Inference: Methods for the Exploration of Posterior Distributions and Likelihood Functions, 2nd edition.
Tong: The Multivariate Normal Distribution.
Vapnik: Estimation of Dependences Based on Empirical Data.
Weerahandi: Exact Statistical Methods for Data Analysis.
West/Harrison: Bayesian Forecasting and Dynamic Models.
Wolter: Introduction to Variance Estimation.
Yaglom: Correlation Theory of Stationary and Related Random Functions I: Basic Results.
Yaglom: Correlation Theory of Stationary and Related Random Functions II: Supplementary Notes and References.

Springer Series in Statistics